Potentially Toxic Elements Pollution in Urban and Suburban Environments II

Potentially Toxic Elements Pollution in Urban and Suburban Environments II

Guest Editor

Ilaria Guagliardi

MDPI

Basel • Beijing • Wuhan • Barcelona • Belgrade • Novi Sad • Cluj • Manchester

Guest Editor
Ilaria Guagliardi
Institute for Agricultural and
Forest Systems in the
Mediterranean (ISAFOM)
National Research Council of
Italy (CNR)
Rende (CS)
Italy

Editorial Office
MDPI AG
Grosspeteranlage 5
4052 Basel, Switzerland

This is a reprint of the Special Issue, published open access by the journal *Toxics* (ISSN 2305-6304), freely accessible at: https://www.mdpi.com/journal/toxics/special_issues/97971P6UV1.

For citation purposes, cite each article independently as indicated on the article page online and as indicated below:

Lastname, A.A.; Lastname, B.B. Article Title. *Journal Name* **Year**, *Volume Number*, Page Range.

ISBN 978-3-7258-5921-4 (Hbk)
ISBN 978-3-7258-5922-1 (PDF)
https://doi.org/10.3390/books978-3-7258-5922-1

Contents

About the Editor

Ilaria Guagliardi

Ilaria Guagliardi is a permanent researcher at the National Research Council of Italy, Institute for Agricultural and Forestry Systems in the Mediterranean (ISAFOM), located in Rende (CS), Italy. Her scientific expertise lies in environmental geochemistry, with a focus on pollution processes affecting agricultural and urban soils, water bodies, and sediments. She applies advanced geostatistical and multivariate techniques, including compositional and multifractal data analysis, to investigate contamination patterns and develop geospatial models for assessing geochemical risks and managing environmental impacts at regional scales.

Her research also extends to water-related topics such as hydrological modeling, watershed management, and landscape ecology. Dr. Guagliardi is actively involved in editorial and peer-reviewing activities, serving as associate editor and reviewer for over 50 international scientific journals. She has authored more than 80 peer-reviewed publications and contributes as an expert panel member for evaluating international research projects. She maintains ongoing collaborations with several universities and research institutions in Italy and abroad, fostering interdisciplinary approaches to environmental science.

toxics

MDPI

Editorial

Potentially Toxic Elements Pollution in Urban and Suburban Environments II

Ilaria Guagliardi

National Research Council of Italy—Institute for Agriculture and Forest Systems in the Mediterranean (CNR-ISAFOM), Via Cavour 4/6, 87036 Rende, Italy; ilaria.guagliardi@cnr.it

1. Introduction

Following the success of the first edition of the Special Issue "Potentially Toxic Elements Pollution in Urban and Suburban Environments", which gathered 13 high-quality contributions from international research groups, this second edition has once again proved to be a scientific success, with an equivalent number of published papers. The renewed interest and engagement of the scientific community clearly reflect the growing importance of monitoring and understanding pollution by potentially toxic elements (PTEs) as a serious and widespread environmental challenge.

Over the past few decades, accelerated population growth, rapid industrialization, urbanization, and changes in land use have dramatically increased the release and redistribution of inorganic contaminants into the environment [1–4]. The resulting contamination by PTEs, such as lead, cadmium, arsenic, chromium, and mercury, has become a global concern because of their persistence, non-biodegradability, and potential to bioaccumulate in living organisms [5–7]. Once introduced into the environment, PTEs can migrate across environmental matrices, air, water, soil, and biota [8–11]. They can be transferred to humans through dermal contact, inhalation of dust, and ingestion of contaminated food or drinking water [12,13]. Their presence, even at trace concentrations, may exert toxic effects on plants, animals, and humans, ultimately threatening the stability of ecosystems and public health [14–16].

Due to their ubiquity and complex environmental behavior, PTEs present substantial challenges for monitoring, risk assessment, and remediation [17,18]. Understanding their sources, transport mechanisms, and interactions within and between environmental compartments requires an interdisciplinary approach combining environmental geochemistry, soil and water sciences, and ecotoxicology [19]. These disciplines, supported by geostatistical and geochemical modeling, play a crucial role in identifying contamination sources, whether geological or anthropogenic, quantifying exposure risks, and informing sustainable management and policy decisions [20,21].

The present Special Issue aims to collect and compare case studies from diverse geographical contexts to advance our understanding of the behavior, transport, fate, and ecotoxicological implications of PTEs in both urban and suburban settings. The selected contributions address a broad spectrum of issues, including the assessment of soil, water, and food contamination, the evaluation of human and ecological risks, the development of analytical and geostatistical tools for source identification, and the use of bioindicators and model organisms to evaluate exposure. Together, they highlight the global dimension of PTE pollution and the need for harmonized strategies to prevent, mitigate, and monitor contamination across environmental matrices.

https://doi.org/10.3390/toxics13110920

2. An Overview of Published Articles

Recent scientific contributions collected in this issue provide a comprehensive and global perspective on the pervasive challenge of environmental contamination by potentially toxic elements (PTEs), agrochemicals, and natural emissions, as well as their multifaceted implications for ecosystems, food safety, and human health. The studies span diverse geographical, environmental, and socio-economic contexts, collectively advancing our understanding of contamination sources, exposure pathways, and analytical methodologies for monitoring and risk assessment.

In Lisbon (Portugal), Arán et al. (contribution 1) examined soils from urban and rural kitchen gardens to assess fertility and contamination by trace metals. Despite elevated concentrations of Cr, Ni, and Cu exceeding regulatory thresholds, the low bioavailable fractions of these elements suggest limited ecological and health risks. The authors highlighted the importance of regular monitoring to ensure the long-term safety of urban agriculture, a growing practice driven by sustainability and food security goals.

In northern Thailand, Santha et al. (contribution 2) investigated the legacy of unregulated zinc mining in the Mae Tao watershed, resulting in widespread soil and water contamination by Cd, Zn, Pb, and Mn. Controlled leaching experiments revealed that ionic strength, rather than pH, plays a dominant role in governing metal mobility, with Cd identified as the only element posing a carcinogenic risk. This study provides valuable insights into the geochemical behavior of heavy metals in tropical agroecosystems, highlighting the need for risk-based remediation and long-term environmental monitoring.

In southwestern China, Yang et al. (contribution 3) applied hydrochemical and isotopic analyses to trace nitrate contamination in urban groundwater. Fertilizers, sewage, and livestock manure were identified as major contributors to NO_3^- pollution, with nearly one-fifth of samples exceeding the national drinking water limit. Nonetheless, the overall water quality remained within acceptable ranges. The study demonstrates the utility of isotopic fingerprinting in disentangling natural and anthropogenic influences on groundwater quality in rapidly urbanizing environments.

Agricultural systems located in volcanic regions illustrate the dual role of natural processes as both stressors and stimulants of plant metabolism. In the Cotopaxi region of Ecuador, Mihai et al. (contributions 4 and 5) found that volcanic ash deposition significantly enhanced antioxidant activity and the accumulation of bioactive compounds in potatoes, corn, and beans. However, these metabolic benefits were accompanied by shifts in elemental composition and the presence of trace heavy metals, raising potential food safety concerns. The findings illustrate how volcanic inputs can simultaneously enrich and challenge agroecosystem sustainability, highlighting the importance of balancing nutraceutical gains with toxicological vigilance.

Analytical rigor remains central to contamination studies, as shown by Guagliardi et al. (contribution 6) in southern Italy, who evaluated the comparative performance of X-ray fluorescence (XRF) and inductively coupled plasma mass spectrometry (ICP-MS) for PTE determination in soils. The study identified systematic differences between methods, especially for elements such as V and Zn, emphasizing that instrument sensitivity, calibration, and matrix effects critically influence measurement reliability. These results reinforce the need for harmonized analytical protocols to ensure data comparability in environmental monitoring.

Several investigations addressed contamination pathways and biological indicators. In Zacatecas (Mexico), Ávila Vázquez et al. (contribution 7) discovered extremely high Pb concentrations in apples cultivated near mine tailings, exceeding international food safety limits by several orders of magnitude. The contamination pathway was identified as foliar absorption rather than root uptake, underscoring the vulnerability of crops in mining-impacted landscapes.

Similarly, in the Comarca Lagunera region (Mexico), Ocampo-Lopez et al. (contribution 8) found elevated Pb levels in pigeon tissues, particularly in bones, confirming their value as bioindicators of urban and industrial pollution.

A complementary methodological study by Zou et al. (contribution 9) in Changchun (China) improved the accuracy of portable XRF data through matrix correction models, providing a cost-effective and rapid approach for large-scale urban soil quality assessments.

Exposure to contaminants within indoor and aquatic environments was also explored. Somsunun et al. (contribution 10) analyzed household dust from Chiang Mai and Lamphun (Thailand) and found higher as concentrations in rural homes and elevated Cd levels in urban ones. Children were identified as the most vulnerable group, facing both carcinogenic and non-carcinogenic risks, thereby calling for stronger indoor air quality management.

In South Africa, Mugudamani et al. (contribution 11) detected herbicides, especially simazine, atrazine, and terbuthylazine, in rivers, reservoirs, and drinking water. The results revealed significant ecological risks and potential carcinogenic effects, underscoring the importance of integrated pesticide management and enhanced water treatment practices.

At a broader toxicological scale, Notario-Barandiaran et al. (contribution 12) conducted a scoping review of 77 epidemiological studies examining the neuropsychological effects of arsenic exposure during pregnancy and childhood. The majority reported adverse cognitive and psychomotor outcomes, although limitations in exposure speciation and study design hinder definitive causal inference. This synthesis highlights critical gaps in current exposure assessment and the need for longitudinal, mechanistic research.

Conversely, Al-Raddadi et al. (contribution 13) provided a reassuring perspective by analyzing the medicinal plant Boerhavia elegans, demonstrating that essential micronutrients such as Fe, Mn, and Zn were abundant. At the same time, toxic elements remained below WHO safety limits, thereby supporting its continued use in traditional medicine.

Collectively, these studies reveal the global interconnectedness of environmental contamination issues, transcending climatic, economic, and geographic boundaries. They collectively emphasize the importance of multidisciplinary approaches combining geochemical, biological, and analytical perspectives. Furthermore, they point to a shared need for improved methodological standardization, continuous environmental monitoring, and the development of science-based policies to safeguard ecosystem integrity, agricultural productivity, and public health.

3. Future Perspectives and Concluding Remarks

The research contributions collected in this Special Issue clearly demonstrate that contamination by potentially toxic elements remains a persistent and complex problem that transcends geographical and socio-economic boundaries. Despite considerable progress in analytical methodologies, environmental monitoring, and risk assessment frameworks, several critical gaps in knowledge still hinder the full understanding of PTE dynamics and their long-term ecological and health consequences.

Future research should focus on improving the spatial and temporal resolution of PTE monitoring through integrated, multi-scale approaches that combine field measurements, remote sensing, and geostatistical modeling. Particular attention should be paid to the interactions between PTEs and emerging contaminants, such as microplastics and nanomaterials, which may influence their mobility, bioavailability, and toxicity. Similarly, understanding the coupling between geogenic and anthropogenic sources in rapidly urbanizing regions will be essential to designing effective prevention and remediation strategies.

Advances in analytical chemistry, isotopic tracing, and in situ sensing technologies offer new opportunities to refine source apportionment and exposure assessment. These should be complemented by experimental and modeling studies addressing biogeochemical transformations and long-term sequestration processes in soils and sediments. In parallel, the integration of human biomonitoring data with environmental measurements will enhance the ability to assess real exposure pathways and cumulative risks, particularly in vulnerable populations such as children and communities living near industrial or mining areas.

From a policy perspective, there is a pressing need to harmonize environmental standards and monitoring protocols across countries, ensuring data comparability and supporting international initiatives for pollution control and health protection. Collaborative networks and open-access data platforms can play a key role in achieving this goal, promoting transparency, reproducibility, and the global sharing of best practices.

Overall, this second edition of the Special Issue "Potentially Toxic Elements Pollution in Urban and Suburban Environments" reinforces the critical importance of continued interdisciplinary research and coordinated policy actions. By combining geochemical insight with technological innovation and public health awareness, the scientific community can contribute to a more accurate understanding of PTE behavior and to the sustainable management of contaminated environments, ultimately safeguarding ecosystem integrity and human well-being.

Data Availability Statement: No new data were created or analyzed in this study. Data sharing is not applicable to this article.

Acknowledgments: As Guest Editors of the Special Issue "Potentially Toxic Elements Pollution in Urban and Suburban Environments II", I would like to express my sincere gratitude to all the authors who contributed their valuable research to this Special Issue and to the reviewers for their constructive evaluations and timely feedback. I also warmly thank the Toxics Editorial Office for their professional support throughout the peer-review and publication process. The success of this second edition reflects the collaborative commitment of the scientific community to addressing the global challenge of potentially toxic element pollution.

Conflicts of Interest: The author declares no conflicts of interest.

List of Contributions

1. Arán, D.; Santos, O.; Feteira-Santos, R.; Benhalima, Y.; Santos, E.S. Soil Quality and Trace Element Risk in Urban and Rural Kitchen Gardens: A Comparative Analysis. *Toxics* **2025**, *13*, 697.
2. Santha, N.; Watcharamai, T.; Benjakul, R.; Saenton, S. Leachability and Health Risk Assessment of Cadmium and Other Heavy Metals in Agricultural Soils from the Mae Tao Watershed, Northern Thailand. *Toxics* **2025**, *13*, 687.
3. Yang, C.; Chen, S.; Dong, J.; Zhang, Y.; Wang, Y.; Kang, W.; Zhang, X.; Liang, Y.; Fu, D.; Yan, Y.; et al. Hydrochemical Characteristics, Controlling Factors, and High Nitrate Hazards of Shallow Groundwater in an Urban Area of Southwestern China. *Toxics* **2025**, *13*, 516.
4. Mihai, R.A.; Vivanco Gonzaga, R.F.; Romero Balladares, N.R.; Catana, R.D. Characterization of Volcanic Ash Influence on the Nutritional Quality and Biological Traits in Potato Crops of the Cotopaxi Region. *Toxics* **2025**, *13*, 453.
5. Mihai, R.A.; Rodríguez Valencia, K.E.; Sivizaca Flores, N.G.; Ramiro Fernando, V.G.; Nelson Santiago, C.I.; Catana, R.D. Consequences of Volcanic Ash on Antioxidants, Nutrient Composition, Heavy Metal Accumulation, and Secondary Metabolites in Key Crops of Cotopaxi Province, Ecuador. *Toxics* **2025**, *13*, 75.

6. Guagliardi, I.; Ricca, N.; Cicchella, D. Comparative Evaluation of Inductively Coupled Plasma Mass Spectrometry (ICP-MS) and X-Ray Fluorescence (XRF) Analysis Techniques for Screening Potentially Toxic Elements in Soil. *Toxics* **2025**, *13*, 314.
7. Ávila Vázquez, V.; Aguilera Flores, M.M.; Veyna Robles, A.N.; Solís Lerma, L.E.; Sánchez Mata, O.; Durón Torres, S.M. Determination of Lead in Fruit Grown in the Vicinity of Tailings Dams of a Mine in Zacatecas, Mexico. *Toxics* **2025**, *13*, 188.
8. Ocampo-Lopez, A.; Puente-Valenzuela, C.O.; Sánchez-Galván, H.; Valenzuela-García, A.A.; Estrada-Arellano, J.R.; Delgado-González, R.A.; Aguirre-Joya, J.A.; Torres-León, C.; Ocampo-Lopez, A.; Aguillón-Gutiérrez, D.R. Lead Concentrations in Tissues of Pigeons (*Columba livia*) in the Urban Area of Comarca Lagunera, Mexico. *Toxics* **2024**, *12*, 830.
9. Zou, X.; Lu, J.; Zhao, X.; Wei, Q.; Gou, Z.; Hou, Y.; Lai, Y. An Investigation into the Viability of Portable Proximal Sensor X-Ray Fluorescence Data for Assessing Heavy Metal Contamination in Urban Soils: A Case Study in Changchun, China. *Toxics* **2024**, *12*, 798.
10. Somsunun, K.; Prapamontol, T.; Kuanpan, T.; Santijitpakdee, T.; Kohsuwan, K.; Jeytawan, N.; Thongjan, N. Health Risk Assessment of Heavy Metals in Indoor Household Dust in Urban and Rural Areas of Chiang Mai and Lamphun Provinces, Thailand. *Toxics* **2023**, *11*, 1018.
11. Mugudamani, I.; Oke, S.A.; Gumede, T.P.; Senbore, S. Herbicides in Water Sources: Communicating Potential Risks to the Population of Mangaung Metropolitan Municipality, South Africa. *Toxics* **2023**, *11*, 538.
12. Notario-Barandiaran, L.; Compañ-Gabucio, L.M.; Bauer, J.A.; Vioque, J.; Karagas, M.R.; Signes-Pastor, A.J. Arsenic Exposure and Neuropsychological Outcomes in Children: A Scoping Review. *Toxics* **2025**, *13*, 542.
13. Al-Raddadi, T.M.; Al-Khateeb, L.A.; Sadaka, M.W.; Bahaffi, S.O. Trace Element Speciation and Nutrient Distribution in Boerhavia elegans: Evaluation and Toxic Metal Concentration Across Plant Tissues. *Toxics* **2025**, *13*, 14.

References

1. Nuruzzaman, M.; Mezbaul Bahar, M.; Ravi Naidu, R. Diffuse soil pollution from agriculture: Impacts and remediation. *Sci. Total Environ.* **2025**, *962*, 178398. [CrossRef] [PubMed]
2. Strokal, M.; Bai, Z.; Franssen, W.; Hofstra, N.; Koelmans, A.A.; Ludwig, F.; Ma, L.; van Puijenbroek, P.; Spanier, J.E.; Vermeulen, L.C.; et al. Urbanization: An increasing source of multiple pollutants to rivers in the 21st century. *Urban Sustain.* **2021**, *1*, 24. [CrossRef]
3. Ale, T.O.; Ogunribido, T.H.; Ademila, O.; Akingboye, A.S. Soil pollution status due to potentially toxic elements in active open dumpsites: Insights from different Nigerian geological environments. *Environ. Earth Sci.* **2024**, *83*, 535. [CrossRef]
4. Ali, M.U.; Liu, G.; Yousaf, B.; Abbas, Q.; Ullah, H.; Munir, M.A.M.; Fu, B. Pollution characteristics and human health risks of potentially (eco)toxic elements (PTEs) in road dust from metropolitan area of Hefei, China. *Chemosphere* **2017**, *181*, 111–121. [CrossRef]
5. Sakan, S.; Mihajlidi-Zelić, A.; Sakan, N.; Frančišković-Bilinski, S.; Kodranov, I.; Đorđević, D. Potentially toxic elements in sediments near mines—A comprehensive approach for the assessment of pollution status and associated risk for the surface water environment. *Environ. Sci. Pollut. Res.* **2024**, *31*, 16613–16628. [CrossRef]
6. Joseph, A.M.; Gandhi, M.S. Assessment of Potentially Toxic Elements Contamination in Water and Sediments from Vembanad Lake, Kerala, India: In Accounts with the Influence of its Major In Flow Rivers. *Int. J. Environ. Res.* **2025**, *19*, 62. [CrossRef]
7. Wang, F.; Xiang, L.; Sze-Yin Leung, K.; Elsner, M.; Zhang, Y.; Guo, Y.; Pan, B.; Sun, H.; An, T.; Ying, G.; et al. Emerging contaminants: A One Health perspective. *Innovation* **2024**, *5*, 100612. [CrossRef]
8. Bineshpour, M.; Payandeh, K.; Nazarpour, A.; Sabzalipour, S. Status, source, human health risk assessment of potential toxic elements (PTEs), and Pb isotope characteristics in urban surface soil, case study: Arak city, Iran. *Environ. Geochem. Health* **2021**, *43*, 4939–4958. [CrossRef]
9. Jiang, F.; Ren, B.; Hursthouse, A.; Deng, R.; Wang, Z. Distribution, source identification, and ecological-health risks of potentially toxic elements (PTEs) in soil of thallium mine area (southwestern Guizhou, China). *Environ. Sci. Pollut. Res.* **2019**, *26*, 16556–16567. [CrossRef]
10. Fakhri, Y.; Limam, I.; Kamali, M.; Zare, A.; Ranaei, V.; Mohamadi, S.; Khaneghah, A.M. A systematic review of potentially toxic elements (PTEs) in river sediments from China: Evaluation of associated non-dietary health risks. *Environ. Monit. Assess.* **2025**, *197*, 269. [CrossRef]

11. Obodai, J.; Amaning Adjei, K.; Duncan, A.E.; Odai, S.N. Potentially Toxic Elements (PTEs) contamination and ecological risk of sediment in the upper course of the Ankobra River, Ghana. *Environ. Monit. Assess.* **2022**, *194*, 446. [CrossRef] [PubMed]
12. Infusino, E.; Guagliardi, I.; Gaglioti, S.; Caloiero, T. Vulnerability to Nitrate Occurrence in the Spring Waters of the Sila Massif (Calabria, Southern Italy). *Toxics* **2022**, *10*, 137. [CrossRef] [PubMed]
13. Mihajlović, D.; Srdić, S.; Benka, P.; Čereković, N.; Ilić, P.; Radanović, D.; Antić-Mladenović, S. Potentially toxic elements in the agricultural soils of northwestern Bosnia and Herzegovina: Spatial and vertical distribution, origin and ecological risk. *Environ. Monit. Assess.* **2025**, *197*, 295. [CrossRef] [PubMed]
14. Ma, Y.; Wang, Q.; Su, W.; Cao, G.; Fu, G.; Du, W. Potential Sources, Pollution, and Ecological Risk Assessment of Potentially Toxic Elements in Surface Soils on the North-Eastern Margin of the Tibetan Plateau. *Toxics* **2022**, *10*, 368. [CrossRef]
15. Mullineaux, S.T.; McKinley, J.M.; Marks, N.J.; Scantlebury, D.M.; Doherty, R. Heavy metal (PTE) ecotoxicology, data review: Traditional vs. a compositional approach. *Sci. Total Environ.* **2021**, *769*, 145246. [CrossRef]
16. Bavi, H.; Gharaie, M.H.M.; Moussavi-Harami, R.; Zand-Moghadam, H.; Mahboubi, A.; Tohidi, M.R. Spatial dispersion hot spots of contamination and human health risk assessments of PTEs in surface sediments of streams around porphyry copper mine, Iran. *Environ. Geochem. Health.* **2023**, *45*, 3907–3931. [CrossRef]
17. Pan, L.; Fang, G.; Wang, Y.; Wang, L.; Su, B.; Li, D.; Xiang, B. Potentially Toxic Element Pollution Levels and Risk Assessment of Soils and Sediments in the Upstream River, Miyun Reservoir, China. *Int. J. Environ. Res. Public Health* **2018**, *15*, 2364. [CrossRef]
18. Ambrosino, M.; Albanese, S.; De Vivo, B.; Guagliardi, I.; Guarino, A.; Lima, A.; Cicchella, D. Identification of Rare Earth Elements (REEs) distribution patterns in the soils of Campania region (Italy) using compositional and multivariate data analysis. *J. Geochem. Expl.* **2022**, *243*, 107112. [CrossRef]
19. Ribeiro, D.R.G.; Faccin, H.; Dal Molin, T.R.; Machado de Carvalho, L.; Amado, L.L. Metal and metalloid distribution in different environmental compartments of the middle Xingu River in the Amazon, Brazil. *Sci. Total Environ.* **2017**, *605–606*, 66–74. [CrossRef]
20. Guagliardi, I.; Astel, A.M.; Cicchella, D. Exploring Soil Pollution Patterns Using Self-Organizing Maps. *Toxics* **2022**, *10*, 416. [CrossRef]
21. Rouhani, A.; Hejcman, M.; Trögl, J. A review of soil pollution by potentially toxic elements and remediation strategies in copper mining areas in Iran. *Int. J. Environ. Sci. Technol.* **2025**, *22*, 9793–9806. [CrossRef]

toxics

MDPI

Article

Herbicides in Water Sources: Communicating Potential Risks to the Population of Mangaung Metropolitan Municipality, South Africa

Innocent Mugudamani [1,*], Saheed A. Oke [2], Thandi Patricia Gumede [1] and Samson Senbore [2]

[1] Department of Life Sciences, Central University of Technology, Free State, Bloemfontein 9301, South Africa; tgumede@cut.ac.za

[2] Department of Civil Engineering, Centre for Sustainable Smart Cities, Central University of Technology, Free State, Bloemfontein 9301, South Africa; soke@cut.ac.za or okesaheed@gmail.com (S.A.O.); senboresamson@gmail.com (S.S.)

* Correspondence: imugudamani@gmail.com or innomugudamani@outlook.com

Abstract: Pesticides are an important tool for maintaining and improving the global population's standard of living. However, their presence in water resources is concerning due to their potential consequences. Twelve water samples from rivers, dams/reservoirs, and treated drinking water were collected from Mangaung Metropolitan Municipality in South Africa. The collected samples were analysed using high-performance liquid chromatography linked to a QTRAP hybrid triple quadrupole ion trap mass spectrometer. The ecological and human health risks were assessed by risk quotient and human health risk assessment methods, respectively. Herbicides, such as atrazine, metolachlor, simazine and terbuthylazine, were analysed in water sources. The average concentrations of simazine in rivers (1.82 mg/L), dams/reservoirs (0.12 mg/L), and treated drinking water (0.03 mg/L) were remarkable among all four herbicides detected. Simazine, atrazine, and terbuthylazine posed high ecological risks for both acute and chronic toxicity in all water sources. Moreover, simazine is the only contaminant in the river water that poses a medium carcinogenic risk to adult. It can be concluded that the level of herbicide detected in water sources may affect aquatic life and human beings negatively. This study may aid in the development of pesticide pollution management and risk reduction strategies within the municipality.

Keywords: pesticide and herbicide; potential risks; water sources; ecological and health risks; Mangaung; residue

1. Introduction

Water resources play an important role in our daily lives as a source of water [1]. They are, however, increasingly exposed to an extensive kind of organic chemical of an anthropogenic source, such as pesticides [2]. Pesticides are chemicals or mixtures of chemicals that are primarily utilised for the protection of agricultural yields from weeds, insects, and pests, and humans from diseases [3,4]. Pesticides' beneficial effects make them an important tool for maintaining and improving the global populations' living standards. Each year, approximately two million tons of pesticides are utilised worldwide to control weeds, insects, and pests. Herbicides, insecticides, rodenticides, fungicides, and other pesticides are traditionally classified based on target species. Herbicides and insecticides are the primary intoxicants among pesticides. Herbicides, on the other hand, are the most commonly used type of pesticides, accounting for 47.5% of total global pesticide consumption [4]. A herbicide, according to Wang et al. [5], is any chemical, alone or in combination, whose purpose is to regulate, terminate, prevent, or alleviate the development of weeds in a crop. It is also used in forestry, community areas, parks, golf courses, and sports fields to control weeds [6].

Pesticides may contaminate water sources as a result of runoff, wastewater discharges, and return flow from agricultural and irrigated areas. They enter water through direct application to control aquatic weeds or indirectly through transportation from treated areas. Leakage and runoff from agricultural areas are the primary sources of conveyance to water sources [7,8]. The detection and quantification of pesticide in water sources together with their potential environmental health risks have been extensively documented in many areas of the world, such as in Argentina [9], Japan [10], the United States [11], India [12], the Czech Republic [13], China [14], and Malaysia [15]. In spite of such significant research outputs, the potential human health risks associated with pesticides found in various water sources essentially remain unrevealed in most of these studies. Moreover, most of these studies focus less on water sources, such as treated drinking water, which is considered the safest drinking water source. As a result of the environmental dynamics and incessant use of a number of pesticides, their presence in the water environment as well as their end products are concerning, as they may trigger possible effects on aquatic ecosystems and human health, particularly via the ingestion of water [1,10]. Pesticide exposure through water ingestion, in particular, can cause gastrointestinal and neurological effects, mimic the human body's hormones, which reduce body immunity, disrupt hormone balance, trigger reproductive issues, pose carcinogenic effects, and reduce intelligence, particularly in children at the body development stage [4]. To protect aquatic ecosystems, regulations for the viable utilisation of pesticides have been implemented in both Europe [16,17] and America [18]. Regrettably, appropriate legislation and inspections are frequently missing in developing countries, such as South Africa, resulting in pesticide mismanagement and misuse, together with the continued use of disqualified chemicals [19].

In South Africa, approximately 26,000 tons of pesticides are used each year, with approximately 700 active chemicals registered for agricultural usage. In the African continent, South Africa adds roughly one third of all pesticides used, rendering it a high-risk country for pesticide contamination [20–23]. Monitoring pesticide levels in water resources is thus critical not only for assessing water quality but also for protecting the health of the ecosystem and South African water consumers. Previous studies on the occurrence of herbicides in South African aquatic environment found these substances in rivers in the Western Cape Province [24], dams in the Free State Province [25], seawater in the Western Cape Province [26], and wastewater influent and effluent in the KwaZulu Natal Province [27]. Hence, they did not shed light on the ecological risks of pesticides. In addition to the paucity of data on risk assessment of pesticides in various water sources, there is sparse information on the human health risks of pesticides in the country, particularly in the Free State Province (home to Mangaung Metropolitan Municipality) where almost fourteen percent (14%) of profitable agri-business in the country occurs [28]. The intensive agricultural activities in this area encourage the use of agricultural chemicals. Trizine herbicides, such as atrazine, simazine, and terbuthylazine, and chloroacetanilides herbicides, such as metolachlor, were targeted in this study because agriculture dominates the landscape of the Mangaung Metropolitan Municipality. These compounds are also the most commonly used around the world and are associated with serious health effects [4].

The widespread utilisation of agricultural chemicals, such as herbicides, in the Mangaung Metropolitan Municipality prompted this study. Here, we report the first assessment of herbicides in water sources around the Mangaung Metropolitan Municipality in the Free State Province of South Africa. To ensure good surface and drinking water quality within the Mangaung Metropolitan Municipality, it was critical to assess potential pesticide adverse effects on aquatic ecosystems and human health. To date, there is a scarcity of studies in the Mangaung Metropolitan Municipality that provide a comprehensive assessment of pesticides and their potential environmental health risks across a variety of water sources (rivers, dams/reservoirs, and treated drinking water) that are used by thousands of people for various purposes. Thus, in this metropolitan, more detailed risk assessment studies of pesticides in rivers, dams/reservoirs, and treated drinking water are desperately needed for water and human health security. Therefore, this project will fill in such a gap by determining the concentrations of herbicides in such water sources. The secondary goal is to assess the ecological, non-carcinogenic, and carcinogenic risks associated with pesticides exposure. The novelty of this study stems from the fact that it is the first to report on the environmental and human health risks of pesticides in rivers, dams/reservoirs, and treated drinking water within the Mangaung Metropolitan Municipality. The findings of the study may help to notify the community and relevant stakeholders on the environmental and public health risks of pesticide exposure. Furthermore, it will guide potential interventions to curtail pesticide pollution and concomitant risks in the region and throughout the country.

2. Materials and Methods

2.1. Study Area

The Mangaung Metropolitan Municipality is classified as a category-A municipality. It is centrally positioned in South Africa's Free State Province (Figure 1). It is located at 29°10′00″ S and 26°1′67″ E, at an area that covers approximately 9899 km^2. It is home to 878,834 people, which accounts for roughly 28% of the provincial population. The temperatures may vary from 1.3 °C to 21 °C during winter and from 13 °C to 30.9 °C during summer. The municipality's average annual rainfall is approximately 476 mm, with February being the wettest month. The Mangaung Metropolitan Municipality has only one biome, grassland, which covers the entire municipality. In the Mangaung Metro, commercial agriculture accounts for 22.96% of land, with pivot irrigation accounting for 1.4% and cultivated orchards accounting for 0.02%. Agriculture is a major economic activity in the Mangaung Metro, with commercial, small-scale, and subsistence farming all practiced. An estimated 46,172 households, or 19.4% of the total households in the municipality, rely on agricultural activities for a living. Crop farming accounts for 67.5% of agricultural activities in the Mangaung Metro, followed by livestock farming, mixed farming, and finally other types of agriculture. Community services, finance, trade, transportation, and manufacturing are some of the other economic sectors in the municipality. As both the Water Services Authority and the Water Service Provider, the Mangaung Metropolitan Municipality is obligated to fulfil its mandate of providing safe and dependable portable water to its consumers. The Mangaung Metro's water resources include dams (reservoirs), rivers, wetlands, and groundwater resources. The city's bulk water supply is currently 31% supplied by its water treatment works and 69% supplied by other providers [29].

2.2. Sample Collection and Analysis

In September 2022, the sampling campaign was launched. Rivers, dams/reservoirs, and water treatment works (WWTW) in the Mangaung Metropolitan Municipality were targeted during the campaign. As shown in Table 1, twelve (12) water samples were collected from rivers (5), dams/reservoirs (5), and treated drinking water (2). When choosing a sampling location, accessibility, site representation, and pollution sources were all factors to consider. To collect grab water samples, 750 millilitre (mL) hygienic glass bottles with screw caps were used. During the sampling campaign, cooler boxes, ice cubes, and tags were also purchased to protect and label the samples. The 750 millilitre (mL) glass bottles were washed several times with clean water or river water prior to sampling. Furthermore, before taking the treated drinking-water sample, the tap was opened and permitted to run freely for a few minutes. Following fieldwork, all collected samples placed in a cooler box packed with ice cubes were conveyed to the laboratory, where they were stored at 4 °C until analysis.

In the laboratory (Bloemfontein, South Africa), the samples were filtered through glass fibre filters to remove particulate matter before being concentrated at a flow rate of 5 mL/minutes (min) onto methanol conditioned C18-6 mL solid phase extraction cartridges (Strata, Phenomenex, Torrance, CA, USA). The bound sample was slowly eluted from the dried cartridges with 2 mL methanol and 2 mL ethyl acetate. The eluant was vacuum dried until almost dry (Thermo Scientific Savant Speedvac, Waltham, MA, USA) and reconstituted in 1 mL purified water.

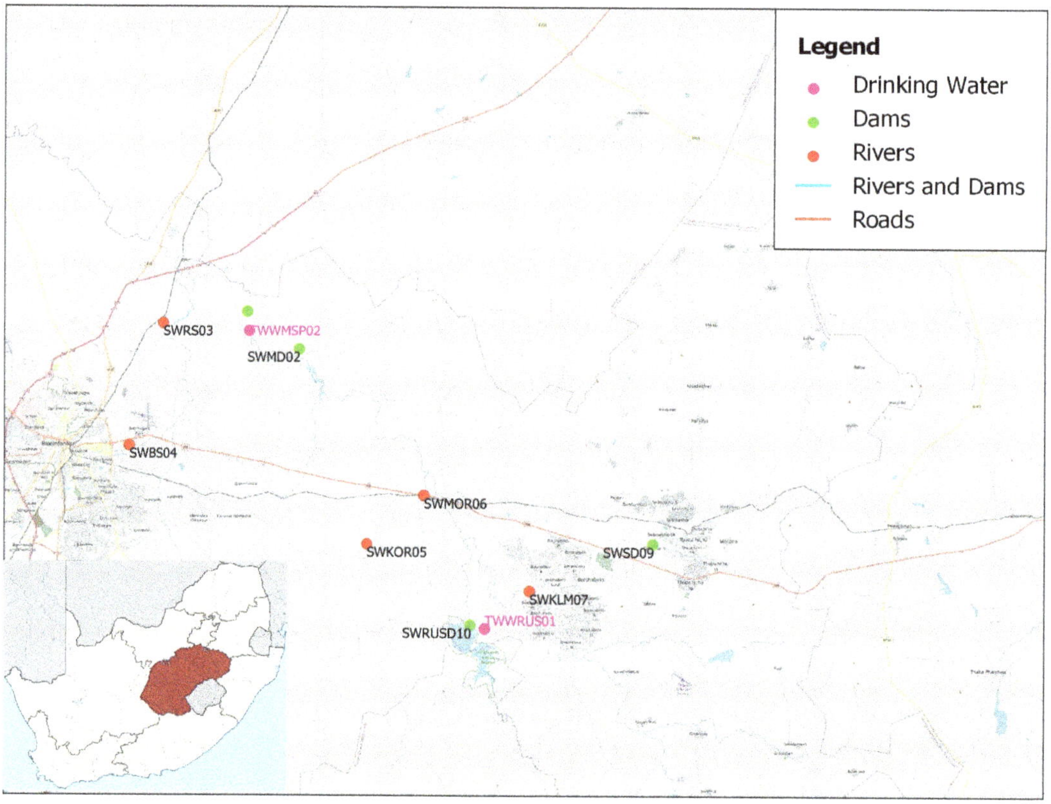

Figure 1. Mangaung Metropolitan Municipality.

Table 1. Location of the sampling points within Mangaung Metropolitan Municipality.

Sample ID	Description	Coordinates		
		Longitude	Latitude	Elevation
River samples				
SWRS03	Near intense agricultural activities	29°06′9.6″	26°19′7.2″	1379 m
SWBS04	Close to residential area and WWTW	29°07′2.4‴	26°17′1.5″	1390 m
SWKOR05	Passes through agricultural farms	29°12′8.3″	25°31′04″	1334 m
SWMOR06	Near farms, railroad, and national road	29°09′39.3″	26°34′20.3″	1327 m
SWKLM07	Passes through a township and near WWTW	29°14′34.3″	26°40′26.2″	1373 m
Dam/Reservoirs samples				
SWMS01	It also serves as a resort	29°01′4.9″	26°24′2.7″	1344 m
SWMD02	Serve as a resort, conference centre	29°02′8.4″	26°27′5.8″	1354 m
SWKD08	Fishing activities, and farms nearby	29°53′03″	25°57′21″	1226 m
SWSD09	Near settlement and industrial activities	29°12′10‴	26°47′38″	1460 m
SWRUSD10	Near farms and residents	29°16′20″	26°37′00″	1370 m
Treated drinking water sample				
TWWRU01	Water treatment works 1	29°16′31″	26°37′51″	-
TWWMSP02	Water treatment works 2	29°01′10.3″	26°24′9.2″	-

High-performance liquid chromatography linked to a QTRAP hybrid triple quadrupole ion trap mass spectrometer was used to analyse the water samples. Analyst 1.5 (AB SCIEX) software was used for all data acquisition and processing. Positive and negative ionisation modes were used to analyse the samples. During analysis, 20 microliter (μL) of each extracted sample was separated on a C18 (150 mm × 4.6 mm, Gemini NX, Phenomenex) column at a flow rate of 300 μL/min using a 5 min gradient from 5% solvent A (H_2O/0.1% formic acid) to 95% solvent B (MeOH/0.1% formic acid) with a total run time of 9 min and 10 min in positive and negative ionisation modes, respectively, to allow for column re-equilibration. Eluting analytes were electrospray ionised in the TurboV ion source with a heater temperature of 500 "°C" to evaporate the excess solvent, 40 psi nebuliser gas, 40 psi heater gas, and a curtain gas of 15 psi. In positive ionisation mode, the ion spray voltage was set at 5500 V, while in the negative ionisation mode, it was set at −4500 V.

Pesticide analyses were carried out using multiple reaction monitoring transitions per analyte. The quantifier was the peak area on the chromatogram generated by the first and most sensitive transition, while the qualifier was the peak area generated by the second transition. The qualifier served as an additional level of confirmation for the analytes' presence. The retention times for these two transitions must be the same as those listed in Table 2.

Table 2. Multiple reaction monitoring transition values for targeted compounds.

Analyte	Positive Ionisation Mode		Negative Ionisation Mode		Retention Time (Minutes)
	Q1 (m/z)	Q3 (m/z)	Q1 (m/z)	Q3 (m/z)	
Atrazine 1	216.049	174.2	216.049	174.2	21.80
Atrazine 2	216.049	68.1	216.049	68.1	21.80
Metolachlor 1	284.347	252	284.347	252	25.30
Metolachlor 2	284.347	176.2	284.347	176.2	25.30
Simazine 1	202.039	132.1	202.039	132.1	19.80
Simazine 2	202.039	104.1	202.039	104.1	19.80
Terbuthylazine 1	230.087	174.3	230.087	174.3	24.00
Terbuthylazine 2	230.087	68	230.087	68	24.00

To validate instrument performance, the selectivity, linearity and limit of quantification (LOQ) were taken into account. Samples were submitted in batches with solvent blank runs between each sample analysed and quality control samples of known concentration interspersed. For each analyte, a four-point calibration curve with a linear fit through the origin was generated, ranging in concentration from 0.001 ppm to 1 ppm. The linear fit yielded a correlation coefficient (r) value above 0.98. Furthermore, the quantification limits ranged from 0.0001 mg/L to 0.01 mg/L as shown in Table 3.

Table 3. Linearity, and limit of quantification of the targeted analytes.

Analyte	Linearity (r-Value)	LOQ
Simazine	0.99	0.0100
Atrazine	0.99	0.0001
Terbuthylazine	0.99	0.0001
Metolachlor	0.99	0.0001

2.3. Environmental Risk Assessment

The potential ecological risks of herbicides in water sources were evaluated by the environmental risk quotient (*RQ*) method. The risks were evaluated based on acute and chronic toxicities, which measure the toxic effects on the most vulnerable organisms, such as algae, invertebrate, and fish [30]. Using Equation (1), the risk quotient values were computed by comparing the measured environmental concentration (*MEC*) and the predicted no effect concentration (*PNEC*):

$$RQ = \frac{MEC}{PNEC} \quad (1)$$

When performing a risk analysis for a specific contaminant and aquatic organism, i.e., algae, invertebrate and fish, it was crucial to estimate the level of contaminant in the water sources as well as the toxicity of a particular contaminant to an organism in the water sources (Table 4). The predicted no effect concentration was calculated by dividing the acute or chronic toxicity value by an assessment factor (AF). The toxicity results were corrected by an assessment factor during the process of determining the predicted no effect concentration of a specific herbicide in water sources as shown in Equation (2). Acute toxicity was considered the median lethal concentration (LC_{50}) or mean effective concentration (EC_{50}), where AF = 1000. Chronic toxicity was determined by the no observable effect concentration ($NOEC$), which can be 100, 50, or 10 mg/L for algae, invertebrates, and fish, respectively [31,32]. The risk quotient was then calculated by comparing the predicted no effect concentration ($PNEC$) with the measured environmental concentrations (MEC) of the herbicide of interest after the toxicity data was corrected by an assessment factor [30]:

$$PNEC = \frac{EC_{50}}{AF} \; or \; PNEC = \frac{NOEC}{AF} \tag{2}$$

where RQ is the risk quotient calculated using the effective concentration (EC_{50}) or no observable effect concentration ($NOEC$). The measured concentrations of the herbicides were represented by the MEC. The predicted no effect concentration ($PNEC$) was the highest concentration of a drug known to have no negative risks on organisms in the environment. The risks of herbicides in water sources were categorised as (i) low ($RQ \leq 0.1$); (ii) medium ($0.1 < RQ < 1$); or (iii) high ($RQ \geq 1$) [31,33].

Table 4. Acute and chronic toxicity data for detected herbicides in algae, invertebrate and fish.

Compound	Taxon	Specie	Acute Toxicity (EC_{50})	Chronic Toxicity ($NOEC$)	References
Atrazine	Algae	*P. Subcapitata*	0.059 mg/L	0.1 mg/L	[34]
	Invertebrate	*Daphnia magma*	6.9 mg/L	0.1 mg/L	[34]
	Fish	*Oncorhynchus mykiss*	4.5 mg/L	2 mg/L	[34]
Metolachlor	Algae	*P. Subcapitata*	57.1 mg/L	3.0 mg/L	[34]
	Invertebrate	*Daphnia magma*	23.5 mg/L	3.0 mg/L	[34]
	Fish	*Oncorhynchus mykiss*	3.9 mg/L	1.0 mg/L	[34]
Simazine	Algae	*P. Subcapitata*	0.04 mg/L	0.6 mg/L	[34]
	Invertebrate	*Daphnia magma*	1.1 mg/L	0.6 mg/L	[34]
	Fish	*Oncorhynchus mykiss*	9.0 mg/L	0.7 mg/L	[34]
Terbuthylazine	Algae	*P. Subcapitata*	0.02 mg/L	-	[35]
	Invertebrate	*Daphnia magma*	39.4 mg/L	0.21 mg/L	[35]
	Fish	*Oncorhynchus mykiss*	3.6 mg/L	0.13 mg/L [a]	[35]

[a] mysid shrimp.

2.4. Health Risk Assessment

The human health risk assessment method aids in appraising the likelihood and severity of adverse risks that a specific herbicide may pose in humans. It consists of various steps, such as hazard identification, exposure assessment, dose–response analysis, and risk characterisation [36–38]. The goal of the hazard identification step is to look into the type, concentration, and distribution of pesticides in a specific area [37]. The dose–response assessment step aids in determining the specific relationship between the contaminant exposure dose and the likelihood of adverse reactions in the exposed population [38]. The exposure assessment step assesses the amount, rate, and time of a human's exposure to a contaminant. Ingestion remains a major risk among the various routes of exposure. The average daily dose (ADD) of pesticides ingested by adults and children is calculated using Equation (3). Because of their behavioural and physiological differences, it is calculated separately [36,37,39]:

$$ADD_{ing} = \frac{C \times IngR \times EF \times ED \times CF}{BW \times AT} \tag{3}$$

$IngR$ denotes the rate of ingestion for adults and children. The concentration of herbicides is C, the exposure duration is ED, the conversion factor is CF, the body weight is BW, the exposure frequency is EF, and the average time is AT (Table 5). The non-carcinogenic risk is then calculated by dividing the average daily doses in Equation (4) by a corresponding reference dose (RfD) in Table 6. Equation (5) is then used to calculate the hazard index (HI) [36,39]:

$$HQ = \frac{ADD}{RfD} \tag{4}$$

$$HI = \sum HQ \tag{5}$$

where if the hazard quotient or index is less than one, it indicates a very low risk; between one and four, it indicates a possible risk; and when greater than four, it indicates a high risk [36,39]. The ADD values in Equation (3) are also used to compute the cancer risk (CR) of each examined contaminant. Using Equation (6), the cancer risk is then assessed by multiplying the ADD by a corresponding slope factor (SF) in Table 6:

$$CR = ADD \times SF \tag{6}$$

The permitted risk ranges are $<10^{-6}$ (very low risk), 10^{-6}–10^{-5} (low risk), 10^{-5}–10^{-4} (medium risk), 10^{-4}–10^{-3} (high risk), and $>10^{-3}$ (very high risk) [36,39].

Table 5. Exposure assessment parameters for ingestion pathway.

Parameters	Description	Unit	Values		Reference
			Adult	Children	
BW	Body weight	kg	70	28	[37,40]
EF	Exposure frequency	d/year	350	350	[36,39]
ED	Exposure duration	years	30	6	[36,39]
$IngR$	ingestion rate	L/day	2	1.5	[37,41]
AT	Average time (cancer)	days	365 × 70	365 × 70	[36,39]
	Average time(non-cancer)	days	365 × ED	365 × ED	[36,39]
CF	Conversion factor	L/cm^3	0.003	0.003	[37,40]
C	Concentration	mg/L	-	-	-

Table 6. Herbicide reference doses and cancer slope factors.

Contaminant	Reference Doses (RfD)	Cancer Slope Factor (CSF)	Reference
Atrazine	0.0035 mg/kg-day	0.23 mg/kg-day	[42]
Metolachlor	0.15 mg/kg-day	0.0092 mg/kg-day	[42]
Simazine	0.005 mg/kg-day	0.12 mg/kg-day	[42]
Terbuthylazine	-	-	-

3. Results and Discussion

3.1. Pesticides in Water Resources

Table 7 displays the rate of detection and concentrations of pesticides found in this study. Minimum, maximum, average, and standard deviation were the summary for descriptive statistics. When a contaminant was detected in only one sample, no standard deviation was calculated. The findings were discussed in terms of their presence in rivers, dams/reservoirs, and treated drinking water. This study targeted four (4) herbicides: atrazine, simazine, terbuthylazine, and metolachlor. The pesticides discovered were classified as triazines (atrazine, simazine, and terbuthylazine) and chloroacetanilides (metolachlor). Triazine herbicides, such as atrazine, simazine, and terbuthylazine, and chloroacetanilide herbicides, such as metolachlor, were targeted in this study because agriculture dominates the landscape of Mangaung Metropolitan Municipality. These compounds are the most commonly used around the world and are associated with serious health effects [4]. The availability of their standards in the laboratory was also one of the influencing factors.

Table 7. Herbicides concentrations in rivers, dams/reservoirs and treated drinking water.

	Concentration (mg/L)		
Compounds	**DF (%)**	**Min-Max**	**Mean ± SD**
	Rivers (n = 5)		
Atrazine	100	0.002–0.06	0.03 ± 0.03
Metolachlor	100	0.003–0.03	0.01 ± 0.01
Simazine	80	<LOQ-5.67	1.82 ± 2.66
Terbuthylazine	100	0.01–0.21	0.06 ± 0.09
	Dams/Reservoirs (n = 5)		
Atrazine	100	0.01–0.03	0.02 ± 0.01
Metolachlor	100	0.002–0.03	0.01 ± 0.01
Simazine	80	<LOQ-0.20	0.12 ± 0.08
Terbuthylazine	100	0.01–0.06	0.03 ± 0.02
	Treated drinking water (n = 2)		
Atrazine	100	0.015–0.02	0.02 ± 0.003
Metolachlor	100	0.009–0.01	0.01 ± 0.001
Simazine	50	<LOQ-0.03	0.03
Terbuthylazine	100	0.019–0.02	0.02 ± 0.001

Notation: DF = detection frequency, n = number of samples, min = minimum concentration, max = maximum concentration, mg/L = milligram per litre, SD = standard deviation.

3.1.1. Pesticides in Rivers

As shown in Table 7, all herbicides had 100% detection frequency, with the exception of simazine, which was detected in 80% of the 5 river samples collected. The mean concentrations of atrazine, metolachlor, simazine, and terbuthylazine were 0.03 mg/L, 0.01 mg/L, 1.82 mg/L, and 0.06 mg/L, respectively. Simazine had the highest concentration among the herbicides, despite being detected in 80% of the collected samples, while metolachlor had the lowest. The detection of pesticides in river water in this study is comparable to other published works, which detected herbicides, such as simazine, in Cape Town, South Africa [26], terbuthylazine in Western Cape, South Africa [24], metolachlor in Hungary [43], and atrazine in Maryland, United States [11]. Most of the rivers in this study pass through the city, industrial areas, and townships. They are also located near the roadside and agricultural fields. Therefore, the presence of herbicides in river water was not unexpected in this study, especially triazine herbicides, which were the most prevalent herbicides in river samples. Triazine herbicides are considered effective and low-cost compounds that are primarily used in crop production [5]. Herbicides are also used to control weeds in industrial areas, along roadsides, and in public squares [6]. The agricultural sector is the backbone of the economy in the Mangaung Metropolitan Municipality. The area is known for producing a lot of maize, soybeans, wheat, sorghum, sunflower, potatoes, groundnuts, and wool. All of these activities necessitate the use of weed control herbicides before, during, and after farming. In addition, the Mangaung Metropolitan Municipality has well-developed roads, public parks, industrial areas, and golf clubs. With all these noticeable areas, significant amounts of herbicides are applied to control weeds in paving, parks, golf courts, roadsides, buildings, and industrial areas. Runoff from agricultural fields, roads, public squares, golf clubs, and industrial areas may increase the concentrations of herbicides in rivers around the municipality. Moreover, trace amounts of herbicides in this study may be introduced into streams by wastewater effluents mostly discharged in rivers.

3.1.2. Pesticides in Dams/Reservoirs

The detection rates of atrazine, simazine, terbuthylazine, and metolachlor in dams/reservoirs were 100%, 80%, 100% and 100%, respectively, as shown in Table 7. Their corresponding average concentrations were 0.02 mg/L, 0.12 mg/L, 0.03 mg/L, and

0.01 mg/L. The concentrations of these herbicides were trending as simazine > terbuthylazine > atrazine = metolachlor. Simazine had the highest mean concentration of any herbicides detected. The present study highlighted the occurrence of herbicides in dams, showing that water pollution by organic compounds is a serious issue. Curchod et al.'s [24] study in Western Cape, South Africa, found concentrations of simazine and terbuthylazine which were lower than the current results. In Eastern Goiás of Brazil [44], atrazine and metolachlor concentrations were lower than the findings of this study, while in the semiarid region of Argentina [9], the detected concentrations of atrazine and metolachlor were higher than the findings of this study. Herbicides are predominantly utilised in agricultural activities, but substantial quantities are also utilised in forestry, industrial areas, public areas, parks, golf courts, and sports grounds for weed control and maintenance [5,6]. Pandey et al. [45] also stated that these herbicides can be used to control invasive plants in water. As a result, the presence of these herbicides in dams/reservoirs may be attributed to their use to control aquatic weeds, such as algae and submerged weeds. Some dams/reservoirs are distinguished by well-kept large open spaces with lawns and turf grasses near bodies of water for picnics. Herbicides may be required to suppress and control annual and perennial broadleaf and grassy weeds in these large open spaces with lawn and turf grasses. As a result, runoff from these lawns and turf grass areas may contaminate dams with simazine, terbuthylazine, atrazine, and metolachlor. According to Wang et al. [5], herbicides can be useful prior to and after cultivation to restrict broadleaf and grassy weeds in agricultural fields. Given that the majority of the dams/reservoirs are surrounded by and located near agricultural fields, runoff from those sites may possibly be a contributing factor. The use of these herbicides in dams may endanger the aquatic life and water consumers [5]. Moreover, rivers that discharge their water in dams may also introduce trace amounts of herbicides in dams. This is because most rivers receive effluents from wastewater treatment works, which receive water from various areas and are a source of many pollutants, including herbicides.

3.1.3. Pesticides in Treated Drinking Water

As presented in Table 7, a 100% detection rate for atrazine, metolachlor, and terbuthylazine was recorded, whereas a 50% detection rate was observed for simazine. The mean concentrations of atrazine, metolachlor, simazine and terbuthylazine were 0.02 mg/L, 0.01 mg/L, 0.03 mg/L and 0.02 mg/L, respectively. Although not detected in all samples, the concentration of simazine was higher than that of all the detected herbicides in treated drinking water. From their measured concentration, these herbicides were trending as simazine > atrazine = tebuthylazine > metolachlor. The occurrence of herbicides such as atrazine and terbuthylazine in treated drinking water was also reported by Odendaal et al. [46] in their study aimed at determining contaminants of emerging concern in drinking water in South Africa. Machete and Shadung [47] also reported the occurrence of atrazine and terbuthylazine in treated drinking water in the Vals and Renoster catchment, South Africa. The use of herbicides to control weeds in dams/reservoirs used as a source of water in water treatment works may be the source of these herbicides. Rivers that discharge their water into these dams may contain traces of herbicides, as they mostly receive wastewater effluents. Atrazine detection in water sources may also be connected to their persistent nature. Almberg et al. [48] reported that herbicides such as atrazine are persistent in soil and their transport to water, making it the most commonly detected pesticide in water sources. Moreover, the presence of pesticides in treated drinking water revealed that the methods used in selected water treatment works are incapable of removing these compounds, which include abstraction, macro/micro sieving, coagulation, flocculation, sedimentation, filtration, and disinfection. Despite the widespread use of pesticides in South Africa, many of the country's registered pesticides products have not been re-evaluated since their initial approval [49]. Among the detected pesticides in this study, South Africa only has water quality guidelines for the protection of human health and aquatic environments against atrazine, which is 0.01 mg/L [49,50]. Atrazine concentration was above the Republic of South African (RSA) acceptable limit. In order to protect public health from serious health effects, the World Health Organisation (WHO) has also implemented pesticides guideline levels in drinking water [51]. The mean concentration of herbicides in this study were above their corresponding World Health Organisation guidelines, except metolachlor, which was equal to its corresponding value. These findings show a negative situation and a serious health concern for the Mangaung community because chronic pesticide exposure through water ingestion can have serious health implications [4]. Table 8 compares herbicide detection to the World Health Organisation and South African guideline values.

Table 8. Concentration of pesticides in treated drinking water with guidelines values [49,51].

Pesticides	Guideline Value		Concentration (mg/L)		Reference
	RSA (mg/L)	WHO (mg/L)	Maximum	Mean	
Atrazine	0.01	0.002	0.019	0.02	[49,51]
Metolachlor	-	0.01	0.01	0.01	[51]
Simazine	-	0.002	0.03	0.03	[51]
Terbuthylazine	-	0.007	0.02	0.02	[51]

3.2. Environmental Risk Assessment

The mean values of measured environmental concentrations were utilised to measure the ecological effects of herbicides on aquatic organisms. The risks of contaminants with concentrations below the limit of quantification (LOQ) were not assessed. In the absence of effective concentration (*EC*) values, lethal concentration (LC_{50}) values were used. In cases where no observed effect concentration (*NOEC*) values were unavailable, the lowest observed effect concentration (*LOEC*) values were used [19,52]. As three representative organisms of the aquatic environment, the predicted non-effect concentrations for algae, invertebrates, and fish were used. In the event of data gaps, different species and endpoints were included. The results of risk assessment based on three representative organisms are shown in Table 9.

Table 9. Ecological risks of pesticides in water resources.

Contaminant	Taxonomic Class	Acute Toxicity		Chronic Toxicity	
		HQ Mean	Risk	HQ Mean	Risk
		River			
Atrazine	Algae [a]	508	High	30	High
	Invertebrate [b]	4.35	High	15	High
	Fish [c]	6.67	High	0.15	Medium
Metolachlor	Algae [a]	0.17	Medium	0.33	Medium
	Invertebrate [b]	0.5	Medium	0.17	Medium
	Fish [c]	2.5	High	0.1	Medium
Simazine	Algae [a]	45,500	High	303	High
	Invertebrate [b]	1820	High	182	High
	Fish [c]	202	High	26	High
Terbuthylazine	Algae [a]	3000	High	-	-
	Invertebrate [b]	1.54	High	14.28	High
	Fish [c]	16.67	High	46.15	High
		Dams/Reservoirs			
Atrazine	Algae [a]	339	High	20	High
	Invertebrate [b]	2.9	High	10	High
	Fish [c]	4.44	High	0.1	Medium
Metolachlor	Algae [a]	0.17	Medium	0.33	Medium
	Invertebrate [b]	0.5	Medium	0.17	Medium
	Fish [c]	2.5	High	0.1	Medium

Table 9. *Cont.*

Contaminant	Taxonomic Class	Acute Toxicity		Chronic Toxicity	
		HQ Mean	Risk	HQ Mean	Risk
Simazine	Algae [a]	2500	High	16.67	High
	Invertebrate [b]	100	High	10	High
	Fish [c]	11	High	1.43	High
Terbuthylazine	Algae [a]	1500	High	-	-
	Invertebrate [b]	0.77	Medium	0.71	Medium
	Fish [c]	8.33	High	2.31	High
		Treated drinking water			
Atrazine	Algae [a]	339	High	20	High
	Invertebrate [b]	2.89	High	10	High
	Fish [c]	4.44	High	0.1	Medium
Metolachlor	Algae [a]	0.17	Medium	0.33	Medium
	Invertebrate [b]	0.5	Medium	0.17	Medium
	Fish [c]	2.5	High	0.1	Medium
Simazine	Algae [a]	750	High	5	High
	Invertebrate [b]	30	High	3	High
	Fish [c]	3.33	High	0.43	Medium
Terbuthylazine	Algae [a]	1000	High	-	-
	Invertebrate [b]	0.77	Medium	4.76	High
	Fish [c,d]	5.55	High	0.15	Medium

Notation: HQ: hazard quotient. Acute and chronic toxicity data of selected species were extracted from data available in the literature. [a] *P. subcapitata;* [b] *Daphnia magna;* [c] *Oncorhynchus mykiss;* [d] *Mysid shrimp* [34,35].

The study looked at both acute and chronic toxicities on aquatic organisms. In all water media studied, the risk of toxicity ranged from medium to high. In terms of acute toxicity, atrazine and simazine posed high risks to all aquatic organisms in all water media. Furthermore, high terbuthylazine risks were noticed in all aquatic organisms in rivers. However, in dams/reservoirs and treated drinking water, it posed a high risk to algae and fish. Metolachlor posed a high risk only to fish in all water sources. Remarkably, simazine was found to pose the greatest ecological risk in rivers, with risk quotient (*RQ*) values of 45,500, 1820, and 202 for algae, invertebrates, and fish, respectively.

In chronic toxicity, simazine showed a high risk to all aquatic organisms in rivers and dams. In treated drinking water, its high risks were noticed for algae and invertebrates. Algae and invertebrates were also sensitive to atrazine, with risk quotient (*RQ*) values greater than one in all water media. Terbuthylazine posed high risks to invertebrates and fish in rivers. In dams/reservoirs, it only posed a high risk to fish, while in treated drinking water, its high risk was observed for invertebrates. Metolachlor did not show high risk to all aquatic life in all water media. Moreover, in treated drinking water, none of the herbicides posed a high risk to fish. The greatest environmental risk was observed in river water for simazine with risk quotient (*RQ*) values of 303, 182, and 26 for algae, invertebrates, and fish, respectively.

The high risks of herbicides, such as simazine, atrazine, and terbuthylazine in water resources show an undesirable condition for the aquatic ecosystem in the Mangaung Metropolitan Municipality. Furthermore, unobserved pesticides and their mixtures may pose greater risks than those observed in this study [19]. These outcomes are expected to have an undesirable effect on aquatic life and human beings. There is evidence showing that exposure to simazine may cause weight changes as well as effects on the serum and thyroid gland [53]. Terbuthylazine may have an effect on carp growth rate, early ontogeny, and antioxidant enzyme [49]. Atrazine exposure may increase the risk of cancer, reproductive problems, and antibiotic resistance. In animals, metolachlor can cause salivation, lacrimation, and convulsions [54,55]. Crop farming, which accounts for 67.5% of all agricultural activities in the Mangaung Metro, may have exacerbated herbicide concentrations in water resources, which lead to their high ecological risks. Therefore, these pesticides should be prioritised and included in municipal environmental monitoring plans. Furthermore, the qualitative screening of unknown pesticides should also be considered in the near future. This will help to obtain a full understanding of the pesticide contaminants and associated risks in water resources within the Mangaung Metro.

3.3. Human Health Risk Assessment

Water sources, such as rivers, dams/reservoirs, and treated drinking water, play an important role in the lives of people within the Mangaung Metropolitan Municipality. In rural areas under the authority of Mangaung, these water sources serve as a source of water for various domestic purposes, including drinking. Therefore, the health risks of pesticides were focused on oral ingestion. This is because ingestion remains a major risk among the various routes of exposure [36,37,39]. The risks of pesticides with unknown cancer slope factors were not assessed. Tables 10 and 11 present the results of non-carcinogenic and carcinogenic risks, respectively.

Table 10. Non-carcinogenic risks of pesticides in water resources.

Compound	Rivers		Dams/Reservoir		Treated Drinking Water	
	ADD	HQ	ADD	HQ	ADD	HQ
			Adults			
Atrazine	8.2×10^{-7}	2.3×10^{-7}	5.5×10^{-7}	1.6×10^{-4}	5.5×10^{-7}	1.6×10^{-4}
Metolachlor	2.7×10^{-7}	1.8×10^{-6}	2.7×10^{-7}	1.8×10^{-6}	2.7×10^{-7}	1.8×10^{-6}
Simazine	5×10^{-5}	1.6×10^{-2}	2.7×10^{-6}	5.4×10^{-4}	8.2×10^{-7}	1.6×10^{-4}
Terbuthylazine	1.6×10^{-6}	-	8.2×10^{-7}	-	5.5×10^{-7}	-
Total	5.3×10^{-5}	1.6×10^{-2}	4.3×10^{-6}	7×10^{-4}	2.2×10^{-6}	3.2×10^{-4}
			Children			
Atrazine	1.5×10^{-6}	4.3×10^{-4}	4.1×10^{-7}	1.2×10^{-4}	4.1×10^{-7}	1.2×10^{-4}
Metolachlor	5×10^{-7}	3.3×10^{-6}	2×10^{-7}	1.3×10^{-6}	2×10^{-7}	1.3×10^{-6}
Simazine	9.3×10^{-5}	1.9×10^{-2}	2.1×10^{-6}	4.2×10^{-4}	6.1×10^{-7}	1.2×10^{-4}
Terbuthylazine	3.1×10^{-6}	-	6.1×10^{-7}	-	4.1×10^{-7}	-
Total	9.8×10^{-5}	1.9×10^{-2}	3.3×10^{-6}	5.4×10^{-4}	1.6×10^{-7}	2.4×10^{-4}

Notation: ADD: average daily dose; HQ: hazard quotient.

Table 11. Carcinogenic risks of pesticides in water resources.

Compound	Rivers		Dams/Reservoir		Treated Drinking Water	
	ADD	CR	ADD	CR	ADD	CR
			Adults			
Atrazine	3.5×10^{-7}	8×10^{-8}	2.3×10^{-7}	5.3×10^{-8}	2.3×10^{-7}	5.3×10^{-8}
Metolachlor	1.2×10^{-7}	1.1×10^{-9}	1.2×10^{-5}	1.1×10^{-7}	1.2×10^{-7}	1.1×10^{-9}
Simazine	2.1×10^{-5}	2.5×10^{-6}	1.2×10^{-6}	1.4×10^{-7}	3.5×10^{-8}	4.2×10^{-9}
Terbuthylazine	7×10^{-7}	-	3.5×10^{-7}	-	2.3×10^{-7}	-
Total	2.2×10^{-5}	2.5×10^{-6}	1.4×10^{-5}	3×10^{-7}	6.5×10^{-7}	5.8×10^{-8}
			Children			
Atrazine	1.3×10^{-7}	2.9×10^{-8}	8.8×10^{-8}	2×10^{-8}	8.8×10^{-8}	2×10^{-8}
Metolachlor	4.3×10^{-8}	3.9×10^{-10}	4.3×10^{-8}	3.9×10^{-10}	4.3×10^{-8}	3.9×10^{-10}
Simazine	8×10^{-6}	9.6×10^{-7}	4.3×10^{-7}	5.2×10^{-8}	1.3×10^{-8}	1.5×10^{-9}
Terbuthylazine	2.6×10^{-7}	-	1.2×10^{-7}	-	4.3×10^{-7}	-
Total	8.4×10^{-6}	9.9×10^{-7}	6.8×10^{-7}	7.2×10^{-8}	5.7×10^{-7}	2.2×10^{-8}

Notation: ADD: average daily dose; CR: cancer risk.

Pesticides in the water resources of the Mangaung Metropolitan Municipality were assessed for human health risks. In terms of non-carcinogenic risk, all pesticides posed low risks to both adults and children, with hazard quotient values less than one (Table 10). These values indicate that the Mangaung Metropolitan Municipality community is free of non-carcinogenic effects caused by the identified herbicides (atrazine, simazine, and metolachlor) exposure in water resources. In terms of carcinogenic risk, all pesticides in the river water showed very low risk to adults and children, with the exception of simazine, which showed medium risk to adults (Table 11). Furthermore, the presence of all pesticides posed a very low risk to adults and low risk to children in dams and treated drinking water. Identified herbicides in this study may have serious health consequences for the community, as the water from dams/reservoirs and rivers in the Mangaung Metropolitan Municipality is used to support farmers and local communities. Treated drinking water is also used for sanitary and household purposes. Furthermore, while the level of risk in this study was low to medium, it may be harmful to the community in the near future due to the continuous introduction of these contaminants. Unintended contact, the bioaccumulation of pesticide residues in fish and locally grown crops, and biomagnification in the food chain can cause considerable risks to the community of the Mangaung Metropolitan Municipality as a result of water polluted by these compounds [19]. Furthermore, herbicide degradation may yield one or more complex transformation products that may be more tenacious or noxious than the original compound [56]. The various health effects associated with herbicide exposure include, but are not limited to, oxidative stress, cytotoxicity, dopaminergic effects, sexual maturation delays, breast cancer, reproductive, and endocrine effects [3]. Human health risk assessment studies of pesticides in water sources are critical in this community for regulatory purposes [19] as well as the protection of public health and the country's limited water resources.

4. Conclusions

The current project was motivated by the dearth of data on the level of herbicides in water sources around the Mangaung Metropolitan Municipality and their associated environmental and health risks. To the best of our knowledge, this was the first study on the Mangaung Metropolitan Municipality to address such issues. The study revealed that herbicides are present in rivers, dams/reservoirs and treated drinking water in this area. The presence of these herbicides shows a possibility to cause high ecological risks and medium carcinogenic risks during spring season, demonstrating that aquatic and human beings may be affected by these contaminants. The presence of pesticides in water sources within the Mangaung Metropolitan Municipality is concerning because South Africa is a water-stressed country, and exposure to pesticides may cause serious effects. Therefore, it is strongly advised that interventions aimed at determining the source of contamination be implemented to safeguard water resources and the health of water consumers and aquatic organisms. The findings of the study may serve as an awareness to the community and relevant stakeholders on the environmental and public health risks of pesticide exposure. It will also aid in the development of municipal pollution management and risk-reduction strategies for herbicides and other toxic compounds. Some limitations of this study include the fact that it only focused on a few pesticide families, which should be expanded in future studies to obtain an overall understanding of the risks associated with them. Furthermore, the predicted non-effect concentration values are based on currently available data and may change as more reliable data become available.

Author Contributions: Conceptualization, I.M., S.A.O., T.P.G. and S.S.; methodology, I.M., S.A.O., T.P.G. and S.S.; software, S.S., I.M. and S.A.O.; validation, I.M., S.A.O. and T.P.G.; formal analysis, I.M.; investigation, I.M., S.A.O. and S.S.; resources, S.A.O.; data curation, I.M.; writing—original draft preparation, I.M., S.A.O. and S.S.; writing—review and editing, T.P.G.; visualization, S.S.; supervision, T.P.G. and S.A.O.; project administration, S.A.O.; funding acquisition, S.A.O. All authors have read and agreed to the published version of the manuscript.

Funding: This work was funded by the Water Research Commission (WRC), Project No.: 2022/2023-00791. It was also financed by research development and postgraduate studies at the Central University of Technology, Free State (CUT). Part of the APC was also received from the DSI-NRF Centre of Excellence (CoE) in Human Development at the University of the Witwatersrand, Johannesburg, South Africa.

Institutional Review Board Statement: Not applicable.

Informed Consent Statement: Appropriate permissions were obtained and followed during this study. Relevant national and institutional guidelines were also followed and complied with.

Data Availability Statement: Available on request from the corresponding author.

Acknowledgments: The authors gratefully acknowledge the financial support from WRC, Project No.: 2022/2023-00791 and the Central University of Technology, Free State. This research was also supported by the DSI-NRF Centre of Excellence (CoE) in Human Development at the University of the Witwatersrand, Johannesburg, South Africa. The content is solely the responsibility of the authors and does not reflect the views of the aforementioned institutions. We acknowledge our research team member Elias K. Mphosho, who contributed to the field data collection.

Conflicts of Interest: The authors declare no conflict of interest.

References

1. Husk, B.; Sanchez, J.S.; Leduc, R.; Takser, L.; Savary, O.; Cabana, H. Pharmaceuticals and pesticides in rural community drinking water of Quebec, Canada—A regional study on the susceptibility to source contamination. *Water Qual. Res. J.* **2019**, *54*, 88–103. [CrossRef]
2. Bai, X.; Lutz, A.; Carroll, R.; Keteles, K.; Dahlin, K.; Murphy, M.; Nguyen, D. Occurrence, distribution, and seasonality of emerging contaminants in urban watersheds. *Chemosphere* **2018**, *200*, 133–142. [CrossRef] [PubMed]
3. Nicolopoulou-Stamati, P.; Maipas, S.; Kotampasi, C.; Stamatis, P.; Hens, L. Chemical pesticides and human health: The urgent need for a new concept in agriculture. *Front. Public Health* **2016**, *4*, 148. [CrossRef]
4. Syafrudin, M.; Kristanti, R.A.; Yuniarto, A.; Hadibarata, T.; Rhee, J.; Al-Onazi, W.A.; Algarni, T.S.; Almarri, A.H.; Al-Mohaimeed, A.M. Pesticides in drinking water—A review. *Int. J. Environ. Res. Public Health* **2021**, *18*, 468. [CrossRef] [PubMed]
5. Wang, M.; Lv, J.; Deng, H.; Liu, Q.; Liang, S. Occurrence and Removal of Triazine Herbicides during Wastewater Treatment Processes and Their Environmental Impact on Aquatic Life. *Int. J. Environ. Res. Public Health* **2022**, *19*, 4557. [CrossRef] [PubMed]
6. Stevanović, M.; Gasic, S. Herbicides in surface water bodies—Behaviour, effects on aquatic organisms and risk assessment. *Pestic. Phytomed.* **2019**, *34*, 157–172. [CrossRef]
7. Goncalves-Filho, D.; Silva, C.C.G.; De Souza, D. Pesticides determination in food and natural waters using solid amalgambased electrodes: Challenges and trends. *Talanta* **2020**, *212*, 120756. [CrossRef] [PubMed]
8. Kruc-Fijalkowska, R.; Dragon, K.; Drozdzynski, D.; Gorski, J. Seasonal variation of pesticides in surface water and drinking water wells in the annual cycle in western Poland, and potential health risk assessment. *Sci. Rep.* **2022**, *12*, 3317. [CrossRef]
9. Mas, L.I.; Aparicio, V.; De Gerónimo, E.; Costa, J.L. Pesticides in water sources used for human consumption in the semiarid region of Argentina. *SN Appl. Sci.* **2020**, *2*, 691. [CrossRef]
10. Chidya, C.G.R.; Abdel-dayem, S.M.; Takeda, K.; Sakugawa, H. Spatio-temporal variations of selected pesticide residues in the Kurose River in Higashi-Hiroshima city, Japan. *J. Environ. Sci. Health Part B* **2018**, *53*, 602–614. [CrossRef]
11. Panthi, S.; Sapkota, A.R.; Raspanti, G.; Allard, S.M.; Bui, A.; Craddock, H.A.; Murray, R.; Zhu, L.; East, C.; Handy, E.; et al. Pharmaceuticals, herbicides, and disinfectants in agricultural water sources. *Environ. Res.* **2019**, *174*, 1–8. [CrossRef]
12. Ganaie, M.I.; Jan, I.; Mayer, A.N.; Dar, A.A.; Mayer, I.A.; Ahmed, P.; Sofi, J.A. Health Risk Assessment of Pesticide Residues in Drinking Water of Upper Jhelum Region in Kashmir Valley-India by GC-MS/MS. *Int. J. Anal. Chem.* **2023**, *27*, 6802782. [CrossRef] [PubMed]
13. Medkova, D.; Hollerova, A.; Riesova, B.; Blahova, J.; Hodkovicova, N.; Marsalek, P.; Doubkova, V.; Weiserova, Z.; Mares, J.; Faldyna, M.; et al. Pesticides and Parabens Contaminating Aquatic Environment: Acute and Sub-Chronic Toxicity towards Early-Life Stages of Freshwater Fish and Amphibians. *Toxics* **2023**, *11*, 333. [CrossRef] [PubMed]
14. Zheng, S.; Chen, B.; Qiu, X.; Chen, M.; Ma, Z.; Yu, X. Distribution and risk assessment of 82 pesticides in Jiulong River and estuary. *Chemosphere* **2016**, *144*, 1177–1192. [CrossRef] [PubMed]
15. Elfikrie, N.; Ho, Y.B.; Zaidon, S.Z.; Juahir, H.; Tan, E.S.S. Occurrence of pesticides in surface water, pesticides removal efficiency in drinking water treatment plant and potential health risk to consumers in Tengi River Basin, Malaysia. *Sci. Total Environ.* **2020**, *712*, 136540. [CrossRef]
16. European Union (EU). Council directive 98/83/EC of 3 November 1998 on the quality of water intended for human consumption. *Off. J. Eur. Union L* **1998**, *330*, 32–54. Available online: https://eur-lex.europa.eu/legal-content/EN/TXT/PDF/?uri=CELEX:01998L0083-20151027&FROM=EN (accessed on 25 December 2022).
17. European Union (EU). European Union Directive 2008/105/EC of the European Parliament and of the Council on environmental quality standards in the field of water policy. *Off. J. Eur. Union L* **2008**, *348*, 84–97.
18. Graefe, J.J. Glyphosate's Fate: Comparing strategies for the precautionary cancellation of glyphosate registrations in the United States and the European Union. *Conn. J. Int. Law* **2019**, *35*, 248.
19. Abera, B.; Van Echelpoel, W.; De Cock, A.; Tytgat, B.; Kibret, M.; Spanoghe, P.; Mengistu, D.; Adgo, E.; Nyssen, J.; Goethals, P.L.M.; et al. Environmental and Human Health Risks of Pesticides Presence in the Lake Tana Basin (Ethiopia). *Sustainability* **2022**, *14*, 14008. [CrossRef]

20. Food and Agricultural organisation (FAO). *OECD-FAO Agricultural Outlook 2018–2027*; OECD Publishing: Paris, France; Food and Agriculture Organization of the United Nations: Rome, Italy, 2018. Available online: https://www.fao.org/3/I9166EN/I9166E (accessed on 24 December 2022).
21. Fuhrimann, S.; Wan, C.; Blouzard, E.; Veludo, A.; Holtman, Z.; Chetty-Mhlanga, S.; Dalvie, M.A.; Atuhaire, A.; Kromhout, H.; Röösli, M.; et al. Pesticides Research on Environmental and Human Exposure and Risks in Sub-Saharan Africa: A Systematic Literature Review. *Int. J. Environ. Res. Public Health* **2022**, *19*, 259. [CrossRef]
22. Tang, F.H.M.; Lenzen, M.; McBratney, A.; Maggi, F. Risk of pesticidespollution at a global scale. *Nat. Geosci.* **2021**, *14*, 206–210. [CrossRef]
23. Degrendele, C.; Prokes, R.; Senk, P.; Jikova, S.R.; Kohoutek, J.; Melymuk, L.; Pribylova, P.; Dalvie, M.A.; Roosli, M.; Klanova, J.; et al. Human exposure to pesticides in dust from two agricultural sites in South Africa. *Toxics* **2022**, *10*, 629. [CrossRef] [PubMed]
24. Curchod, L.; Oltramare, C.; Junghans, M.; Stamm, C.; Dalvie, M.A.; Roosli, M.; Fuhrimann, S. Temporal variation of pesticides mixtures in rivers of three agricultural watersheds during a major drought in the Western Cape, South Africa. *Water Res.* **2020**, *6*, 100039. [CrossRef]
25. Horn, S.; Pieters, R.; Bohn, T. A first assessment of glyphosate, 2,4-D and Cry proteins in surface water of South Africa. *S. Afr. J. Sci.* **2019**, *115*, 1–7. [CrossRef]
26. Ojemaye, C.Y. Identification and Quantification of Chemicals of Emerging Concern (Persistent Organic and Inorganic Pollutants) in Some Selected Marine Environments of Cape Town, South Africa. Ph.D. Thesis, University of the Western Cape, South Africa, 2020.
27. Spath, J.; Arumugam, P.; Lindberg, R.H.; Abafe, A.O.; Jansson, S.; Fick, J.; Buckly, C.A. Biochar for the removal of detected micropollutants in South African domestic wastewater: A case study from a demonstration-scale decentralised wastewater treatment system in eThekwini. *Water SA* **2021**, *47*, 396–416.
28. Free State Development Corporation. Agriculture and Agro Processing: Annual Report 2017/208. 2022. Available online: Https://www.fdc.co.za/index.php/addons/fs.economic-sectors/agriculture-agro-processing (accessed on 22 December 2022).
29. Mangaung Metropolitan Municipality Integrated Development Plan (MMM). "Final Integrated Plan 2022/2027". 2022. Available online: http://www.mangaung.co.za/wp-content/uploads/2022/05/2-Final-IDP-2022-2027-24.05.2022-Signed.pdf (accessed on 30 August 2022).
30. Archer, E.; Wolfaardt, G.M.; Van Wyk, J.H. Pharmaceutical and personal care products (PPCPs) as endocrine disrupting contaminants (EDCs) in South African surface waters. *Water SA* **2017**, *43*, 1–23. [CrossRef]
31. Vasilachi, I.C.; Asiminicesei, D.M.; Fertu, D.I.; Gavrilescu, M. Occurrence and Fate of Emerging Pollutants in Water Environment and Options for Their Removal. *Water* **2021**, *13*, 181. [CrossRef]
32. Nannou, C.; Kaprara, E.; Psaltou, S.; Salapasidou, M.; Palasantza, P.A.; Diamantopoulos, P.; Lambropoulou, D.A.; Mitrakas, M.; Zouboulis, A. Monitoring of a broad set of pharmaceuticals in wastewaters by high-resolution mass spectrometry and evaluation of heterogenous catalytic ozonation for their removal in a pre-industrial level unit. *Analytica* **2022**, *3*, 195–212. [CrossRef]
33. Pei, S.; Li, B.; Wang, B.; Liu, J.; Song, X. Distribution and Ecological Risk Assessment of Pharmaceuticals and Personal Care Products in Sediments of North Canal, China. *Water* **2022**, *14*, 1999. [CrossRef]
34. Kock-Schulmeyer, M.; Ginebreda, A.; Gonzalez, S.; Cortina, J.L.; De Alda, M.L.; Barcelo, D. Analysis of the occurrence and risk assessment of polar pesticides in the Llobregat River Basin (NE Spain). *Chemosphere* **2012**, *86*, 8–16. [CrossRef]
35. Australian National Registration Authority for Agricultural and Veterinary chemicals (NRA). *Evaluation of the Active Terbuthylazine in the Product Swimcareot Swimming Pool Algaecide*; National Registration Authority for Agricultural and Veterinary Chemicals: Armidale, NSW, Australia, 2001. Available online: https://apvma.gov.au/sites/default/files/publication/14061-prs-terbuthylazine.pdf (accessed on 24 December 2022).
36. USEPA (United State Environmental Protection Agency). *Risk Assessment Guidance for Superfund Volume I: Human Health Evaluation Manual (Part E, Supplemental Guidance for Dermal Risk Assessment)*; USEPA: Washington, DC, USA, 2004. Available online: https://www.epa.gov/risk/risk-assessment-guidance-superfund-rags-part-e (accessed on 10 November 2022).
37. Kamunda, C.; Mathuthu, M.; Madhuku, M. Potential human risk of dissolved heavy metals in gold mine waters of Gauteng Province, South Africa. *J. Toxicol. Environ. Health Sci.* **2018**, *10*, 56–63.
38. Tian, M.; Li, W.; Ruan, M.; Wei, J.; Ma, W. Water Quality Pollutants and Health Risk assessment for Four Different Drinking Water Sources. In *E3S Web of Conferences*; EDP Sciences: Les Ulis, France, 2019; Volume 78, p. 03004.
39. Mugudamani, I.; Oke, S.A.; Gumede, T.P. Influence of Urban Informal Settlements on Trace Element Accumulation in Road Dust and Their Possible Health Implications in Ekurhuleni Metropolitan Municipality, South Africa. *Toxics* **2022**, *10*, 253. [CrossRef] [PubMed]
40. USEPA (United States Environmental Protection Agency). *Supplemental Guidance for Developing Soil Screening Levels for Superfund Sites*; Office of Emergency and Remedial Response: Washington, DC, USA, 2002. Available online: https://epa.gov/superfund/superfund-soil-screening-guidance (accessed on 3 November 2022).
41. USEPA (United States Environmental Protection Agency). *Chemical Search (IRIS)*; USEPA: Washington, DC, USA, 2016. Available online: https://cfpub.epa.gov/ncea/iris/search/index.cfm?keyword=atrazine (accessed on 17 November 2022).
42. USEPA (United State Environmental Protection Agency). *Regional Screening Levels (RSLs)*; USEPA: Washington, DC, USA, 2021. Available online: https://www.epa.gov/risk/regional-screening-levels-rsls-generic-tables (accessed on 10 April 2022).
43. Székács, A.; Mörtl, M.; Darvas, B. Monitoring PesticidesResidues in Surface and Ground Water in Hungary: Surveys in 1990–2015. *J. Chem.* **2015**, *2015*, 717948. [CrossRef]

44. Correia, M.; Carbonari, C.A.; Velini, E.D. Detection of herbicides in water bodies of the Samambaia River sub-basin in the Federal District and eastern Goiás. *J. Environ. Sci. Health Part B* **2020**, *55*, 574–582. [CrossRef] [PubMed]
45. Pandey, P.; Caudill, J.; Lesmeister, S.; Zheng, Y.; Wang, Y.; Stillway, M.; Hoffmann, K.; Gilbert, P.; Kwong, M.; Conrad, L.; et al. Assessing Glyphosate and Fluridone Concentrations in Water Column and Sediment Leachate. *Front. Environ. Sci.* **2019**, *7*, 22. [CrossRef]
46. Odendaal, C.; Seaman, M.T.; Kemp, G.; Patterton, H.E.; Patterton, H.G. An LC-MS/MS based survey of contaminants of emerging concern in drinking water in South Africa. *S. Afr. J. Sci.* **2015**, *111*, 2014-0401. [CrossRef]
47. Machete, M.; Shadung, J.M. Detection of selected agricultural pesticides in river and tap water in Letsitele, Lomati and Vals–Renoster catchments, South Africa. *Water SA* **2019**, *45*, 716–720. [CrossRef]
48. Almberg, K.S.; Mary, E.M.E.; Jones, R.M.; Rankin, K.; Freels, S.; Stayner, L.T. Atrazine Contamination of Drinking Water and Adverse Birth Outcomes in Community Water Systems with Elevated Atrazine in Ohio, 2006–2008. *Int. J. Environ. Res. Public Health* **2018**, *15*, 1889. [CrossRef]
49. Department of Water Affairs and Forestry (DWAF). South African Water Quality Guidelines Volume 1: Domestic Water Use, Second Edition. 1996. Available online: https://www.dws.gov.za/Groundwater/documents/Pol_saWQguideFRESHDomesticusevol.pdf (accessed on 24 December 2022).
50. Horak, I.; Horn, S.; Pieters, R. Agrochemicals in freshwater systems and their potential as endocrine disrupting chemicals: A South African context. *Environ. Pollut.* **2021**, *268*, 115718. [CrossRef]
51. Hamilton, D.; Dieterle, R.; Felsot, A.; Harris, C.; Holland, P.; Katayama, A.; Kurihara, N.; Linders, J.; Unsworth, J.; Wong, S.S. Regulatory limits for pesticides residues in water (IUPAC Technical Report). *Pure Appl. Chem.* **2003**, *75*, 1123–1155. [CrossRef]
52. Lewis, K.A.; Tzilivakis, J.; Warner, D.J.; Green, A. An international database for pesticides risk assessments and management. *Hum. Ecol. Risk Assess. Int. J.* **2016**, *22*, 1050–1064. [CrossRef]
53. Mendes, K.F.; Régo, A.P.J.; Takeshita, V.; Tornisielo, V.L. Water Resource Pollution by Herbicide Residues. In *Biochemical Toxicology—Heavy Metals and Nanomaterials*; Ince, M., Ince, O.K., Ondrasek, G., Eds.; IntechOpen: London, UK, 2019; pp. 1–16.
54. Velisek, J.; Stara, A.; Koutnik, D.; Machova, J. Effects of terbuthylazine-2-hydroxy at environmental concentration on early stages of common carp (*Cyprinus carpio* L.). *Biomed. Res. Int.* **2014**, *2014*, 621304. [CrossRef] [PubMed]
55. Manahil, M. Endocrine-Disrupting Properties of Pharmaceuticals and Personal Care Products (PPCPS): An Evaluation Using Aquatic Model Organisms. Master's Thesis, Wayne State University, Detroit, MI, USA, 2017.
56. Ccanccapa, A.; Masi, A.; Navarro-Ortega, A.; Pico, Y.; Barcelo, D. Pesticides in the Ebro River basin: Occurrence and risk assessment. *Environ. Pollut.* **2016**, *211*, 414–424. [CrossRef] [PubMed]

toxics

MDPI

Article

Health Risk Assessment of Heavy Metals in Indoor Household Dust in Urban and Rural Areas of Chiang Mai and Lamphun Provinces, Thailand

Kawinwut Somsunun [1,2], Tippawan Prapamontol [1,*], Todsabhorn Kuanpan [1], Teetawat Santijitpakdee [1], Kanyapak Kohsuwan [1], Natwasan Jeytawan [1] and Nathaporn Thongjan [1]

1 Environment and Health Research Group, Research Institute for Health Sciences (RIHES), Chiang Mai University, Chiang Mai 50200, Thailand; kawinwut_s@cmu.ac.th (K.S.); todkuanpan@gmail.com (T.K.); teetawat_san@cmu.ac.th (T.S.); kanyapak_koh@cmu.ac.th (K.K.); natwasan_jeytawan@cmu.ac.th (N.J.); nathaporn_t@cmu.ac.th (N.T.)

2 PhD Degree Program in Environmental Science, Environmental Science Research Center, Faculty of Science, Chiang University, Chiang Mai 50200, Thailand

* Correspondence: tippawan.prapamontol@cmu.ac.th

Abstract: Indoor exposure to heavy metals poses human health risks worldwide, but study reports from Thailand are still limited, particularly in rural and urban areas. We measured the heavy metals in a hundred indoor household dust samples collected from urban and rural areas in Chiang Mai and Lamphun provinces and found a significantly higher concentration of As in rural areas and Cd in urban areas with industrial activities. The source identification of the heavy metals showed significant enrichment from traffic emissions, paint, smoking, and mixed sources with natural soil. From health risk assessment models, children were more vulnerable to noncarcinogenic risks (HI = 1.45), primarily via ingestion (HQ = 1.39). Lifetime cancer risks (LCRs) due to heavy metal exposure were found in adults (LCR = 5.31×10^{-4}) and children (LCR = 9.05×10^{-4}). The cancer risks from As were higher in rural areas via ingestion, while Cr and Ni were higher in urban areas via inhalation and ingestion, respectively. This study estimated that approximately 5 out of 10,000 adults and 9 out of 10,000 children among the population may develop cancer in their lifetime from exposure to indoor heavy metals in this region.

Keywords: indoor household dust; heavy metals; health risk assessment; source identification; urban area; rural area; cancer risk

1. Introduction

Since the start of the COVID-19 pandemic, people around the world have been staying at home progressively more, frequently spending up to 90% of their time indoors [1–3]. Therefore, domestic indoor environments may represent the greatest risk factor for human exposure to indoor contaminants, especially in children, the elderly, and vulnerable people who have lived primarily indoors. According to the World Health Organization [4], in 2020, exposure to residential air pollution resulted in approximately 3.2 million deaths per year. This exposure was associated with the development of noncommunicable diseases, including stroke, ischemic heart disease, and chronic obstructive pulmonary disease (COPD). Additionally, it was determined that approximately 6% of all lung cancer deaths could be attributable to exposure to carcinogens originating from indoor household air pollution [4]. Hence, the identification, characterization, and mitigation of indoor household pollutants are important.

Indoor environments contain many pollutants, including carbon and sulfur oxides, volatile organic compounds, particulate matter, biological particles, radon, and chemicals emitted by furniture and interior decorations [5,6]. Household dust is an important indoor pollutant that contains a variety of organic and inorganic contaminants, including trace heavy metals, which can enter and harm the human body [7].

Over the past few decades, numerous studies have demonstrated that dust samples exhibit elevated concentrations of elemental species and organic pollutants [8–10]. The dispersion of settled dust into the atmosphere can occur through the influence of wind and various natural or human activities that have significant implications for both air quality and human health [11–14]. Furthermore, there is a correlation between the resuspension of settled dust and the presence of particulate matter indoors [15,16]. Heavy metal contamination in household dust arises from various internal and external sources of human activities. These sources include mining, vehicle emissions and transportation, fossil fuel combustion and heating methods, cooking, smoking, painting, agricultural and industrial activities, and natural sources [17,18].

The transfer of heavy metals present in dust to humans can occur through various routes of exposure, including ingestion, inhalation, and dermal absorption [19,20]. The potential health risks associated with high concentrations of toxic metals found in settled dust are a matter of concern due to their acute and chronic toxicity, especially for children and vulnerable individuals, who are more susceptible [21–23]. Heavy metals exhibit significant levels of persistence and biotoxicity, hence, exerting detrimental effects on several organs such as the lungs, kidneys, and other systems in the body associated with cardiovascular, cerebrovascular, skeletal, and neurological problems [24,25]. Arsenic (As) is a highly toxic substance that has been linked to the development of numerous complications in various organ systems of the body, as well as many types of cancers [26,27]. Exposure to cadmium (Cd) has been identified as a potential immunotoxicant [28], with established associations with many cancers and pronounced toxic effects on the liver and kidneys [29]. Chronic exposure and bioaccumulation of chromium (Cr) have been found to induce allergic responses, anemia, and toxicity in the male reproductive system [30]. Inhalation of manganese (Mn) has been found to have detrimental effects on the respiratory, renal, and hepatic systems, ultimately, resulting in the development of a neurological disorder known as manganism [31]. Nickel (Ni) has been associated with several health concerns, including allergies, cardiovascular and kidney problems, and lung fibrosis, as well as an increased risk of developing lung and nasal cancers [32]. Exposure to lead (Pb) has the potential to impact multiple systems inside the body, and it is especially detrimental to the neurological development of young children, as it alters the functioning of the brain and central nervous system [33]. Many studies have indicated that As, Cd, Cr, Pb, and Ni may be linked to cancer development, resulting in their classification as carcinogens [22,23,34–36]. Furthermore, in the body, some heavy metals exhibit a biological half-life exceeding 10–35 years, hence, presenting enduring health risks [28].

Chiang Mai and Lamphun provinces are important areas in upper Northern Thailand, which consist of urban areas, rural areas with agricultural activities, and industrial areas. These urban, industrial, and agricultural areas can be significant sources of heavy metal pollution in the form of dust from various activities. The process of urbanization has a substantial influence on the environmental quality in the area. Therefore, the primary objective of this study was to assess the concentrations of eight heavy metals in the indoor household dust in Chiang Mai and Lamphun provinces, representing urban areas in the City of Chiang Mai, industrial areas near the City of Lamphun, and agricultural areas in the rural areas of both provinces. The second objective was to identify the sources of the heavy metals in the indoor household dust of Chiang Mai and Lamphun provinces. Finally, the third objective was to evaluate a health risk assessment in terms of the carcinogenic and noncarcinogenic risks.

2. Materials and Methods

2.1. Study Area

This study was carried out within the geographical boundaries of Chiang Mai and Lamphun provinces (Figure 1). Chiang Mai and Lamphun cities are located in one of the most developed regions in Thailand, which has been undergoing rapid urbanization over the last few decades [37]. This study was conducted in urban, industrial, and rural areas with agriculture. The urban area in this study consisted of Mueang Chiang Mai district, Nai Mueang Lamphun subdistrict, and some parts of Hang Dong, San Sai, Mae Rim, and San Kamphaeng districts. The industrial area was the Industrial Estate in Lamphun province, the largest industrial area in upper Northern Thailand. These industries have primarily been engaged in the electronics, automotive components, agricultural, and manufacturing sectors. Meanwhile, the other areas of this region remain abundant in agricultural land [38]. Regardless of the forest and mountain areas, except for the urban and industrial areas, the other areas in this region were assumed to be rural areas that could be indicated as agricultural areas in this study.

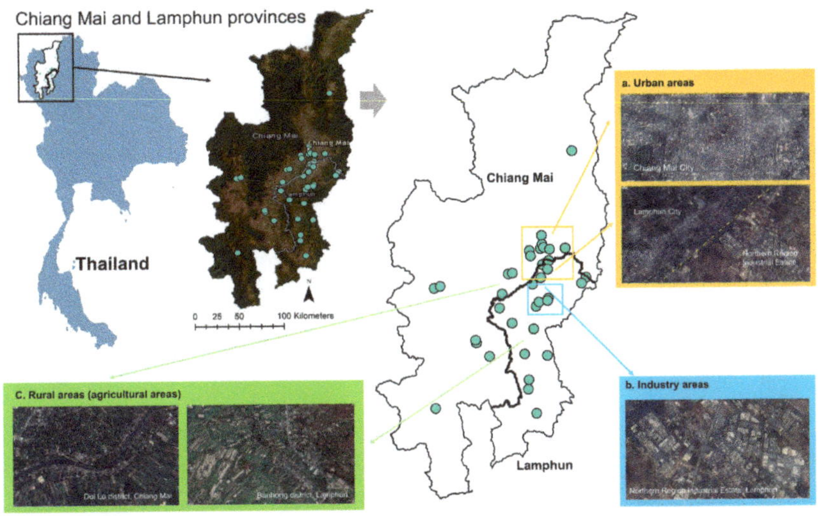

Figure 1. Map of the study site and the locations of collected indoor household dust.

2.2. Samples and Data Collection

Potential locations of houses were geographically mapped and house owners were informed of the study and invited to participate with their informed consent. This study was approved by the Human Experimentation Committee, Research Institute for Health Sciences (study code: Project No. 1/59, approved on: 10 May 2016). Totally, 100 settled indoor household dust samples were collected in the houses of participants. The land use types of urban, industrial, and agricultural areas of the dust-collected houses were identified within a 5 km radius during the period of dust collection. The settled dust was collected using a vacuum cleaner (HITACHI CV-SF18 220C) from surface areas of at least 1 m above the ground level of the living room and bedroom of all participants (and on the surface of furniture). Dust samples from the vacuum cleaner were dried, transferred, and filtered using a 63-micron sieve. After being sieved, approximately 0.5–2 g of dust samples was kept in a clean polyethylene Ziplock bag at −20 °C until analysis. Household characteristics and household activities were recorded during the period of dust collection.

2.3. Sample Analysis

To determine the heavy metal concentrations of As, Cd, Cr, Cu, Mn, Ni, Pb, and Zn in indoor settled house dust, the method of Falciani et al. was modified [39]. Briefly, 150 mg of each dust sample was digested with 5 mL of HNO_3 and 5 mL of HCl in an ETHOS UP High-Performance Microwave Digestion System (Milestone Inc., Sorisole, Italy) using TFM microwave digestion vessels at 210 °C for 40 min. After cooling, the digested samples were diluted to 25 mL with Milli-Q water (18.2 MΩ-cm) and then filtered through a 0.2 μm nylon filter. Digestions and quality controls were analyzed for concentrations of heavy metals in triplicate with an inductively coupled plasma optical spectrometer, or ICP-OES (Agilent 5800, Agilent Scientific Technology Ltd., Santa Clara, California, USA). To prepare the calibration curve, a certified reference material, ICP Multi-Element Standard (CPA chem. Stara Zagora, Bulgaria), was used. The radiofrequency (RF) power was 1.2 kW, the nebulization flow was 0.7 L min^{-1}, and the argon plasma flow was 12.0 L min^{-1}. The blank experiments were conducted by repeating the steps in the sample preparation. The composition of the blank was compared with the sample solution to identify the elemental composition of the heavy metals in the dust. The calibration curves for absorbance and concentration were used to determine the concentrations of heavy metals in every sample with a straight line of $r > 0.999$. To ensure the method's correctness, each heavy metal analysis was carried out in triplicate using standard reference material (SRM2584). The range of the results for the percentage recovery of standard heavy metals was 81.8% to 97.6% (Table S3).

2.4. Enrichment Factor (EF) Calculation

The enrichment factor (*EF*) is usually used to estimate the degree of enrichment of an element in soil and dust samples compared with its abundance in the Earth's crust [40,41]. To determine the anthropogenic input, or the impact of human activities, in the metal values in the settled house dust, the *EF* can identify the origin of each element in the dust sample and distinguish whether that element originated from anthropogenic activity or natural sources [42] based on a reference element that is considered to be stable in its performance and not susceptible to environmental and analysis processes [43]. The *EF* was calculated using the following Equation (1).

$$EF = (C_x/C_{ref})_{sample}/(C_x/C_{ref})_{background} \quad (1)$$

where C_x is the examined metal concentration, and C_{ref} is the reference metal concentration for normalization. In this study, Mn was applied as the reference metal [14]. The background values of the chemical elements in the continental crust followed Taylor's report [44].

2.5. Health Risk Assessment

The exposure dose to the heavy metals in the house dust was estimated via three routes of exposure [45]. The model developed by the US Environmental Protection Agency [46] was used to calculate the exposure to metals in settled house dust. The average daily doses (*ADDs*, mg kg^{-1} day^{-1}) of heavy metals in household dust via ingestion, dermal contact, and inhalation as exposure pathways were calculated separately using Equations (2)–(4) for noncarcinogenic risk and (5)–(7) for carcinogenic risk [14,47]. The average daily doses (*ADDs*, mg kg^{-1} day^{-1}) for noncarcinogenic risk were calculated as follows:

$$ADDingest = C \times \frac{IngR \times ExF \times ED}{BW \times AT} \times CF \quad (2)$$

$$ADDinhal = C \times \left[\frac{InhR \times ExF \times ET \times ED}{PEF \times BW \times AT}\right] \quad (3)$$

$$ADDdermal = C \times \left[\frac{SA \times SL \times ABS \times ExF \times ED}{BW \times AT}\right] \times CF \quad (4)$$

The average daily doses (*ADDs*, mg kg^{-1} day^{-1}) for carcinogenic risk were calculated as follows:

$$ADDingest_{carcinogenic} = C \times \left[\frac{IR \times ExF}{AT}\right] \times CF \quad (5)$$

$$ADDinhal_{carcinogenic} = C \times \left[\frac{ExF \times ET \times ED}{PEF \times 24 \times AT} \right] \times 1000 \quad (6)$$

$$ADDdermal_{carcinogenic} = C \times \left[\frac{ABS \times ExF \times DFS}{AT} \right] \times CF \quad (7)$$

where C is the measured concentration of heavy metals in household dust (mg kg^{-1}), and *IngR* and *InhR* are the ingestion and inhalation rates, respectively; *ExF* is the exposure frequency (day $year^{-1}$); *ED* is the exposure duration, represented by years of stay in the house (years), obtained from the questionnaire; *BW* is the body weight (kg); *AT* is the average time stay in the house per day; *PEF* is the particle emission factor; *SL* is the skin adherence factor; *SA* is the exposed skin area; and *ABS* is the dermal absorption factor. The values of the parameters are listed in Table S1.

Subsequently, the hazard index and cancer risk methods were used to assess the health risks of heavy metal exposure to household dust. The calculated doses of each metal for the three exposure pathways were compared to their corresponding reference dose (*RfD*) (mg kg^{-1} day^{-1}) to yield a hazard quotient (*HQ*), as shown in Equation (8), and then summed to obtain the total noncarcinogenic risk of all pathways using the hazard index (*HI*), as shown in Equation (9) [46]. For noncarcinogenic risk, if the *HQ* or *HI* < 1, this indicates that there is no significant risk effect. In contrast, if the *HQ* or $HI \geq 1$, there is possibly a noncarcinogenic effect, which tends to increase in effect when the *HQ* or *HI* increases [46].

$$HQ = \frac{ADD}{RfD} \quad (8)$$

$$HI = \sum HQ_{Ingest} + HQ_{Inhal} + HQ_{Dermal} \quad (9)$$

While the carcinogenic risk, or cancer risk (*CR*), estimates the carcinogenic effects over the lifetime, the dose was multiplied by the corresponding cancer slope factor (*SF*) $[(mg\ kg^{-1}\ day^{-1})]^{-1}$. Total cancer risks, or lifetime cancer risk (*LCR*), were assessed by summing the cancer risk of each exposure route [13,46], which is defined as follows:

$$CR = ADD \times SF \quad (10)$$

$$LCR = \sum CR_{Ingest} + CR_{Inhal} + CR_{Dermal} \quad (11)$$

For the carcinogenic risk, if the *CR* or *LCR* is in a range between 1×10^{-6} and 1×10^{-4}, this can indicate an acceptable or tolerable risk; if higher than 1×10^{-4}, this suggests that, at least, 1 in 10,000 people may develop any cancer from lifetime exposure [14]. The reference values for the noncarcinogenic and carcinogenic risk assessment are shown in Table S2.

2.6. Source Apportionment Model

In environmental research, positive matrix factorization (PMF) is typically utilized to identify and allocate pollution sources according to the composition of the pollutants measured at a certain location. It is frequently used to evaluate complex combinations of contaminants and to determine how various sources contribute to the total contamination. To evaluate the source apportionment of heavy metals in indoor settled household dust, PMF was employed. According to the US EPA, the PMF 5.0 user manual (version 5.0.14) software was used. Source identification was established according to representative elements of base run factors that were selected by sample concentration. To determine the source contributions and component profiles of the pollutant sources under non-negative limitations, the concentration of the elements and uncertainty values were employed in PMF [48], as per the following Equation (12):

$$x_{ij} = \sum_{k=1}^{p} \left(g_{ik} f_{kj} + e_{ij} \right) \quad (12)$$

where X_{ij} is the concentration of a species, g_{ik} is the factor contribution, f_{kj} is the factor profile, e_{ij} is the residual matrix, p is the factor number, i is the sample number, and j is the species of element.

To produce the ideal matrices of G and F by repeatedly breaking down the heavy metal concentration matrix X, PMF minimizes the objective function Q, as per the following Equation (13):

$$Q = \sum_{i=1}^{n}\sum_{j=1}^{m}\left[\frac{e_{ij}}{u_{ij}}\right]^2 \quad (13)$$

The uncertainty is calculated using a fixed fraction of the method detection limit (*MDL*) according to the EPA PMF 5.0 user guide, as shown in Equations (14) and (15).

When concentration less than the *MDL*:

$$u_{ij} = \frac{5}{6} \times MDL \quad (14)$$

When concentration more than *MDL*:

$$u_{ij} = \sqrt{error\ fraction \times c + MDL} \quad (15)$$

The numbers of the F (factor profile) and G (factor contribution) matrices are examined with FPEAK and classified as "rotational ambiguity", which can be used to calculate the minimum Q value. PMF could be performed on five FPEAK runs in this dataset. Every variable had a strong signal-to-noise ratio and no missing values or outliers in the samples. Thirty runs of the model yielded five possible source assumptions. Model residual analysis and model diagnostics were utilized to ascertain the five possible components. The mapping of the bootstrap factors to the base factors was over 80%, indicating that the bootstrap uncertainties could be interpreted and appropriated to the number of factors.

2.7. Statistical Analysis

According to the normality distribution test, most of the heavy metal levels showed a non-normal distribution ($p < 0.001$, except for Mn). Therefore, nonparametric tests, including the Kruskal–Wallis and Mann–Whitney tests, were used to compare the median values of the datasets. A p-value of <0.05 shows significant differences between the median of the compared groups.

3. Results and Discussion

3.1. Heavy Metal Concentrations in Indoor Household Dust

The median and IQR concentrations (mg kg^{-1}) of eight heavy metals in household dust are shown in Table 1. We found that Mn (542.7, 270.3 mg kg^{-1}) exhibited the highest concentration in indoor household dust followed by Zn (352.7, 246.9 mg kg^{-1}), >Cu (82.5, 56.3 mg kg^{-1}), >Pb (44.8, 27.7 mg kg^{-1}), >Cr (32.4, 18.3 mg kg^{-1}), >Ni (28.9, 17.9 mg kg^{-1}), >As (10.3, 6.1 mg kg^{-1}), and Cd (0.9, 1.5 mg kg^{-1}), which was the lowest concentration. The concentration ranges of As, Cd, Cr, Cu, Mn, Ni, Pb, and Zn in indoor household dust in the present study were from 4.2 to 62.1 mg kg^{-1}, 0.2 to 20.2 mg kg^{-1}, 17.2 to 148.0 mg kg^{-1}, 25.1 to 401.5 mg kg^{-1}, 204.8 to 1318.1 mg kg^{-1}, 11.2 to 146.2 mg kg^{-1}, 18.0 to 426.4 mg kg^{-1}, and 123.8 to 1527.2 mg kg^{-1}, respectively. The distribution map of the heavy metal concentrations in indoor household dust in Chiang Mai and Lamphun provinces is shown in Figure 2.

Table 1. The concentrations of heavy metals (mg kg^{-1}) in indoor household dust in this study.

Heavy Metals	Concentration (mg kg^{-1})					
	Mean	SD	Median	IQR	Min.	Max.
As	12.5	8.8	10.3	6.1	4.2	62.1
Cd	2.2	3.9	0.9	1.5	0.2	20.2
Cr	38.5	20.1	32.4	18.3	17.2	148.0
Cu	107.3	78.2	82.5	56.3	25.1	401.5
Mn	577.6	222.5	542.7	270.3	204.8	1318.1
Ni	34.9	23.6	28.9	17.9	11.2	146.2
Pb	62.1	57.8	44.8	27.7	18.0	426.4
Zn	408.7	246.9	352.7	210.1	123.8	1527.2

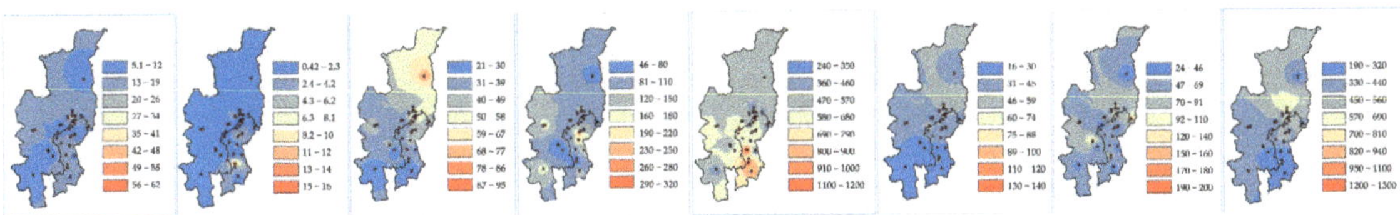

As Cd Cr Cu Mn Ni Pb Zn

Figure 2. Distribution map of heavy metal concentrations (mg kg^{-1}) in indoor household dust using the inverse distance weighting (IDW) interpolation method.

The concentrations of heavy metals in this study were within the range of previous studies worldwide (Table 2). However, in contrast to other studies, only Mn in this study was different, changing the order and showing a higher concentration than Zn in this region. Regardless of the highest value of Mn, the orders of the heavy metal concentrations in this region are likely similar to those found in China, the UK, a meta-analysis of 35 countries [17], and other countries, with slight changes in the order for Cu, Pb, Cr, and Ni (Table 2). The higher concentration of Mn than Zn in this study may be due to the origin of the household dust from contamination by outdoor air and soil resuspension, which is highly abundant in Mn. While Zn-rich content was related to industrial emissions and urban areas [14,17], there was also found a higher concentration of Zn in urban areas than in rural areas with a borderline significant difference in the present study (Table 3). Additionally, in the case of this study, the geological background might be a crucial factor in the metal content. Some studies revealed that detached homes had higher levels of Mn [17,49], and most houses in this study were detached single houses (data are not shown). Moreover, most houses in Northern Thailand are open-air houses, which allows natural ventilation throughout by keeping the doors and windows open [6]. Soil dust that contains abundant Mn can easily enter houses. In addition, agricultural chemicals are a potential source of Mn [50], and most people in this region are engaged in agricultural work that easily transfers contaminated soil and agricultural chemicals onto the body and into homes.

Regarding the differences in urbanization, the concentrations of heavy metals in urban areas, industrial areas, and agricultural areas, such as rural areas, were investigated. The comparison of the heavy metal concentrations (mg kg^{-1}) in indoor household dust in the different areas is shown in Table 3. On the basis of the results, a significantly higher concentration of As was found in rural areas than urban areas, and a borderline significantly higher concentration of Zn was found in urban areas than rural areas. Meanwhile, the concentrations of Cd, Cr, Cu, Mn, Ni, and Pb showed no significant differences between urban and rural areas. It was previously reported that the As level decreases with an increase in the distance from the city center [17] and is related to pesticide use in agricultural lands [35,51]. Moreover, the higher concentration of As in rural areas could be linked to the contamination of outdoor soil [40], while Zn was mostly linked to activities in urban areas, such as vehicle transportation [14] and industrial activities [17].

In the industrial area, there was a significantly higher concentration of only Cd in the indoor household dust of the houses located near the industrial area than in the houses far from the industrial area. While for the other metals, there were found no significant differences. It has previously been reported that emissions of Cd mainly originate from various industrial activities [18,52]. In addition, electronic waste recycling in electronic industries also played a significant part in Cd contamination [52]. The differences in the heavy metal concentrations in the different areas suggests that anthropogenic activities can influence a variety of heavy metals.

Table 2. A comparison of heavy metals in indoor dust with previous studies.

Countries (*n*)	Heavy Metal Concentrations in Indoor Dust (mg kg^{-1})								Ref.
	As	Cd	Cr	Cu	Mn	Ni	Pb	Zn	
35 Countries (*n* = 2265) [a]	25.3		128	264	333	77.6	224	1470	[17]
35 Countries (*n* = 2265) [b]	13.3	0.76	86	176	257	39	94	1110	[17]
Australia (*n* = 1310) [a]	31.9		105	232	336	49.4	305	1680	[17]
China (*n* = 111) [a]	17.3		134	242	245	70.6	161	1190	[17]
Ghana (*n* = 54) [a]	6.2		43.4	74.9	185	48	163	252	[17]
UK (*n* = 148) [a]	6.9		93.6	136	269	35.8	131	532	[17]
USA (*n* = 345) [a]	20.5		207	549	385	165.4	93.6	1785	[17]
Turkey (*n* = 85) [b]	4.41	0.35	23.8	65.7	65.9	32.3	27.5	263	[53]
Australia (*n* = 224) [a]	20.2		99.8	298	247	56.7	364	2437	[47]
Across China (*n* = 3392) [b]	15.6	2.73	85.9	136.2		40.7	161.5	602.7	[18]
Canada (*n* = 125) [a]	13	11	92	1900	250	60	4500	14,000	[54]
UK (*n* = 32) [c]		1.2		301	524	53.1	150	622	[55]
Japan (*n* = 100) [c]		1.02	67.8	304	226	59.6	57.9	920	[56]
Sukhothai, Thailand (*n* = 16) [a]		9		3			226	1051	[57]
Ubon Ratchathani, Thailand (*n* = 56) [a]			0.99			0.92	3.00		[58]
This study (*n* = 100) [a]	12.5	2.2	38.5	107.3	577.6	34.9	62.1	408.7	
This study (*n* = 100) [b]	10.3	0.9	32.4	82.5	542.7	28.9	44.8	352.7	

[a] Mean. [b] Median. [c] Geomean.

Table 3. A comparison of the heavy metal concentrations (mg kg^{-1}) in indoor household dust in urbanized areas, agricultural areas, and industrial areas in Chiang Mai and Lamphun provinces.

Areas	Median, IQR Concentration, (mg kg^{-1})							
	As	Cd	Cr	Cu	Mn	Ni	Pb	Zn
Urbanized								
Rural (agriculture) (79)	10.97, 5.89	0.89, 1.31	30.97, 18.00	79.18, 57.25	577.41, 284.14	27.37, 18.93	45.25, 29.18	332.33, 193.00
Urban (21)	8.80, 4.14	1.08, 1.53	41.89, 15.12	84.13, 57.58	461.83, 288.92	30.94, 17.36	44.16, 41.19	420.46, 234.74
p-Value	0.042	0.308	0.137	0.557	0.102	0.245	0.632	0.053
Industrial area nearby								
Yes (21)	10.97, 6.20	1.37, 9.49	34.61, 16.33	81.83, 34.49	519.73, 205.02	29.86, 24.55	49.29, 31.13	384.10, 199.74
No (74)	10.25, 6.50	0.83, 1.18	31.41, 19.91	84.16, 63.33	573.72, 281.19	27.92, 16.89	42.18, 30.28	344.83, 214.65
p-Value	0.784	0.028	0.654	0.438	0.510	0.319	0.247	0.654

3.2. Source Identification of Heavy Metal in Indoor Household Dust

Source identification was conducted using source modeling of the enrichment factor (EF) and positive matrix factorization (PMF). The *EF* values can indicate the sources of influence, whether anthropogenic activities or natural sources, on dust and soil concentrations when compared to Mn as the crush metal [42].

The mean value of the *EF* for each heavy metal in the indoor household dust is shown in Table 4. The *EF* values of the heavy metals in the indoor household dust are ranked in the order of Cd (18.2) followed by As (11.1) > Zn (10.8) > Pb (8.0) > Cu (3.8) > Ni (0.8) > Cr (0.7). The *EF* can indicate the natural conditions when the value is less than 2 and suggest an anthropogenic influence when it is greater than 2 [59]. On the basis of the results, the mean *EF* values indicated an extremely high enrichment of Cd, As, and Zn, which were higher than 10, suggesting a significant enrichment level derived from anthropogenic activity [59,60]. The *EF* values of Pb and Cu were between 2 and 10, indicating that these metals were lightly enriched and mainly originated from human activities. The *EFs* of Ni and Cr were lower than 1, therefore, indicating that they certainly originated from natural sources and were not affected by human activities. Regarding anthropogenic activities and heavy metals, As, Cd, and Zn have been reported as potential metals from agricultural chemical sources [50,51] and industrial activities [18,52]. Moreover, As, Pb, and Zn are reportedly related to traffic emissions and motor vehicles [14,17,61]. Paint is associated with increased As, Cu, Pb, and Zn in household dust [14,18,62,63]. Regarding natural sources, Cr and Ni, in this study, were identified as contaminants from a natural origin, like natural soil. Soil is the dominant contributor of Ni and Cr to household dust [40,56]. Nevertheless, some heavy metals can originate from both natural and anthropogenic sources [64]. Moreover, the simultaneous accumulation of metals from soil and anthropogenic sources can enhance the enrichment of heavy metals in household dust, which may lead to increased *EF* values.

Table 4. The enrichment factor of heavy metals in indoor household dust.

Heavy Metals	*n*	Enrichment Factor					
		Mean	SD	Median	IQR	Min.	Max.
As	100	11.13	5.66	10.13	4.93	3.82	47.88
Cd	96	18.18	27.69	9.43	13.41	1.47	141.17
Cr	100	0.70	0.48	0.61	0.39	0.21	4.15
Cu	97	3.76	3.24	2.87	2.14	0.72	19.77
Ni	99	0.84	0.61	0.65	0.50	0.18	3.60
Pb	100	7.99	5.14	6.48	5.03	2.41	27.99
Zn	99	10.84	7.68	9.21	8.77	1.85	51.82

For further source identification, the PMF model was employed. The results show the heavy metals in the house dust in this study mainly came from five sources. The factor fingerprints plot (Figure 3) illustrates the distribution of the metal species by the various sources and provides the percentage contribution of the metal species according to the various sources. Factor profiles are shown in Figure 3. Factor 1 explains 19.01%, factor 2 explains 15.83%, factor 3 explains 21.64%, factor 4 explains 13.24%, and factor 5 explains 30.28% of all data for the heavy metals in the indoor household dust.

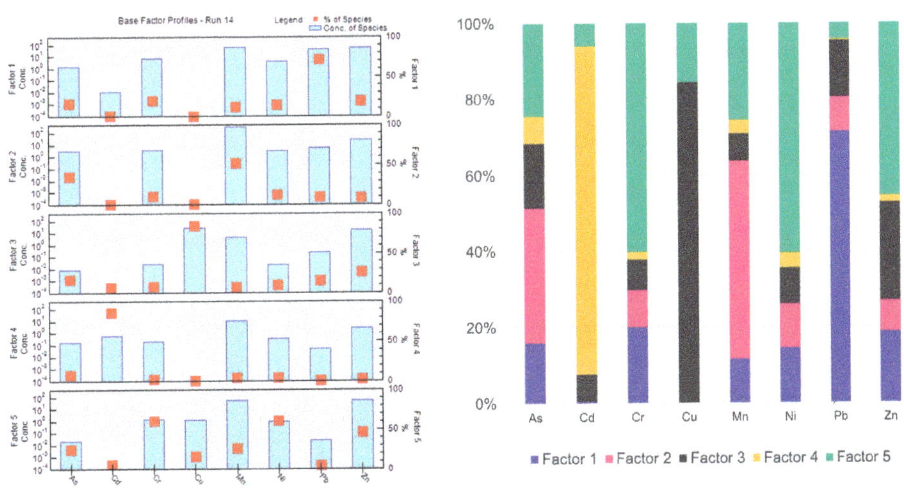

(a) (b)

Figure 3. Factor profiles and factor fingerprints plot of 8 heavy metals in the indoor household dust in this study: (**a**) factor fingerprints plot; (**b**) factor profiles.

Factor 1 was dominated by Pb, which was attributed to traffic emissions and the painting of houses. It has been reported that Pb is primarily related to leaded gasoline usage, and it is associated with house dust in homes with a high traffic density nearby [17,40,61,65,66]. Paint and coating materials were also related to the higher concentration of Pb in house dust [14,18,66,67].

As and Mn were high in concentration, as explained by factor 2, representing natural soil and agriculture. Soil and the Earth's crust are the most important sources of these metals [17,40,64].

Factor 3, represented by Cu, was caused by paint pigment in houses. The As, Cu, Pb, and Zn in the household dust were also associated with colored paints for houses, building materials, and stainless steel used in household furniture [14,17,62]. Wall paint and coating materials may lead to the accumulation of Cu in house dust, especially green paint [63].

Factor 4 was dominated by Cd, which is attributed to cigarette smoking. Several studies have reported tobacco smoking to be an important factor of elements enriched in house dust, particularly Cd, Ni, Pb, and Zn [14,68], as cigarettes contain these elements [69,70]. However, Cd is also used in color pigments for paints and coating materials [14,18], and they originate from agricultural chemicals, traffic emissions, and wildfires [14,17,18,50]. Probably, the sources of Cd are not only contamination from tobacco smoke but also from other mixed sources.

Factor 5 may be identified as multiple sources or a complex mixture of outdoor sources including natural sources of Cr and Ni and anthropogenic sources of Zn. It has been reported that Cr and Ni can originate from soil [17,64]. Moreover, Zn in household dust has been related to outdoor agriculture [50], motor vehicles [14], and paint [18]. Additionally, the very high enrichment of Zn content might not be of natural origin and must come from either industrial emissions [17] or from the geological background of this area.

3.3. Health Risk Assessment

The human health risks of carcinogenic and noncarcinogenic risks were calculated as a result of exposure to heavy metals in indoor household dust via inhalation, ingestion, and dermal contact routes. The calculated carcinogenic and noncarcinogenic risks of the heavy metals are shown in Tables 5 and 6, respectively. For the noncarcinogenic risk, the hazard quotient (*HQ*) exhibits the main risk of the heavy metals in the household dust from ingestion, followed by the dermal contact and inhalation exposure pathways. Except for only Cu in adults and Mn in children, the highest exposure pathway was ingestion followed by inhalation and dermal contact. The hazard index (*HI*) or overall health risk attributable to exposure to heavy metals in household dust via all tree pathways in adults and children were in the same order as As > Mn > Pb > Cr > Cd > Cu > Ni > Zn. On the basis of these results, in rural and urban areas, the overall *HI* values of all heavy metals in adults were lower than 1, suggesting that the heavy metals in household dust were within tolerable limits or had no effect on noncarcinogenic risk in this study. However, the *HI* in children exhibited noncarcinogenic risk primarily through ingestion. Suggesting that there was a possibility of noncarcinogenic effects in children in this area from heavy metal exposure. The finding of noncancer risk only in children may be due to children exhibiting a greater vulnerability to indoor heavy metal contamination than adults. This susceptibility is attributed to their lower body weight and higher dust ingestion rates resulting from increased physical activity [17,18]. Consequently, children face a higher health risk from indoor heavy metals than adults.

Table 5. Noncarcinogenic risk or hazard quotient (*HQ*) of 8 heavy metals in household dust in Chiang Mai and Lamphun provinces.

Heavy Metal	Noncarcinogenic Risk							
	Adult				Child			
	HQ_{ing}	HQ_{inh}	HQ_{dermal}	HI	HQ_{ing}	HQ_{inh}	HQ_{dermal}	HI
As	6.20×10^{-2}	1.49×10^{-4}	1.81×10^{-2}	8.02×10^{-2}	5.21×10^{-1}	2.91×10^{-4}	1.81×10^{-2}	5.39×10^{-1}
Urban	5.79×10^{-2}	1.39×10^{-4}	1.69×10^{-2}	7.49×10^{-2}	4.86×10^{-1}	2.72×10^{-4}	1.69×10^{-2}	5.03×10^{-1}
Rural	6.31×10^{-2}	1.51×10^{-4}	1.84×10^{-2}	8.17×10^{-2}	5.30×10^{-1}	2.96×10^{-4}	1.84×10^{-2}	5.49×10^{-1}
Cd	1.09×10^{-2}	2.29×10^{-4}	1.74×10^{-3}	1.28×10^{-2}	9.14×10^{-2}	4.48×10^{-4}	8.68×10^{-3}	1.01×10^{-1}
Urban	3.23×10^{-3}	6.79×10^{-5}	5.15×10^{-4}	3.81×10^{-3}	2.71×10^{-2}	1.33×10^{-4}	2.57×10^{-3}	2.98×10^{-2}
Rural	1.29×10^{-2}	2.72×10^{-4}	2.06×10^{-3}	1.53×10^{-2}	1.09×10^{-1}	5.32×10^{-4}	1.03×10^{-2}	1.19×10^{-1}
Cr	1.94×10^{-2}	6.99×10^{-5}	3.88×10^{-3}	2.34×10^{-2}	1.63×10^{-1}	1.37×10^{-4}	1.94×10^{-2}	1.83×10^{-1}
Urban	2.09×10^{-2}	7.50×10^{-5}	4.16×10^{-3}	2.51×10^{-2}	1.75×10^{-1}	1.47×10^{-4}	2.08×10^{-2}	1.96×10^{-1}
Rural	1.90×10^{-2}	6.85×10^{-5}	3.80×10^{-3}	2.29×10^{-2}	1.60×10^{-1}	1.34×10^{-4}	1.90×10^{-2}	1.79×10^{-1}
Cu	4.99×10^{-3}	3.47×10^{-5}	6.64×10^{-5}	5.09×10^{-3}	4.19×10^{-2}	6.79×10^{-5}	3.32×10^{-4}	4.23×10^{-2}
Urban	3.88×10^{-3}	2.70×10^{-5}	5.16×10^{-5}	3.96×10^{-3}	3.26×10^{-2}	5.28×10^{-5}	2.58×10^{-4}	3.29×10^{-2}
Rural	5.29×10^{-3}	3.68×10^{-5}	7.04×10^{-5}	5.40×10^{-3}	4.44×10^{-2}	7.20×10^{-5}	3.51×10^{-4}	4.49×10^{-2}
Mn	3.64×10^{-2}	2.10×10^{-3}	1.90×10^{-3}	4.04×10^{-2}	3.06×10^{-1}	4.10×10^{-3}	9.47×10^{-3}	3.20×10^{-1}
Urban	3.19×10^{-2}	1.83×10^{-3}	1.66×10^{-3}	3.53×10^{-2}	2.68×10^{-1}	3.59×10^{-3}	8.28×10^{-3}	2.79×10^{-1}
Rural	3.77×10^{-2}	2.17×10^{-3}	1.96×10^{-3}	4.18×10^{-2}	3.16×10^{-1}	4.24×10^{-3}	9.79×10^{-3}	3.30×10^{-1}
Ni	2.69×10^{-3}	3.13×10^{-7}	3.97×10^{-5}	2.73×10^{-3}	2.26×10^{-2}	6.12×10^{-7}	1.98×10^{-4}	2.28×10^{-2}
Urban	2.94×10^{-3}	3.42×10^{-7}	4.35×10^{-5}	2.98×10^{-3}	2.47×10^{-2}	6.70×10^{-7}	2.17×10^{-4}	2.49×10^{-2}
Rural	2.62×10^{-3}	3.05×10^{-7}	3.87×10^{-5}	2.66×10^{-3}	2.20×10^{-2}	5.97×10^{-7}	1.93×10^{-4}	2.22×10^{-2}
Pb	2.68×10^{-2}	3.19×10^{-6}	7.12×10^{-4}	2.75×10^{-2}	2.25×10^{-1}	6.25×10^{-6}	3.56×10^{-3}	2.28×10^{-1}
Urban	2.81×10^{-2}	3.34×10^{-6}	7.46×10^{-4}	2.88×10^{-2}	2.36×10^{-1}	6.55×10^{-6}	3.73×10^{-3}	2.39×10^{-1}
Rural	2.64×10^{-2}	3.15×10^{-6}	7.03×10^{-4}	2.71×10^{-2}	2.22×10^{-1}	6.17×10^{-6}	3.51×10^{-3}	2.25×10^{-1}
Zn	2.11×10^{-3}	2.53×10^{-7}	4.22×10^{-5}	2.16×10^{-3}	1.78×10^{-2}	4.96×10^{-7}	2.11×10^{-4}	1.80×10^{-2}
Urban	2.66×10^{-3}	3.18×10^{-7}	5.30×10^{-5}	2.71×10^{-3}	2.23×10^{-2}	6.24×10^{-7}	2.65×10^{-4}	2.26×10^{-2}
Rural	1.97×10^{-3}	2.36×10^{-7}	3.93×10^{-5}	2.01×10^{-3}	1.65×10^{-2}	4.62×10^{-7}	1.96×10^{-4}	1.67×10^{-2}
Total	1.65×10^{-1}	2.58×10^{-3}	2.65×10^{-2}	1.94×10^{-1}	1.39	5.06×10^{-3}	5.99×10^{-2}	1.45
Urban	1.51×10^{-1}	2.15×10^{-3}	2.41×10^{-2}	1.78×10^{-1}	1.27	4.20×10^{-3}	5.30×10^{-2}	1.33
Rural	1.69×10^{-1}	2.70×10^{-3}	2.71×10^{-2}	1.99×10^{-1}	1.42	5.28×10^{-3}	6.17×10^{-2}	1.49

Table 6. Carcinogenic risk (*CR*) and lifetime cancer risk (*LCR*) of 8 heavy metals in household dust in Chiang Mai and Lamphun provinces.

Heavy Metals	Carcinogenic Risk (CR)							
	Adult				Child			
	CR_{ing}	CR_{inh}	CR_{dermal}	LCR	CR_{ing}	CR_{inh}	CR_{dermal}	LCR
As	5.02×10^{-5}	1.27×10^{-8}	2.73×10^{-6}	5.30×10^{-5}	3.18×10^{-4}	3.18×10^{-9}	2.73×10^{-6}	3.21×10^{-4}
Urban	4.69×10^{-5}	1.19×10^{-8}	2.55×10^{-6}	4.94×10^{-5}	2.97×10^{-4}	2.96×10^{-9}	2.55×10^{-6}	2.99×10^{-4}
Rural	5.11×10^{-5}	1.29×10^{-8}	2.78×10^{-6}	5.39×10^{-5}	3.24×10^{-4}	3.23×10^{-9}	2.78×10^{-6}	3.26×10^{-4}
Cd	7.44×10^{-6}	1.22×10^{-5}	5.39×10^{-7}	2.02×10^{-5}	4.71×10^{-5}	3.05×10^{-6}	5.39×10^{-7}	5.07×10^{-5}
Urban	2.21×10^{-6}	3.61×10^{-6}	1.60×10^{-7}	5.98×10^{-6}	1.40×10^{-5}	9.04×10^{-7}	1.60×10^{-7}	1.50×10^{-5}
Rural	8.84×10^{-6}	1.45×10^{-5}	6.41×10^{-7}	2.39×10^{-5}	5.60×10^{-5}	3.62×10^{-6}	6.41×10^{-7}	6.02×10^{-5}
Cr	5.25×10^{-5}	3.89×10^{-4}	9.50×10^{-8}	4.41×10^{-4}	3.32×10^{-4}	9.72×10^{-5}	9.50×10^{-8}	4.29×10^{-4}
Urban	5.63×10^{-5}	4.18×10^{-4}	1.02×10^{-7}	4.74×10^{-4}	3.57×10^{-4}	1.04×10^{-4}	1.02×10^{-7}	4.61×10^{-4}
Rural	5.14×10^{-5}	3.81×10^{-4}	9.32×10^{-8}	4.33×10^{-4}	3.26×10^{-4}	9.53×10^{-5}	9.32×10^{-8}	4.21×10^{-4}
Ni	1.49×10^{-5}	2.22×10^{-9}	1.59×10^{-7}	1.51×10^{-5}	9.43×10^{-5}	5.55×10^{-10}	1.59×10^{-7}	9.45×10^{-5}
Urban	1.63×10^{-5}	2.43×10^{-9}	1.75×10^{-7}	1.65×10^{-5}	1.03×10^{-4}	6.07×10^{-10}	1.75×10^{-7}	1.03×10^{-4}
Rural	1.45×10^{-5}	2.16×10^{-9}	1.55×10^{-7}	1.47×10^{-5}	9.20×10^{-5}	5.41×10^{-10}	1.55×10^{-7}	9.21×10^{-5}
Pb	1.43×10^{-6}	1.79×10^{-10}	6.42×10^{-8}	1.50×10^{-6}	9.08×10^{-6}	4.46×10^{-11}	6.42×10^{-8}	9.14×10^{-6}
Urban	1.50×10^{-6}	1.87×10^{-10}	6.73×10^{-8}	1.57×10^{-6}	9.52×10^{-6}	4.68×10^{-11}	6.73×10^{-8}	9.58×10^{-6}
Rural	1.42×10^{-6}	1.76×10^{-10}	6.33×10^{-8}	1.48×10^{-6}	8.96×10^{-6}	4.41×10^{-11}	6.33×10^{-8}	9.03×10^{-6}
Total	1.26×10^{-4}	4.01×10^{-4}	3.59×10^{-6}	5.31×10^{-4}	8.01×10^{-4}	1.00×10^{-4}	3.59×10^{-6}	9.05×10^{-4}
Urban	1.23×10^{-4}	4.21×10^{-4}	3.05×10^{-6}	5.47×10^{-4}	7.80×10^{-4}	1.05×10^{-4}	3.05×10^{-6}	8.89×10^{-4}
Rural	1.27×10^{-4}	3.96×10^{-4}	3.73×10^{-6}	5.27×10^{-4}	8.06×10^{-4}	9.89×10^{-5}	3.73×10^{-6}	9.09×10^{-4}

For the carcinogenic risks, As, Cd, Cr, Ni, and Pb are known as carcinogenic metals [35]. Hence, the carcinogenic risks of these metals in the indoor household dust were calculated. The overall total risk or lifetime cancer risk (*LCR*) of carcinogenic heavy metal exposure was highest in the inhalation pathway, followed by ingestion and dermal contact pathways, respectively. In adults, the *LCR* values via the three pathways decreased in the following order: Cr > As > Cd > Ni > Pb. From the calculation, Cr exhibited a carcinogenic risk through inhalation in rural and urban areas. The total carcinogenic risks of the eight metals were found in the ingestion and inhalation pathways, while there was no cancer risk in the dermal contact pathway. On the basis of these results, the inhalation exposure pathway showed a higher cancer risk than the ingestion pathway. The higher inhalation risk may be due to the inhalation cancer risk from Cr exposure (3.89×10^{-4}) that was more significant than 1×10^{-4}, an unacceptable level for carcinogenic risk, while Cd, As, Ni, and Pb were between 1×10^{-4} and 1×10^{-6}, which are regarded as acceptable or tolerable risks. Hence, this study indicates that inhaled Cr in indoor household dust poses a cancer risk and suggests the inhalation of Cr in household dust is the dominant exposure pathway to an attributable risk to cancer in adults in this region. An unacceptable level of Cr in indoor dust was found in most cities in China, Turkey, Australia, and New Caledonia and was most dominant through the inhalation pathway [17,18,47,71]. In epidemiologic studies, Cr was reported to be correlated with lung cancer, usually in chromate-related occupational workers [72–74]. Chromium (VI) compounds are also classified as group 1 carcinogenic to humans, and there is sufficient evidence causing lung cancer [35]. Even though there are two common forms of Cr, including Cr(VI) and Cr(III), Cr(VI) was the primary toxicity of Cr exposure. In addition, the proportions of Cr(VI) to Cr(III) can vary widely, ranging from 0.5 to 2.5 [75], and are influenced by factors such as pH, chemical reactions, biological processes, and environmental conditions [76]. Hence, the parameters for the health risk assessment of Cr in household dust in this study were from Cr(VI) due to the greater toxicity that might be higher than the actual risk. In addition, cancer risk through inhalation was higher in urban areas than rural areas, while the carcinogenic risk via ingestion was higher in rural areas than urban areas. This may be due to the higher concentration of Cr in urban areas that positively correlated to the city area [17,77]. Consequently, this study suggests that 5.3 in 10,000 adults (4 from inhalation and 1.3 from ingestion exposure) may develop any cancer from lifetime exposure to these heavy metals in indoor household dust. It is important to note that the carcinogenic risk from inhalation of Cr may be the crucial factor for the high incidence of respiratory diseases, such as COPD, asthma, and lung cancer, in this region of Northern Thailand [78]. Moreover, lung cancer is reported to have a high incidence in Northern Thailand and continues to have significantly higher incidence and mortality annually compared to the other parts [78–80]. Interestingly, heavy metals such as Cr in household dust might be another cause of respiratory diseases development in this region, particularly lung cancer.

In children, the overall *LCR* value of the three pathways decreased in the following order: Cr > As > Ni > Pb > Cd. On the basis of the results, As ($LCR = 3.21 \times 10^{-4}$), Cr ($LCR = 4.29 \times 10^{-4}$), and Ni ($CR_{ing} = 1.03 \times 10^{-4}$) exhibited carcinogenic risks in children in this study. Through ingestion, As and Cd showed cancer risk in urban and rural areas, while Ni revealed a cancer risk only in urban areas. Likewise, Cr via inhalation was found only to show a cancer risk in urban areas. This might be due to urban areas having higher levels of these metals [17], as well as the correlation with vehicle emissions in the city [14,61,81]. The carcinogenic risks in children were different from those in adults, and it was found that the dominant risk was greater through ingestion than in the inhalation pathway. This is due to children's behaviors, which lead to higher dust ingestion rates, and lower body weights than adults [17,18]. From the overall *LCR*, this study suggests that 9 in 10,000 children (8 from ingestion and 1 from inhalation exposure) may develop any cancer from lifetime exposure to these heavy metals in indoor household dust.

4. Conclusions

This study investigated indoor pollutants, with a particular focus on heavy metal contamination in household dust as a significant health risk factor. The study was conducted in Chiang Mai and Lamphun provinces in Northern Thailand, encompassing urban, industrial, and agricultural areas and contributions to heavy metal pollution through various activities. The research found elevated concentrations of heavy metals in indoor household dust, with Mn showing the highest concentration, followed by Zn > Cu > Pb > Cr > Ni > As > Cd. Using the *EFs* and PMF, we identified the sources of the heavy metals in the indoor dust, which indicated that Cd, As, and Zn were significantly enriched by anthropogenic activities, while Pb and Cu showed moderate enrichment. Factors such as traffic emissions, painting, agriculture, and industrial activities were identified as key contributors to heavy metal contamination in the household dust. Health risk assessments were conducted for both the carcinogenic and noncarcinogenic risks associated with the heavy metals' exposure. The results suggest that the noncarcinogenic risks, primarily from ingestion, could pose potential health effects in children. For adults, the noncarcinogenic risks were generally within tolerable limits. Regarding the carcinogenic risks, Cr exposure through inhalation was identified as the most significant risk for adults. Urban areas exhibited higher inhalation risks due to elevated Cr concentrations and associated vehicle emissions. In children, the carcinogenic risks were observed primarily through ingestion, with significant risks associated with As, Cr, and Ni. This study demonstrates the health risks from heavy metals exposure via indoor household dust in both children and adults from Chiang Mai and Lamphun provinces, Northern Thailand. These findings support the previous scientific knowledge that heavy metals in indoor house dust can be a risk to human health, particularly cancer risks. To the best of our knowledge, our study is the first of its kind with a sample size of hundred and eight heavy metals from Thailand. However, the exposure assessment among both adults and children in high-risk areas is worth being further explored.

Supplementary Materials: The following supporting information can be downloaded at: https://www.mdpi.com/article/10.3390/toxics11121018/s1, Table S1: The values of the parameters for the noncarcinogenic and carcinogenic risk assessment; Table S2: The reference values for the noncarcinogenic and carcinogenic risk assessment; Table S3: The quality control values of the heavy metals' measurement [17,46,47,82–88].

Author Contributions: K.S. and T.P., methodology, formal analysis, and writing—review and editing; K.S., data curation and writing—original draft preparation; T.P., investigation, funding acquisition, and supervision; T.K., T.S., K.K., N.J., and N.T., sample collection and analysis. All authors have read and agreed to the published version of the manuscript.

Funding: This research was funded by the Royal Golden Jubilee (RGJ) PhD Scholarship to K.S. (PHD/0169/2559) and by Chiang Mai University, T.P., CMU 2558-2559.

Institutional Review Board Statement: The study was conducted in accordance with the Declaration of Helsinki, and approved by the Human Experimentation Committee, Research Institute for Health Sciences (Study code: Project No. 1/59, approved on 10 May 2016).

Informed Consent Statement: Informed consent was obtained from all subjects involved in the study.

Data Availability Statement: The data that support the findings of this study are available from the corresponding author on reasonable request.

Acknowledgments: We are grateful to Chaicharn Pothirat and Chalerm Liwsrisakun from Department of Internal Medicine, Chiang Mai University for their kind advice on data analyses and discussion. We appreciate and deeply respect the time and effort of all participants for household dust collection.

Conflicts of Interest: The authors declare no conflict of interest.

References

1. Klepeis, N.E.; Nelson, W.C.; Ott, W.R.; Robinson, J.P.; Tsang, A.M.; Switzer, P.; Behar, J.V.; Hern, S.C.; Engelmann, W.H. The National Human Activity Pattern Survey (NHAPS): A resource for assessing exposure to environmental pollutants. *J. Expo. Sci. Environ. Epidemiol.* **2001**, *11*, 231–252. [CrossRef] [PubMed]
2. Nwanaji-Enwerem, J.C.; Allen, J.G.; Beamer, P.I. Another invisible enemy indoors: COVID-19, human health, the home, and United States indoor air policy. *J. Expo. Sci. Environ. Epidemiol.* **2020**, *30*, 773–775. [CrossRef] [PubMed]

3. González-Martín, J.; Kraakman, N.J.R.; Pérez, C.; Lebrero, R.; Muñoz, R. A state–of–the-art review on indoor air pollution and strategies for indoor air pollution control. *Chemosphere* **2021**, *262*, 128376. [CrossRef] [PubMed]
4. World Health Organization. Household Air Pollution and Health. Available online: https://www.who.int/news-room/fact-sheets/detail/household-air-pollution-and-health (accessed on 28 August 2022).
5. Cincinelli, A.; Martellini, T. Indoor Air Quality and Health. *Int. J. Environ. Res. Public. Health* **2017**, *14*, 1286. [CrossRef] [PubMed]
6. Somsunun, K.; Prapamontol, T.; Pothirat, C.; Liwsrisakun, C.; Pongnikorn, D.; Fongmoon, D.; Chantara, S.; Wongpoomchai, R.; Naksen, W.; Autsavapromporn, N.; et al. Estimation of lung cancer deaths attributable to indoor radon exposure in upper northern Thailand. *Sci. Rep.* **2022**, *12*, 5169. [CrossRef] [PubMed]
7. Barrio-Parra, F.; De Miguel, E.; Lázaro-Navas, S.; Gómez, A.; Izquierdo, M. Indoor Dust Metal Loadings: A Human Health Risk Assessment. *Expo. Health* **2018**, *10*, 41–50. [CrossRef]
8. Al-Khashman, O.A. The investigation of metal concentrations in street dust samples in Aqaba city, Jordan. *Environ. Geochem. Health* **2007**, *29*, 197–207. [CrossRef]
9. Lu, X.; Wang, L.; Li, L.Y.; Lei, K.; Huang, L.; Kang, D. Multivariate statistical analysis of heavy metals in street dust of Baoji, NW China. *J. Hazard. Mater.* **2010**, *173*, 744–749. [CrossRef]
10. Hejami, A.A.; Davis, M.; Prete, D.; Lu, J.; Wang, S. Heavy metals in indoor settled dusts in Toronto, Canada. *Sci. Total Environ.* **2020**, *703*, 134895. [CrossRef]
11. Charlesworth, S.; De Miguel, E.; Ordóñez, A. A review of the distribution of particulate trace elements in urban terrestrial environments and its application to considerations of risk. *Environ. Geochem. Health* **2011**, *33*, 103–123. [CrossRef]
12. Turner, A. Oral bioaccessibility of trace metals in household dust: A review. *Environ. Geochem. Health* **2011**, *33*, 331–341. [CrossRef] [PubMed]
13. Lu, X.; Zhang, X.; Li, L.Y.; Chen, H. Assessment of metals pollution and health risk in dust from nursery schools in Xi'an, China. *Environ. Res.* **2014**, *128*, 27–34. [CrossRef] [PubMed]
14. Cheng, Z.; Chen, L.-J.; Li, H.-H.; Lin, J.-Q.; Yang, Z.-B.; Yang, Y.-X.; Xu, X.-X.; Xian, J.-R.; Shao, J.-R.; Zhu, X.-M. Characteristics and health risk assessment of heavy metals exposure via household dust from urban area in Chengdu, China. *Sci. Total Environ.* **2018**, *619–620*, 621–629. [CrossRef]
15. Rasmussen, P.E.; Levesque, C.; Chénier, M.; Gardner, H.D. Contribution of metals in resuspended dust to indoor and personal inhalation exposures: Relationships between PM10 and settled dust. *Build. Environ.* **2018**, *143*, 513–522. [CrossRef]
16. Deering, K.; Spiegel, E.; Quaisser, C.; Nowak, D.; Schierl, R.; Bose-O'Reilly, S.; Garí, M. Monitoring of arsenic, mercury and organic pesticides in particulate matter, ambient air and settled dust in natural history collections taking the example of the Museum für Naturkunde, Berlin. *Environ. Monit. Assess.* **2019**, *191*, 375. [CrossRef] [PubMed]
17. Isley, C.F.; Fry, K.L.; Liu, X.; Filippelli, G.M.; Entwistle, J.A.; Martin, A.P.; Kah, M.; Meza-Figueroa, D.; Shukle, J.T.; Jabeen, K.; et al. International Analysis of Sources and Human Health Risk Associated with Trace Metal Contaminants in Residential Indoor Dust. *Environ. Sci. Technol.* **2022**, *56*, 1053–1068. [CrossRef]
18. Wang, M.; Lv, Y.; Lv, X.; Wang, Q.; Li, Y.; Lu, P.; Yu, H.; Wei, P.; Cao, Z.; An, T. Distribution, sources and health risks of heavy metals in indoor dust across China. *Chemosphere* **2023**, *313*, 137595. [CrossRef]
19. Shi, G.; Chen, Z.; Xu, S.; Zhang, J.; Wang, L.; Bi, C.; Teng, J. Potentially toxic metal contamination of urban soils and roadside dust in Shanghai, China. *Environ. Pollut.* **2008**, *156*, 251–260. [CrossRef]
20. Gurung, A.; Bell, M.L. The state of scientific evidence on air pollution and human health in Nepal. *Environ. Res.* **2013**, *124*, 54–64. [CrossRef]
21. Kong, S.; Lu, B.; Ji, Y.; Zhao, X.; Chen, L.; Li, Z.; Han, B.; Bai, Z. Levels, risk assessment and sources of PM10 fraction heavy metals in four types dust from a coal-based city. *Microchem. J.* **2011**, *98*, 280–290. [CrossRef]
22. Kim, H.S.; Kim, Y.J.; Seo, Y.R. An Overview of Carcinogenic Heavy Metal: Molecular Toxicity Mechanism and Prevention. *J. Cancer Prev.* **2015**, *20*, 232–240. [CrossRef] [PubMed]
23. Carver, A.; Gallicchio, V. *Heavy Metals and Cancer*; IntechOpen: London, UK, 2018.
24. Lin, Y.; Ma, J.; Zhang, Z.; Zhu, Y.; Hou, H.; Zhao, L.; Sun, Z.; Xue, W.; Shi, H. Linkage between human population and trace elements in soils of the Pearl River Delta: Implications for source identification and risk assessment. *Sci. Total Environ.* **2018**, *610–611*, 944–950. [CrossRef] [PubMed]
25. Yang, S.; Zhao, J.; Chang, S.X.; Collins, C.; Xu, J.; Liu, X. Status assessment and probabilistic health risk modeling of metals accumulation in agriculture soils across China: A synthesis. *Environ. Int.* **2019**, *128*, 165–174. [CrossRef] [PubMed]
26. Kapaj, S.; Peterson, H.; Liber, K.; Bhattacharya, P. Human health effects from chronic arsenic poisoning--a review. *J. Environ. Sci. Health A Tox Hazard. Subst. Environ. Eng.* **2006**, *41*, 2399–2428. [CrossRef] [PubMed]
27. Mohammed Abdul, K.S.; Jayasinghe, S.S.; Chandana, E.P.S.; Jayasumana, C.; De Silva, P.M.C.S. Arsenic and human health effects: A review. *Environ. Toxicol. Pharmacol.* **2015**, *40*, 828–846. [CrossRef] [PubMed]
28. Wang, Z.; Sun, Y.; Yao, W.; Ba, Q.; Wang, H. Effects of Cadmium Exposure on the Immune System and Immunoregulation. *Front. Immunol.* **2021**, *12*, 695484. [CrossRef] [PubMed]
29. Genchi, G.; Sinicropi, M.S.; Lauria, G.; Carocci, A.; Catalano, A. The Effects of Cadmium Toxicity. *Int. J. Environ. Res. Public. Health* **2020**, *17*, 3782. [CrossRef]

30. Hossini, H.; Shafie, B.; Niri, A.D.; Nazari, M.; Esfahlan, A.J.; Ahmadpour, M.; Nazmara, Z.; Ahmadimanesh, M.; Makhdoumi, P.; Mirzaei, N.; et al. A comprehensive review on human health effects of chromium: Insights on induced toxicity. *Environ. Sci. Pollut. Res.* **2022**, *29*, 70686–70705. [CrossRef]
31. Centers for Disease Control and Prevention, The National Institute for Occupational Safety and Health (NIOSH). Mangnese. Available online: https://www.cdc.gov/niosh/topics/manganese/default.html (accessed on 8 October 2023).
32. Genchi, G.; Carocci, A.; Lauria, G.; Sinicropi, M.S.; Catalano, A. Nickel: Human Health and Environmental Toxicology. *Int. J. Environ. Res. Public. Health* **2020**, *17*, 679. [CrossRef]
33. World Health Organization. Lead Poisoning. Available online: https://www.who.int/news-room/fact-sheets/detail/lead-poisoning-and-health (accessed on 8 October 2023).
34. International Agency for Research on Cancer (IARC). *IARC Monographs on the Evaluation of Carcinogenic Risks to Humans*; IARC: Lyon, France, 2012.
35. IARC. Arsenic, metals, fibres, and dusts. *IARC Monogr. Eval. Carcinog. Risks Hum.* **2012**, *100*, 11–465.
36. Pietrzak, S.; Wójcik, J.; Baszuk, P.; Marciniak, W.; Wojtyś, M.; Dębniak, T.; Cybulski, C.; Gronwald, J.; Alchimowicz, J.; Masojć, B.; et al. Influence of the Levels of Arsenic, Cadmium, Mercury and Lead on Overall Survival in Lung Cancer. *Biomolecules* **2021**, *11*, 1160. [CrossRef] [PubMed]
37. Geoinformatics Center. Impact Assessment of Urbanization on Environmental Quality in Chiang Mai-Lumphun Valley, Northern Thailand. Available online: https://geoinfo.ait.ac.th/impact-assessment-of-urbanization-on-environmental-quality-in-chiang-mai-lumphun-valley-northern-thailand/ (accessed on 10 October 2023).
38. Kermel-Torrès, D. Atlas of Thailand: Spatial Structures and Development. Available online: https://books.openedition.org/irdeditions/32537 (accessed on 10 October 2023).
39. Falciani, R.; Novaro, E.; Marchesini, M.; Gucciardi, M. Multi-element analysis of soil and sediment by ICP-MS after a microwave assisted digestion method. *J. Anal. At. Spectrom.* **2000**, *15*, 561–565. [CrossRef]
40. Zhao, X.; Li, Z.; Wang, D.; Tao, Y.; Qiao, F.; Lei, L.; Huang, J.; Ting, Z. Characteristics, source apportionment and health risk assessment of heavy metals exposure via household dust from six cities in China. *Sci. Total Environ.* **2021**, *762*, 143126. [CrossRef] [PubMed]
41. Zhao, X.; Lin, L.; Zhang, Y. Contamination and human health risks of metals in indoor dust from university libraries: A case study from Qingdao, China. *Hum. Ecol. Risk Assess. Int. J.* **2021**, *27*, 152–161. [CrossRef]
42. Liu, E.; Yan, T.; Birch, G.; Zhu, Y. Pollution and health risk of potentially toxic metals in urban road dust in Nanjing, a mega-city of China. *Sci. Total Environ.* **2014**, *476–477*, 522–531. [CrossRef]
43. Rubio, B.; Nombela, M.A.; Vilas, F. Geochemistry of Major and Trace Elements in Sediments of the Ria de Vigo (NW Spain): An Assessment of Metal Pollution. *Mar. Pollut. Bull.* **2000**, *40*, 968–980. [CrossRef]
44. Taylor, S.R. Abundance of chemical elements in the continental crust: A new table. *Geochim. Et. Cosmochim. Acta* **1964**, *28*, 1273–1285. [CrossRef]
45. Cao, S.; Chen, X.; Zhang, L.; Xing, X.; Wen, D.; Wang, B.; Qin, N.; Wei, F.; Duan, X. Quantificational exposure, sources, and health risks posed by heavy metals in indoor and outdoor household dust in a typical smelting area in China. *Indoor Air* **2020**, *30*, 872–884. [CrossRef]
46. United States Environmental Protection Agency. Exposure Factors Handbook 2011 Edition (Final Report). Available online: https://cfpub.epa.gov/ncea/risk/recordisplay.cfm?deid=236252 (accessed on 4 September 2023).
47. Doyi, I.N.Y.; Isley, C.F.; Soltani, N.S.; Taylor, M.P. Human exposure and risk associated with trace element concentrations in indoor dust from Australian homes. *Environ. Int.* **2019**, *133*, 105125. [CrossRef]
48. Zhou, L.; Liu, G.; Shen, M.; Hu, R.; Liu, Y. Source identification of heavy metals and stable carbon isotope in indoor dust from different functional areas in Hefei, China. *Sci. Total Environ.* **2020**, *710*, 135599. [CrossRef]
49. Hunt, A.; Johnson, D.L. Suspension and resuspension of dry soil indoors following track-in on footwear. *Environ. Geochem. Health* **2012**, *34*, 355–363. [CrossRef] [PubMed]
50. Naccarato, A.; Tassone, A.; Cavaliere, F.; Elliani, R.; Pirrone, N.; Sprovieri, F.; Tagarelli, A.; Giglio, A. Agrochemical treatments as a source of heavy metals and rare earth elements in agricultural soils and bioaccumulation in ground beetles. *Sci. Total Environ.* **2020**, *749*, 141438. [CrossRef] [PubMed]
51. Martin, A.P.; Turnbull, R.E.; Rissmann, C.W.; Rieger, P. Heavy metal and metalloid concentrations in soils under pasture of southern New Zealand. *Geoderma Reg.* **2017**, *11*, 18–27. [CrossRef]
52. He, C.-T.; Zheng, X.-B.; Yan, X.; Zheng, J.; Wang, M.-H.; Tan, X.; Qiao, L.; Chen, S.-J.; Yang, Z.-Y.; Mai, B.-X. Organic contaminants and heavy metals in indoor dust from e-waste recycling, rural, and urban areas in South China: Spatial characteristics and implications for human exposure. *Ecotoxicol. Environ. Saf.* **2017**, *140*, 109–115. [CrossRef]
53. Gul, H.K.; Gullu, G.; Babaei, P.; Nikravan, A.; Kurt-Karakus, P.B.; Salihoglu, G. Assessment of house dust trace elements and human exposure in Ankara, Turkey. *Environ. Sci. Pollut. Res.* **2023**, *30*, 7718–7735. [CrossRef] [PubMed]
54. Dingle, J.H.; Kohl, L.; Khan, N.; Meng, M.; Shi, Y.A.; Pedroza-Brambila, M.; Chow, C.-W.; Chan, A.W.H. Sources and composition of metals in indoor house dust in a mid-size Canadian city. *Environ. Pollut.* **2021**, *289*, 117867. [CrossRef]
55. Turner, A.; Simmonds, L. Elemental concentrations and metal bioaccessibility in UK household dust. *Sci. Total Environ.* **2006**, *371*, 74–81. [CrossRef]

56. Yoshinaga, J.; Yamasaki, K.; Yonemura, A.; Ishibashi, Y.; Kaido, T.; Mizuno, K.; Takagi, M.; Tanaka, A. Lead and other elements in house dust of Japanese residences—Source of lead and health risks due to metal exposure. *Environ. Pollut.* **2014**, *189*, 223–228. [CrossRef]
57. Srithawirat, T.; Chimjan, O. Contamination assessment of heavy metals in dust deposited on residential building walls in agricultural areas of Sukhothai. *J. Environ. Manag.* **2017**, *13*, 20–33. [CrossRef]
58. Kuntawee, C.; Tantrakarnapa, K.; Limpanont, Y.; Lawpoolsri, S.; Phetrak, A.; Mingkhwan, R.; Worakhunpiset, S. Exposure to Heavy Metals in Electronic Waste Recycling in Thailand. *Int. J. Environ. Res. Public. Health* **2020**, *17*, 2996. [CrossRef]
59. Barbieri, M. The Importance of Enrichment Factor (EF) and Geoaccumulation Index (Igeo) to Evaluate the Soil Contamination. *J. Geol. Geophys.* **2016**, *5*, 1–4. [CrossRef]
60. da Rosa Couto, R.; Faversani, J.; Ceretta, C.A.; Ferreira, P.A.A.; Marchezan, C.; Basso Facco, D.; Garlet, L.P.; Silva, J.S.; Comin, J.J.; Bizzi, C.A.; et al. Health risk assessment and soil and plant heavy metal and bromine contents in field plots after ten years of organic and mineral fertilization. *Ecotoxicol. Environ. Saf.* **2018**, *153*, 142–150. [CrossRef] [PubMed]
61. Zhou, L.; Liu, G.; Shen, M.; Liu, Y. Potential ecological and health risks of heavy metals for indoor and corresponding outdoor dust in Hefei, Central China. *Chemosphere* **2022**, *302*, 134864. [CrossRef] [PubMed]
62. Huang, S.L.; Yin, C.-Y.; Yap, S.Y. Particle size and metals concentrations of dust from a paint manufacturing plant. *J. Hazard. Mater.* **2010**, *174*, 839–842. [CrossRef] [PubMed]
63. Li, Y.; Pi, L.; Hu, W.; Chen, M.; Luo, Y.; Li, Z.; Su, S.; Gan, Z.; Ding, S. Concentrations and health risk assessment of metal(loid)s in indoor dust from two typical cities of China. *Environ. Sci. Pollut. Res.* **2016**, *23*, 9082–9092. [CrossRef] [PubMed]
64. Davis, H.T.; Marjorie Aelion, C.; McDermott, S.; Lawson, A.B. Identifying natural and anthropogenic sources of metals in urban and rural soils using GIS-based data, PCA, and spatial interpolation. *Environ. Pollut.* **2009**, *157*, 2378–2385. [CrossRef] [PubMed]
65. Marx, S.K.; Rashid, S.; Stromsoe, N. Global-scale patterns in anthropogenic Pb contamination reconstructed from natural archives. *Environ. Pollut.* **2016**, *213*, 283–298. [CrossRef] [PubMed]
66. Taylor, M.P.; Isley, C.F.; Fry, K.L.; Liu, X.; Gillings, M.M.; Rouillon, M.; Soltani, N.S.; Gore, D.B.; Filippelli, G.M. A citizen science approach to identifying trace metal contamination risks in urban gardens. *Environ. Int.* **2021**, *155*, 106582. [CrossRef]
67. O'Connor, D.; Hou, D.; Ye, J.; Zhang, Y.; Ok, Y.S.; Song, Y.; Coulon, F.; Peng, T.; Tian, L. Lead-based paint remains a major public health concern: A critical review of global production, trade, use, exposure, health risk, and implications. *Environ. Int.* **2018**, *121*, 85–101. [CrossRef]
68. Shi, T.; Wang, Y. Heavy metals in indoor dust: Spatial distribution, influencing factors, and potential health risks. *Sci. Total Environ.* **2021**, *755*, 142367. [CrossRef]
69. Rasmussen, P.E.; Levesque, C.; Chénier, M.; Gardner, H.D.; Jones-Otazo, H.; Petrovic, S. Canadian House Dust Study: Population-based concentrations, loads and loading rates of arsenic, cadmium, chromium, copper, nickel, lead, and zinc inside urban homes. *Sci. Total Environ.* **2013**, *443*, 520–529. [CrossRef] [PubMed]
70. Böhlandt, A.; Schierl, R.; Diemer, J.; Koch, C.; Bolte, G.; Kiranoglu, M.; Fromme, H.; Nowak, D. High concentrations of cadmium, cerium and lanthanum in indoor air due to environmental tobacco smoke. *Sci. Total Environ.* **2012**, *414*, 738–741. [CrossRef] [PubMed]
71. Kurt-Karakus, P.B. Determination of heavy metals in indoor dust from Istanbul, Turkey: Estimation of the health risk. *Environ. Int.* **2012**, *50*, 47–55. [CrossRef] [PubMed]
72. Luippold, R.S.; Mundt, K.A.; Austin, R.P.; Liebig, E.; Panko, J.; Crump, C.; Crump, K.; Proctor, D. Lung cancer mortality among chromate production workers. *Occup. Environ. Med.* **2003**, *60*, 451–457. [CrossRef] [PubMed]
73. Park, R.M.; Bena, J.F.; Stayner, L.T.; Smith, R.J.; Gibb, H.J.; Lees, P.S. Hexavalent chromium and lung cancer in the chromate industry: A quantitative risk assessment. *Risk Anal.* **2004**, *24*, 1099–1108. [CrossRef] [PubMed]
74. Mahiout, S.; Kiilunen, M.; Vermeire, T.; Viegas, S.; Woutersen, M.; Santonen, T. Occupational exposure to Cr(VI) in Finland in 1980–2016 and related lung cancer risk assessment. *Regul. Toxicol. Pharmacol.* **2022**, *136*, 105276. [CrossRef] [PubMed]
75. Catrambone, M.; Canepari, S.; Perrino, C. Determination of Cr(III), Cr(VI) and total chromium in atmospheric aerosol samples. *E3S Web Conf.* **2013**, *1*, 07005. [CrossRef]
76. Ukhurebor, K.E.; Aigbe, U.O.; Onyancha, R.B.; Nwankwo, W.; Osibote, O.A.; Paumo, H.K.; Ama, O.M.; Adetunji, C.O.; Siloko, I.U. Effect of hexavalent chromium on the environment and removal techniques: A review. *J. Environ. Manag.* **2021**, *280*, 111809. [CrossRef]
77. Fry, K.L.; Gillings, M.M.; Isley, C.F.; Gunkel-Grillon, P.; Taylor, M.P. Trace element contamination of soil and dust by a New Caledonian ferronickel smelter: Dispersal, enrichment, and human health risk. *Environ. Pollut.* **2021**, *288*, 117593. [CrossRef]
78. Aungkulanon, S.; Tangcharoensathien, V.; Shibuya, K.; Bundhamcharoen, K.; Chongsuvivatwong, V. Post universal health coverage trend and geographical inequalities of mortality in Thailand. *Int. J. Equity Health* **2016**, *15*, 190. [CrossRef]
79. Imsamran, W.; Chaiwerawattana, A.; Wiangnon, S.; Pongnikorn, D.; Suwanrungrung, K.; Sangrajrang, S.; Buasom, R. *Cancer in Thailand Vol. VIII, 2010–2012*; National Cancer Institute: Bangkok, Thailand, 2015.
80. Pongnikorn, D.; Daoprasert, K.; Waisri, N.; Laversanne, M.; Bray, F. Cancer incidence in northern Thailand: Results from six population-based cancer registries 1993-2012. *Int. J. Cancer* **2018**, *142*, 1767–1775. [CrossRef] [PubMed]
81. Olujimi, O.; Steiner, O.; Goessler, W. Pollution indexing and health risk assessments of trace elements in indoor dusts from classrooms, living rooms and offices in Ogun State, Nigeria. *J. Afr. Earth Sci.* **2015**, *101*, 396–404. [CrossRef]

82. Hou, S. Pollution characteristics, sources, and health risk assessment of human exposure to Cu, Zn, Cd and Pb pollution in urban street dust across China between 2009 and 2018. *Environ. Int.* **2019**, *128*, 430–437. [CrossRef] [PubMed]
83. United States Environmental Protection Agency. US EPA Regional Screening Levels (RSL) Calculator. Available online: https://epa-prgs.ornl.gov/cgi-bin/chemicals/csl_search (accessed on 4 September 2023).
84. Wang, S.; Wang, L.; Huan, Y.; Wang, R.; Liang, T. Concentrations, spatial distribution, sources and environmental health risks of potentially toxic elements in urban road dust across China. *Sci. Total Environ.* **2022**, *805*, 150266. [CrossRef]
85. US EPA. Supplemental Guidance for Developing Soil Screening Levels for Superfund Sites. Available online: https://www.epa.gov/superfund/superfund-soil-screening-guidance (accessed on 4 September 2023).
86. Ferreira-Baptista, L.; De Miguel, E. Geochemistry and risk assessment of street dust in Luanda, Angola: A tropical urban environment. *Atmos. Environ.* **2005**, *39*, 4501–4512. [CrossRef]
87. Wignall, J.A.; Muratov, E.; Sedykh, A.; Guyton, K.Z.; Tropsha, A.; Rusyn, I.; Chiu, W.A. Conditional Toxicity Value (CTV) Predictor: An In Silico Approach for Generating Quantitative Risk Estimates for Chemicals. *Environ. Health Perspect.* **2018**, *126*, 057008. [CrossRef]
88. Office of Environmental Health Hazard Assessment (OEHHA). Technical Support Doc.ument for Cancer Potency Factors 2009. Available online: https://oehha.ca.gov/air/crnr/technical-support-document-cancer-potency-factors-2009 (accessed on 5 September 2023).

toxics

MDPI

Article

An Investigation into the Viability of Portable Proximal Sensor X-Ray Fluorescence Data for Assessing Heavy Metal Contamination in Urban Soils: A Case Study in Changchun, China

Xiaoxiao Zou, Jilong Lu *, Xinyun Zhao, Qiaoqiao Wei, Zhiyi Gou, Yaru Hou and Yawen Lai

College of Geo-Exploration Science and Technology, Jilin University, Changchun 130026, China; zouxx22@mails.jlu.edu.cn (X.Z.); zhaoxinyun@jlu.edu.cn (X.Z.); weiqiaoqiao@jlu.edu.cn (Q.W.); gouzy22@mails.jlu.edu.cn (Z.G.); houyr20@mails.jlu.edu.cn (Y.H.); laiyw@jlu.edu.cn (Y.L.)

* Correspondence: lujl@jlu.edu.cn

Abstract: In order to validate the applicability of pXRF for rapid in situ detection of heavy metals in urban soils and to accurately obtain an assessment of soil quality in Changchun, a city in northeast China, 164 soil samples from within the main urban area of Changchun were collected for pXRF analysis. The main stable elements Si and Ti were used to establish a matrix effect correction model, and the values of Cr (64.2 mg·kg^{-1}), Cu (43.8 mg·kg^{-1}), Zn (96.2 mg·kg^{-1}), As (20.9 mg·kg^{-1}), and Pb (57.4 mg·kg^{-1}) were predicted. The empirical findings indicate that the quality of soil data from the pXRF was improved to different degrees under the correction model, and it became a relatively reliable dataset; the order of improvement was Cu > Pb > Cr > Zn > As. A comprehensive assessment indicated that Changchun City is primarily contaminated by the heavy metals As, Pb, and Cu, with the main sources being automobile manufacturing and pharmaceutical chemical production. These findings align with previous studies and have produced favorable outcomes in practical applications. This rapid, non-destructive and economical detection method is very applicable and economical for the sustainable monitoring and control of heavy metals in large cities. This study provides a basis for rapid large-scale prediction of urban soil safety and protection of local human health.

Keywords: portable instruments; correction quality; soil investigation of quality; pollution index; Leopold Matrix

1. Introduction

Urban soil is an organic component of the urban ecological system and the part of the pedosphere most affected by human activities. Large urban populations and community activities result in heavy metal contaminants entering the ground from various sources [1]. These sources include industrial activities, the use of wastewater for irrigation, vehicle emissions, and the extraction of metallic mineral resources [1–3]. In addition to these sources, the heavy metal content in urban areas is also influenced by background sedimentation and soil-forming matrices [4]. Soil organic matter exhibits a strong capacity for adsorbing heavy metals [5]. In general, heavy metal concentrations are higher in roadside soils compared to urban parks, and in urban areas compared to agricultural soils of the same parent material in remote suburban regions. Numerous studies have indicated that Cu and Zn levels are particularly significant in residential and mixed-use areas, while Pb and Cd levels tend to be elevated near industries and businesses associated with heavy metals [6]. Various human activities, such as heating and waste disposal, exert differing impacts on the concentrations of these heavy metals. Solid particulate matter from industrial activities enters the atmosphere, where it can be deposited and subsequently migrate to water bodies through rainfall runoff [7]. This process facilitates the accumulation of heavy metals in urban environments. The limited capacity of urban soils, combined with the slow rate of internal material cycling, further impedes the degradation of heavy metals, thereby posing significant risks to human health. Exposure can occur through oral ingestion, skin contact, and inhalation, leading to a range of health issues. For instance, As and Cr are highly carcinogenic, while Pb, Cu, and Zn are associated with blood disorders and nervous system diseases [8]. This presents a significant risk to the well-being and existence of the urban community. Given its central role in the population and that it is closely related to human survival and health, it is crucial to promptly detect and address soil contamination in urban regions [9,10].

However, traditional laboratory analyses, such as Inductively Coupled Plasma Mass Spectrometry (ICP-MS), have long detection times, complicated operations, and require experimental consumables such as acid reagents [3]. This not only incurs high economic costs but also easily leads to further pollution. Portable energy dispersive X-ray fluorescence (pXRF) instruments are known for their rapid detection and low-cost, non-destructive analysis, making them suitable for in situ heavy metal detection in the field and real-time acquisition of soil geochemical data [11–14]. The accuracy of pXRF analysis is limited by matrix effects that are inherent and cannot be completely eradicated, making it challenging to directly compare its accuracy to that of conventional laboratory analysis techniques [15,16]. By increasing the sample size and implementing rigorous calibration methods, the precision of pXRF analysis technology can be enhanced to fulfil the requirements of large-scale urban pollution assessment [17]. Currently, there are two main types of calibration method: one involves experimental operations, such as the standard built-in method and dilution method for suppression, while the other utilizes mathematical modelling, including linear regression and neural network methods [18]. The practical implementation of the experimental method is complex and time-consuming, especially when working with a large number of samples. Accordingly, a new matrix effect correction mathematical model method for calibration has been developed, which is more suitable and convenient for the timely implementation of experimental decisions, such as the design of an encrypted sampling grid and the conducting of repeated experiments.

PXRF technology has significant potential for application in alloy analysis, archaeological research, and mining ore detection [19]. Although urban soils have been studied by experts in the past, they tend to be mostly in suburban areas and forested farmland near cities, etc., and there are relatively few examples of rapid contamination monitoring using pXRF in medium- to large-scale major built-up areas of cities [20]. In the last five decades, Changchun, an important city in northeast China in terms of its urbanized area, has experienced urbanization fueled by industrial development [21]. Many factories have been established that may discharge a large number of heavy metals, among which the main pollutants associated with urbanization are chromium (Cr), copper (Cu), zinc (Zn), arsenic (As), and lead (Pb) in Changchun, and these five metals pollute the soil more commonly than other heavy metals [22,23]. We hypothesized that pXRF data after correction could accurately represent the heavy metal content in the soil, with an accuracy comparable to that of traditional elemental measurements, thereby meeting the standards for quantitative analysis. Changchun, a typical industrialized city in China, is affected by various urban activities, which may lead to heavy metal concentrations in the soil that exceed safety standards and potentially jeopardize the health of residents. Therefore, timely monitoring to detect contamination is essential. The use of pXRF effectively addresses these monitoring needs, making it an appropriate solution for this purpose. To test this hypothesis, we applied pXRF in situ analysis, predicted the estimates under the constructed correction model, and used multiple evaluation factors to judge the accumulation level of elements. Additionally, the research provides a reference for analyzing other heavy metal elements and includes technical details on implementing this method in fieldwork. This was conducted with the objective of ensuring the safety and health of the local population. In light of the aforementioned findings, recommendations have been formulated pertaining to this matter. Therefore, the objectives of this paper are 1. to assess the applicability of pXRF for analyzing heavy metal content in urban soils; and 2. to evaluate the soil quality of Changchun City based on the precise data obtained.

2. Materials and Methods

2.1. Overview of the Study Area

This research was conducted in the urban area of Changchun (124°18′–127°02′ E, 43°05′–45°15′ N), which is the capital city of Jilin province, located in the hinterland of the Songnen Plain and which serves as the geographical core of northeast China (Figure 1) [24]. Changchun has a temperate continental semi-humid climate, with an average annual temperature of 4.8 °C, annual precipitation ranging from 600–700 mm, and a freeze period lasting for 5 months annually [25,26]. According to the WRB standards and the China Soil Database, the main inherent soil types in Changchun are black soils and meadow black soils. According to the National Standard of the People's Republic of China (HJ962-2018), the potentiometric method is mainly used for the determination of soil pH, and the water–soil ratio is 2.5:1, while the method for the determination of soil organic matter is mainly the potassium dichromate oxidation method [27–29]. The soil type in the eastern part of Changchun City is alfisol, primarily characterized by a clay loam texture, a pH level generally around 7.0, and low organic matter content. In the central area of Changchun, the soil is phaeozem, which has a loamy clay texture, with a pH range of 7.1 to 7.5 and an organic matter content exceeding 2.0%. In the western region, the soil type is Chernozem, featuring an organic matter content greater than 2.0% in the topsoil. This soil typically exhibits weak alkalinity, with a topsoil pH of approximately 7.0, and has a texture of either loamy clay or sandy clay. Before the development of the urban area, the predominant soil types included black soil, meadow soil, and meadow black soil. The soil pH in the Changchun urban area ranged from 7.5 to 8.1, indicating a weakly alkaline condition. Additionally, the soil texture was primarily medium loamy, characterized by a deficiency in organic matter. The city covers an area of built districts of approximately 551.38 km^2. It has a permanent population of 9,087,200, consisting of 3,575,300 urban inhabitants, with an urbanization rate of 66.8% in 2022. Changchun, recognized as the cradle of the automobile manufacturing industry, has produced and assembled a total of 1.56 million automobiles as of 2023. Additionally, it serves as a biomedical hub, boasting an industrial scale of 97.8 billion yuan and a comprehensive industrial chain encompassing medical devices and chemical pharmaceuticals [30]. The city boasts a comprehensive array of industrial facilities, including machinery manufacturing, biomedical equipment, chemical manufacturing, bus manufacturing, power and heat production, and other related sectors [31,32].

2.2. Sampling and Analysis

2.2.1. Sampling and Analysis with Sensors (pXRF)

Quality Assurance and Quality Control (QA/QC)

To ensure the usability of the data for decision-making purposes, this study conducted tests on international standard samples to verify the accuracy and precision of the data. The international standard geochemical analysis sample number used for this analysis is the GSS for Soil Geochemical Standard. A total of 18 international standard geochemical samples were used in this study for quality control purposes (Table 1). In order to ensure temporal and spatial consistency in the data analyses, standards are analyzed concurrently with the soil samples. The calibration of these standards is conducted approximately every ten soil samples. The standard samples are placed in a slightly flattened cylindrical container, covered with a film, and subsequently tested using pXRF.

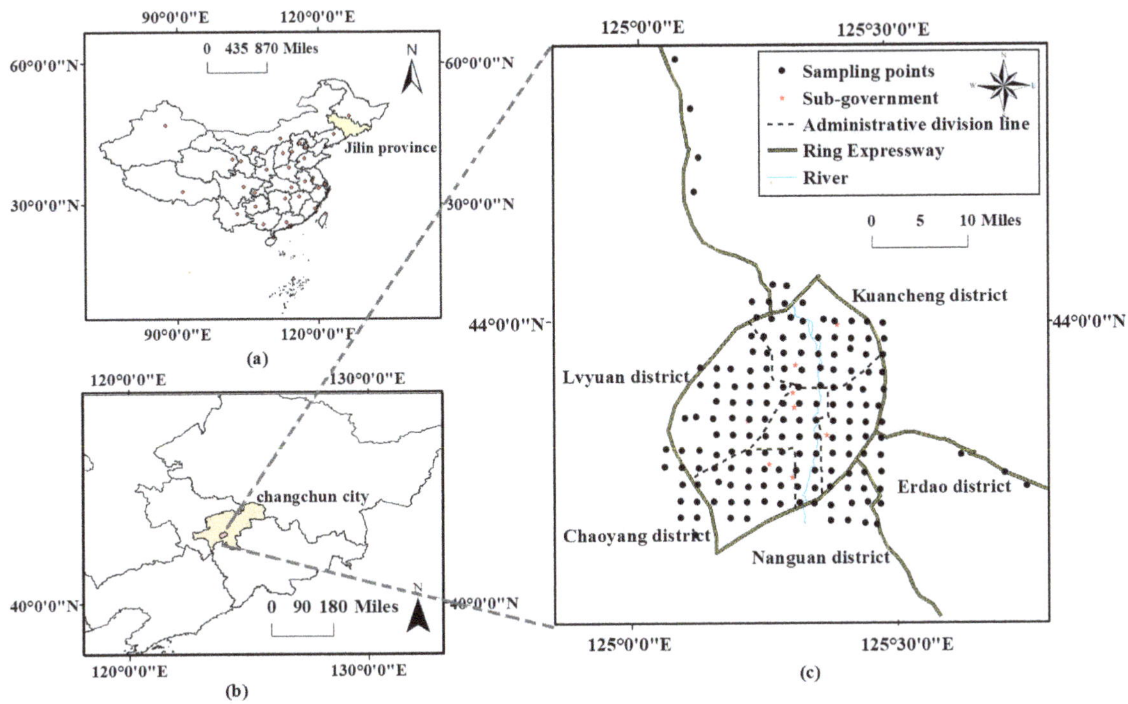

Figure 1. Map of the study area and distribution of soil sampling points. (**a**): Map of Administrative Areas in China, (**b**): Map of Administrative Areas in Jilin Province, (**c**): Changchun main urban area map.

Table 1. Results of certified values of national standard soil geochemical samples (St) and test values of pXRF (pXRF).

	Cr (pXRF)	Cr (St)	Cu (pXRF)	Cu (St)	Zn (pXRF)	Zn (St)	As (pXRF)	As (St)	Pb (pXRF)	Pb (St)	Si % (pXRF)	Si (St)	Ti (pXRF)	Ti% (St)
GSS-1a	51.2	44.00	55.0	42.00	677.0	475.00	48.6	33.00	421.8	339.00	30.62	26.46	3956.4	0.326
GSS-1	45.0	47.00	21.60	21	863.8	680	44.4	34	102.6	98	30.47	29.263	5885.2	0.483
GSS-2	370.8	410.00	30.0	16.30	45.2	42.00	16.6	13.70	-	20.00	33.87	34.29	2954.2	0.271
GSS-3	70.8	70.00	14.00	11.4	19.6	31	-	4.4	20.0	26	34.88	34.929	2378.2	0.224
GSS-4	31.6	25.00	18.00	40	214.4	210	73.0	58	62.8	58	23.42	23.817	13,798.2	1.080
GSS-5	36.8	43.00	118.0	144	605.4	494	530.8	412	690.6	552	24.04	24.575	7633.4	0.629
GSS-6	50.4	57.00	478.60	390	101.2	97	302.2	220	371.0	314	25.09	26.613	5556.	0.439
GSS-7	63.4	82.00	91.4	97.00	149.6	142.00	-	4.80	18.8	14.00	16.25	15.28	25,629.0	2.020
GSS-8	61.0	62.00	18.00	24.3	60.4	68	17.0	12.7	-	21	28.65	27.398	4510.20	0.380
GSS-14	51.2	66.00	29.0	27.40	109.4	96.00	-	6.50	32.6	31.00	30.28	30.16	4700.0	0.406
GSS-17	45.0	61.00	20.4	12.60	28.2	29.00	11.6	6.20	-	17.40	36.17	36.60	1874.0	0.191
GSS-20	80.0	92.00	25.4	28.00	54.6	61.00	11.2	8.70	-	13.40	24.09	22.10	3392.6	0.330
GSS-22	62.4	62	19.8	18.30	65.6	59.00	13.8	7.80	25.0	26.00	32.61	31.90	4395.8	0.380
GSS-23	28.20	32	40.4	32.00	122.4	97.00	20.8	11.80	-	28.00	29.954	27.95	5946.4	0.500
GSS-24	351.40	370	31.6	28.00	96.2	81.00	24.0	15.80	30.2	40.00	32.130	32.31	5361.4	0.450
GSS-25	166.40	118	24.0	23.60	75.8	66.00	17.6	12.90	-	22.00	30.894	28.48	4494.2	0.390
GSS-26	122.20	75	21.8	19.10	68.2	62.00	12.5	8.90	-	21.00	32.830	30.92	4539.0	0.410
GSS-27	64.60	68	52.8	54.00	153.4	127.00	17.7	13.30	44.0	41.00	29.324	27.52	7457.8	0.640

Notes: Unless otherwise indicated, units are in mg·kg^{-1}.

For the testing, a portable energy-dispersive X-ray fluorescence spectrometer (X-Met 7500) manufactured by the Oxford Instruments Group in the UK was used. The device operated at a voltage of 40 KV, a current of 60 mA, and possessed a maximum power output of 2.4 kW [13]. Additionally, the instrument was equipped with a fourth-generation Silicon Drift Detector (SDD) and a Rhodium (Rh) anode target [33,34]. It also offered the option to select the soil test modes Soil-Mining_fp and Mining_LE_FP, which were based on the basic parameter method and empirical coefficient method (they are usually inbuilt algorithms carried out by the manufacturer). The standard sample was placed in the pXRF scan 5 times; each time the analysis duration was 60 s, and the final average value was obtained as the pXRF analysis result. We performed a simple linear fit using the standard values of the standard samples (quality guaranteed by the quality analysis of the national production unit) and the results of the pXRF test and found that the fit was in accordance with the test ($R^2 > 0.95$). The results of the analyses of the international standard geochemical samples are shown in Table 1 below. The calculated R^2 values for the elements Cr, Cu, Zn, As, and Pb are 0.963, 0.978, 0.992, 0.999, and 0.999, respectively. The recovery rate is 100%, except Pb, which is also 62%. This is considered an acceptable level of reporting rate for general research [33]. This quality control and assurance demonstrate that the pXRF data from this study are credible and comparable to studies by other researchers in our experimental group for both conventional laboratory XRF and the pXRF used in the experiments [35]. The detection limit and RSD of portable XRF is Cu (10 $mg{\cdot}kg^{-1}$, 6.3%), Cr (10 $mg{\cdot}kg^{-1}$, 8.6%), Zn (5 $mg{\cdot}kg^{-1}$, 15.4%), As (4 $mg{\cdot}kg^{-1}$, 12.5%), and Pb (5 $mg{\cdot}kg^{-1}$, 7.4%).

Analysis of Soil Samples

During the sample collection process, sampling points were systematically distributed based on a grid pattern, with each point spaced at 2 km intervals. Surface soil within the study area was excavated, and soil samples were obtained at a depth of 0–20 cm using a stainless-steel spade. Additionally, soil specimens were gathered from 3 to 5 different locations surrounding each sampling point, combined, and thoroughly mixed to create a composite sample after the removal of any extraneous materials such as branches, twigs, and leaves. A total of 164 samples were acquired for this study, consisting of 156 samples from the primary urban zone and 8 samples as the background soil from Ring Express (Figure 1c).

For the pXRF test, after removing stones, wood splinters, or some extraneous matter, placing it in a cloth bag and tapping it evenly with a wooden stick, and air-drying it in the sun for about ten minutes, then the air-dried sample was placed into a short cylindrical plastic cup with a diameter of 50 mm and a depth of 20 mm and flattened evenly with paper to facilitate scanning [36]. The experiment was replicated five times at distinct and uniformly distributed locations within a sample. Each test lasted for 60 s, resulting in a total of 600 s for one sample measurement. Finally, the mean value of the five times was calculated as the final result. Based on previous data-processing experience, we filled in the missing data for undetected samples with half the minimum value of the test [37]. To obtain the necessary data, we stored the remaining soil samples and sent them to a laboratory for conventional testing, which included ICP-MS test methods. The study measured the Cr, Cu, Zn, As, and Pb contents using pXRF, in addition to the major elements selected based on element correlation analysis, which included Si and Ti. The experimental data satisfy the precision criteria for data adequacy. The findings from the study of variance (ANOVA) revealed a statistically significant disparity ($p < 0.05$) between the in situ portable XRF measurements and the conventional laboratory data obtained (ICP-MS).

2.2.2. Data Processing and Methods

Calculation of the Correction for Matrix Effects

In theory, the matrix effect arises from various factors. In the context of chemical analysis, the matrix effect refers to the influence on an analytical method caused by all other components of the sample except the specific compound being quantified; the content of an element is affected by all elements except the element being measured [38]. In the actual calculation process, due to the problems of calculation workload and accuracy, it is complicated to calculate all elements one by one. Simplification of the original Sherman equation can help minimize spectral noise and address spectral interferences when analyzing the multi-variable XRF spectra (Equation (1)).

According to the stability tested by the instrument, the main elements such as Si and Ti could be selected as the primary correction indicators in this paper. They have a high content, are minimally influenced by other elements, and are relatively stable in the crust. Multivariate linear matrix effect correction was then carried out on the remaining elements to be measured. On one hand, the impact of the outlier on the correction results diminished compared to the traditional linear regression method (LR); on the other hand, the calculated quantity was simplified.

$$C_i' = \alpha_i C_i + \beta_j Si + \gamma_z Ti + \mu_i \tag{1}$$

where C_i' is the corrected content of heavy metals, (element i) is the predicted value, C_i is the content of elements established by pXRF, Si and Ti is the test content established by pXRF, α_i is the regression coefficient of heavy metal element i, β_j and γ_z are the influence coefficients of the major element Si and element Ti, and μ_i is the regression intercept of heavy metal element i (Equations (2)–(5)) [39,40], all of which are employed by multivariate partial least squares regression and multiple linear regression.

$$M_i = C_i - \sum_k \frac{C_{ik}}{k} \tag{2}$$

$$\alpha_i = \frac{(\sum_k M'_{ik} M_{ik})\left(\sum_k M_{jk}^2\right) - (\sum_k M'_{ik} M_{jk})(\sum_k M_{ik} M_{jk})}{(\sum_k C_{ik}^2)(\sum_k C_{jk}) - (\sum_k C_{ik} C_{jk})^2} \tag{3}$$

$$\beta_j = \frac{\left(\sum_k M'_{ik} M_{jk}\right)(\sum_k M_{ik}^2) - (\sum_k M'_{ik} M_{ik})(\sum_k M_{ik} M_{jk})}{(\sum_k C_{ik}^2)(\sum_k C_{jk}^2) - (\sum_k C_{ik} C_{jk})^2} \tag{4}$$

$$\mu_i = \frac{(\sum_k C'_{ik}) - \alpha_i(\sum_k C_{ik}) - \beta_j(\sum_k C_{jk})}{k} \tag{5}$$

where n is the number of samples and C_{ik} and C_{jk} are the testing values of elements i and j in sample k, respectively [41]. The testing values of i and j are zero-centered to M_{ik} and M_{jk} (Equations (6) and (7)) [42]. Correlation coefficients can be calculated by testing standard samples before testing due to instrument commissioning. Given the complexity of these calculations, this study employs software (such as IBM SPSS Statistics 21.0) that provides such analyses through a streamlined one-step process, as opposed to manual calculations. This approach not only alleviates computational challenges but also enhances accuracy.

$$M_{ik} = C_{ik} - \frac{\sum_{k=1}^{n} C_{ik}}{k} \tag{6}$$

$$M_{ij} = C_{jk} - \frac{\sum_{k=1}^{n} C_{jk}}{k} \tag{7}$$

Performance statistics were used to evaluate the correction results, which included the coefficient of determination (R^2), mean absolute error (MAE) and root mean squared error (RMSE), the ratio of performance to interquartile distance (RPIQ) (Formulas (8)–(11)). Both the RMSE and R^2 metrics evaluate the performance of the linear regression model on the dataset. MAE and RMSE assess the regression model's capability to predict the absolute value of the response variable. In addition, R^2 assesses the predictor's ability to elucidate the change in the response variable [43]. The relationship between the pXRF testing value and an element's predicted value (corrected value) could be estimated using the R^2 values. A stronger correlation is indicated by a larger R^2 value. As R^2 approaches 1, it indicates that the regression model is better fitted. In general, a higher RPIQ indicates a better predictive ability for the model.

$$R^2 = \frac{\sum_N (C_m - \overline{C_n})^2}{\sum_N (C_n - \overline{C_n})^2} \tag{8}$$

$$RPIQ = \frac{Q3 - Q1}{RMSE} \tag{9}$$

$$RMSE = \sqrt{\frac{\Sigma_N (C_m - C_N)^2}{N}} \tag{10}$$

$$MAE = \frac{\Sigma_N |C_m - C_n|}{N} \tag{11}$$

where N is the number of samples, C_m is the predicted values, C_n is the ICP-MS value (represented true values), $\overline{C_n}$ is the mean value of C_n, Q3 is the third quartile, and Q1 is the first quartile distance.

Pollution Evaluation Methods

The overall concentration of heavy metals and the statistical methods do not provide comprehensive information on the extent of soil contamination. They only give a rough indication of the potential for contamination [44]. The pollution index can be used as a tool for comprehensive geochemical assessment of soil environmental status [45]. Single contamination indices may occasionally overestimate the level of contamination at a site due to the methodologies employed in calculations and the selection of background values. Conversely, more aggregated contamination factors may be influenced by varying choices of contaminant element types or may neglect the impact of natural geochemical variability, resulting in inaccurate evaluation outcomes. Therefore, this study employs a diverse array of pollution indicators, encompassing both traditional metrics such as single pollution factors and more comprehensive methodologies for pollution assessment [46] (Table 2). The geochemical background values used in this paper are the soil geochemical background of Jilin province (Table 3). Pollution evaluation is essential for understanding the extent of pollution. This paper employs absolute principal component analysis–multiple linear regression to identify the contributions of various pollution sources [47]. The analysis was primarily conducted using relevant software for calculations.

Table 2. Pollution indices used and description of the corresponding parameters.

Index	Formula	Explain
Single pollution index (PI)	$PI = C_n/B_n$	C_n—the content of the heavy mental element [48]. B_n—the geochemical background value of Jilin province.
Geo-accumulation index (I-$_{geo}$)	$I_{geo} = Log(C_n/k \times B_n)$	C_n—the measured levels of the heavy metal "n" in the soil sample, B_n—used the same way as PI, and K is the correction factor, which was chosen as 1.5 [44].
Contamination factor (CF)	$CF = C_n^i/C_p^i$	C_n—the content of heavy metal from at least five samples of individual metals, C_p—pre-industrial reference value for the substances [49].
Enrichment factor (EF)	$EF = (C_n/C_{ref})/(B_n/B_{ref})$	C_n—content of analyzed heavy metal, C_{ref}—one of the following metals, Ti [50]. B_n—reference content of the analyzed heavy metal, B_{ref}—one of the following metals, Ti in the background [51].
The pollution load index (PLI)		CF_n—the contamination factors of the element n [52].
Risk factor (RI)	$PLI = (CF_1 + CF_2 + \cdots + CF_n)^{\frac{1}{n}}$ $E_r^i = T_r^i \times C_f^i$ $RI = \Sigma_{i=1}^n E_r^i$	n—the number of heavy metals, Tr—the toxicity response coefficient of an individual metal, C_f—contamination factor, Er—single index of the ecological risk factor [53].

Table 2. *Cont.*

Index	Formula	Explain
Degree of contamination (C-deg)	$C_{deg} = \Sigma_{i=1}^{n} CF_i$	CFi—the contamination factor for each element [54].
Nemerow pollution index (PI-Nemerow)	$PI_{Nemerow} = \sqrt{\frac{\left(\frac{1}{n}\Sigma_{i=1}^{n} PI\right)^2 + (PI_{max})^2}{n}}$	n—the total number of elements, PI—the value of the single index, PI_{max}—the maximum value of the PI [55].

Notes: The corresponding soil contamination classifications are given in Appendix A.

Table 3. Soil background values and toxicity factors for Jilin province were used in the calculations.

Elements	Soils Background Values (mg·kg^{-1})	Toxicity Factor
Cr	46.7	2
Cu	17.1	1
Zn	80.4	5
As	8.38	10
Pb	28.8	5

Environmental Impacts

Determining the sources of pollution is essential for conducting a thorough and accurate environmental impact assessment [56]. The impact assessment tool utilized in this study is the Leopold Matrix. This is a two-dimensional interactive matrix, where the horizontal axis represents the environmental factors affected by pollution, including the physical environment and the social environment, among others [8]. The vertical axis denotes the environmental impact factors, which primarily consist of activities that may cause pollution, such as industrial processes and human activities [57]. To present the results more effectively, we utilized a fractional representation (M/I), where M denotes the magnitude of the change in the impacting activity (with values ranging from −5 to +5; positive values indicate positive impacts, while negative values signify negative impacts) and I represents the importance of the change to the environment (on a scale from 1 to 10) [58]. Ultimately, the environmental impact is quantified as the product of M and I, resulting in the environmental impact score [59].

3. Results and Discussion

3.1. Evaluation of Matrix Effect Correction

The Si–Ti matrix effect correction method and the traditional linear regression (LR) method were used to correct the test results and fit the image linearly. A black diagonal line going through the origin is a 1:1 line as we assumed that the ICP-MS data represented the true values (Tables 4 and 5 and Figures 2 and 3). The mean concentrations of five element concentrations determined by in situ pXRF which were relatively lower than data obtained by ICP-MS data. It is noteworthy that the mean of the in situ Zn pXRF (90.1 mg·kg^{-1}) is very close to the mean of the ICP-MS (91.12 mg·kg^{-1}) analyses. These results show that pXRF underestimates soil metal concentrations. Validation indices of MAE and RMSE are shown in Figure 2. The traditional LR generated a lower correction quality than the Si–Ti matrix effect correction method. Although the values for MAE and RMSE remain high within reasonable limits, a significant improvement is evident in the corrected data, which may be due to outliers and model fitness [60].

To better demonstrate the applicability of the model, we also performed validation. The RPIQ values of the element Cr (RPIQ = 1.572) and As (RPIQ = 1.168) are all greater than 1, showing excellent accuracy reliability, and the RPIQs of Cu (RPIQ = 0.968) and Pb (RPIQ = 0.828) are both greater than 0.8, with high accuracy, but the RPIQ of Zn (RPIQ = 0.792) is less than 0.8 but greater than 0.7, which is still in the acceptable range. The above shows that the calibration results for the five heavy metal elements are reasonable and reliable.

Table 4. Basic statistics of the concentrations measured by in situ pXRF, ICP-MS, and corrected pXRF in the urban area of Changchun City, China.

In Situ pXRF (mg·kg^{-1})	Mean	Max	Min
Cr	38.9	203.6	5.3
Cu	21.3	136.6	5.0
Pb	15.8	302.2	6.0
Zn	90.1	1511.6	27.8
As	11.2	21.8	4.0
ICP-MS (mg·kg^{-1})			
Cr	49.00	146.85	15.20
Cu	56.95	592.16	22.45
Pb	43.94	653.32	16.98
Zn	91.12	758.25	18.77
As	18.80	48.13	5.17
The corrected PXRF (mg·kg^{-1})			
Cr	64.22	167.68	38.93
Cu	43.78	392.10	3.90
Pb	57.43	565.72	13.44
Zn	96.23	795.20	39.66
As	20.90	39.86	6.40

Table 5. Correction equations and comparison of R-square before and after calibration.

Elements	Multiple Linear Regression Equations	R^2 Before Correcting	R^2 After Correcting	Degree of Advancement
Cr	Y = 0.652X + 0.001Ti + 0.954Si − 8.381	0.612	0.755	23.37%
Cu	Y = 1.196X + 0.009Ti − 15.030Si + 399.033	0.355	0.768	116.64%
Zn	Y = 0.493X − 0.017Ti − 9.932Si + 395.925	0.688	0.803	16.77%
As	Y = 1.785X + 0.001Ti − 0.569Si + 16.283	0.699	0.761	8.87%
Pb	Y = 1.795X − 0.008Ti − 4.254Si + 179.420	0.586	0.863	47.29%

Notes: where X denotes the content of pXRF of the element to be tested. Y represents the regression values of the element to be corrected. Silicon (Si) in units of 10^{-2} (%) and elemental titanium (Ti) in mg·kg^{-1}.

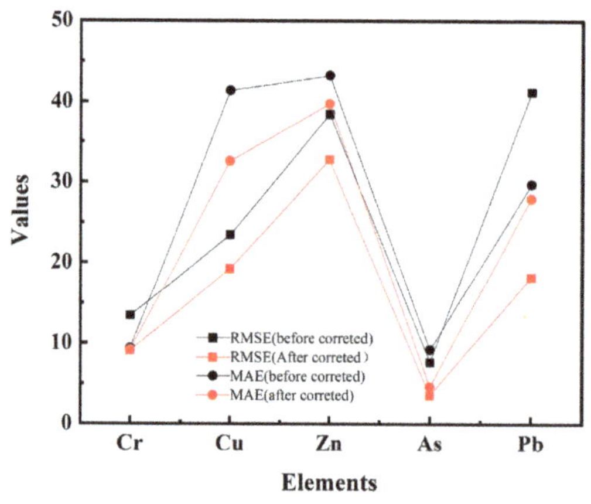

Figure 2. The MAE and RMSE values for the correction results of the improved methods.

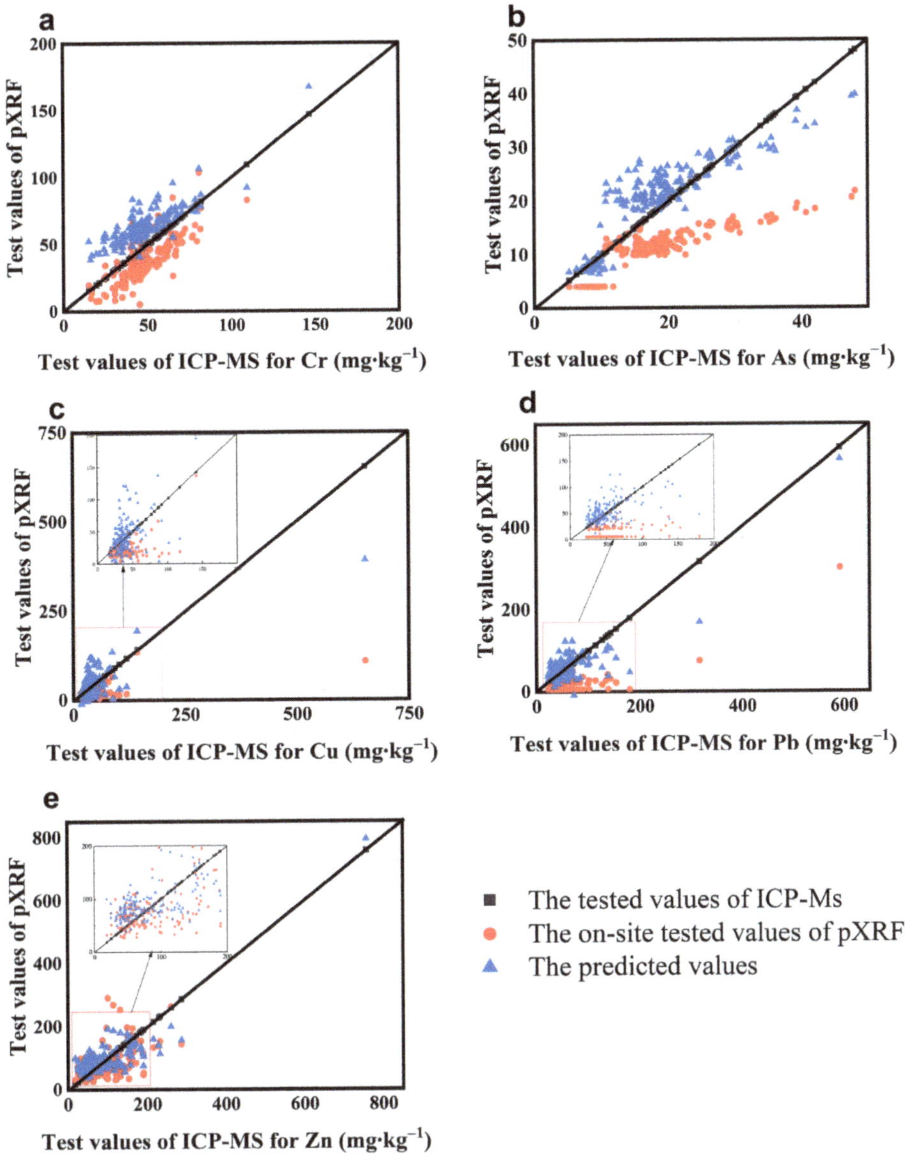

Figure 3. Comparison of scatter plots before and after correction for heavy metal elements (**a**–**e**).

We can see that the corrected data for the elements Cu and Pb are better fitted to the true values (ICP-MS test values) than the uncorrected ones. Element Zn tests higher than the true value at data points with high content, but elements Cu and Zn test lower than the true values. The corrections are closer to the actual values for the high-value ranges of elemental As, but for elemental Cr, the predictions are slightly larger than the true values, but still within reasonable limits, which may be related to the instrumental detection limits (Figure 3).

Compared to traditional and linear regression corrections, the R^2 values for the Cr, Cu, Pb, and Zn elements significantly increased. Cu showed the most significant improvement, followed by Pb, Cr, Zn, and finally As. The R^2 values were greater than 0.75 after calibration, indicating a strong correlation with laboratory testing values obtained through ICP-MS. These corrections have improved the data quality to some extent and weakened the influence of the matrix effect. The comprehensive comparison of elements should be ranked as Cu > Pb > Cr > Zn > As (Table 4). This ranking could be attributed to the constituents' amalgamation or their intrinsic characteristics.

3.2. Assessment of Pollution Degree

The pollution index statistics for the built-up area of Changchun City are shown in Figure 4 below. These box plots illustrate the distribution of the final calculated values for each evaluation method. Additionally, various levels of contamination can be inferred based on the different ranges of contamination value divisions. In the box plot illustrating the distribution of the PI calculation results (Figure 4—Graph 01), the majority of samples exhibit values below three, indicating slight to moderate contamination. However, a few sample points demonstrate severe contamination. In the calculations of the enrichment factor index (Figure 4—Graph 02), most distributions are below five, and none exceed twenty. This indicates a slight to moderate enrichment of the element. Similarly, in the ground accumulation index distribution (Figure 4—Graph 03), most of the samples are in the range of 1–2, showing slight cumulative contamination, and in the final distribution of the ecological risk factor, the data show that most of them are also below 20, but As has a higher ecological risk compared to the other elements (Figure 4—Graph 04). However, considering the toxicity of As and the background level of soils in Jilin province, this could be linked to the selection of background values for the element, which may amplify its contamination effects. The average value of the Nemerow index is equal to 1.20, indicating that the soil standards in the built-up area of Changchun are mildly polluted (Table 6). When calculating the PLI, it can be observed that As has the greatest contribution among the five elements, followed by Cu, Pb, Cr, and Zn.

Table 6. Statistical results of the comprehensive pollution index for the urban area of Changchun.

Index	Cr	Cu	Zn	As	Pb
PI_{Max}	3.59	22.93	9.89	5.02	19.64
PI_{avg}	1.38	2.56	1.20	2.68	1.99
CF	0.71	0.88	0.55	1.39	0.82
Nemerow index			1.20		

Figure 5 illustrates the spatial distribution of heavy metal pollution in the built-up area of Changchun. Each small square in the figure represents the evaluation value of a specific pollution assessment method, with the degree of pollution indicated by a gradient of red colors. Four squares together form a larger square that represents the pollution level at a given sampling point. The greater the number of red squares at a sampling point, the higher the degree of pollution [61]. The distribution of significant contamination for the elements lead, copper, and zinc are similar in the southeast corner, with contamination in this area showing large areas of red color, representing significant contamination regardless of the method of calculation, while the distribution of elemental copper in the northwest corner changes. The combination of several evaluation methods indicates that the level of Cr contamination was low and that the background values of Cr in the soil were not significantly elevated compared to the test values. However, it is important to note that Cr is relatively toxic and requires careful monitoring even at low levels of contamination [62]. The evaluations conducted using various methods indicate differing levels of pollution, which must be assessed carefully and comprehensively. Notably, most heavily polluted sample sites are concentrated around residential areas and industrial parks.

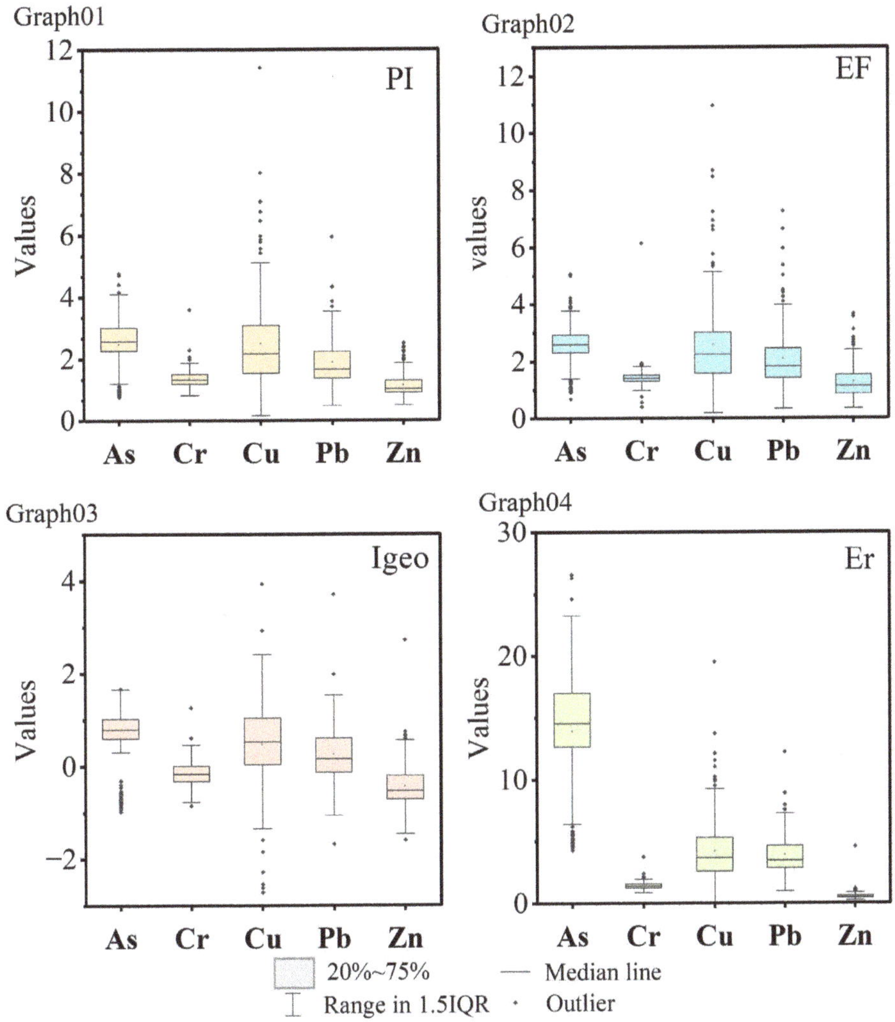

Figure 4. Box plot of pollution index distribution (Graphs 01, 02, 03, 04 correspond to pollution indices PI, EF, Igeo, Er, respectively).

The correlation heat map of heavy metal concentrations in Changchun soil indicates that, at a significance level of $p = 0.05$, the concentrations of Cu, Pb, and Zn are highly correlated, with correlation coefficients exceeding 0.8 [63]. In contrast, the elements of As and Cr show a weak correlation, approximately 0.5 (Figure 6). Combined with the contour plots of these elements, it is evident that there may be homology between these subgroups [64]. Principal component analysis similarly showed that a total of two principal components emerged between the five elements [65]. By employing Kaiser's standard orthogonal rotation method, the eigenvalues of the two factors were determined to be 2.68 and 1.44, respectively [66]. Analyzing the coefficients of the rotated component matrices revealed that Source 1 exhibited higher loadings for Cu, Pb, and Zn, while Source 2 showed greater loadings for As and Cr. These findings were consistent with the results of the correlation analyses (Table 7). The cumulative variance contribution of the principal components reached 82.37%, with the first two principal components explaining 53.59% and

28.77% of the total variance, respectively [67,68]. An absolute principal component multiple linear regression analysis was conducted, revealing that all R-squared values exceeded 0.65, indicating the model's applicability. The identified pollution sources comprised two known sources and one unidentified source.

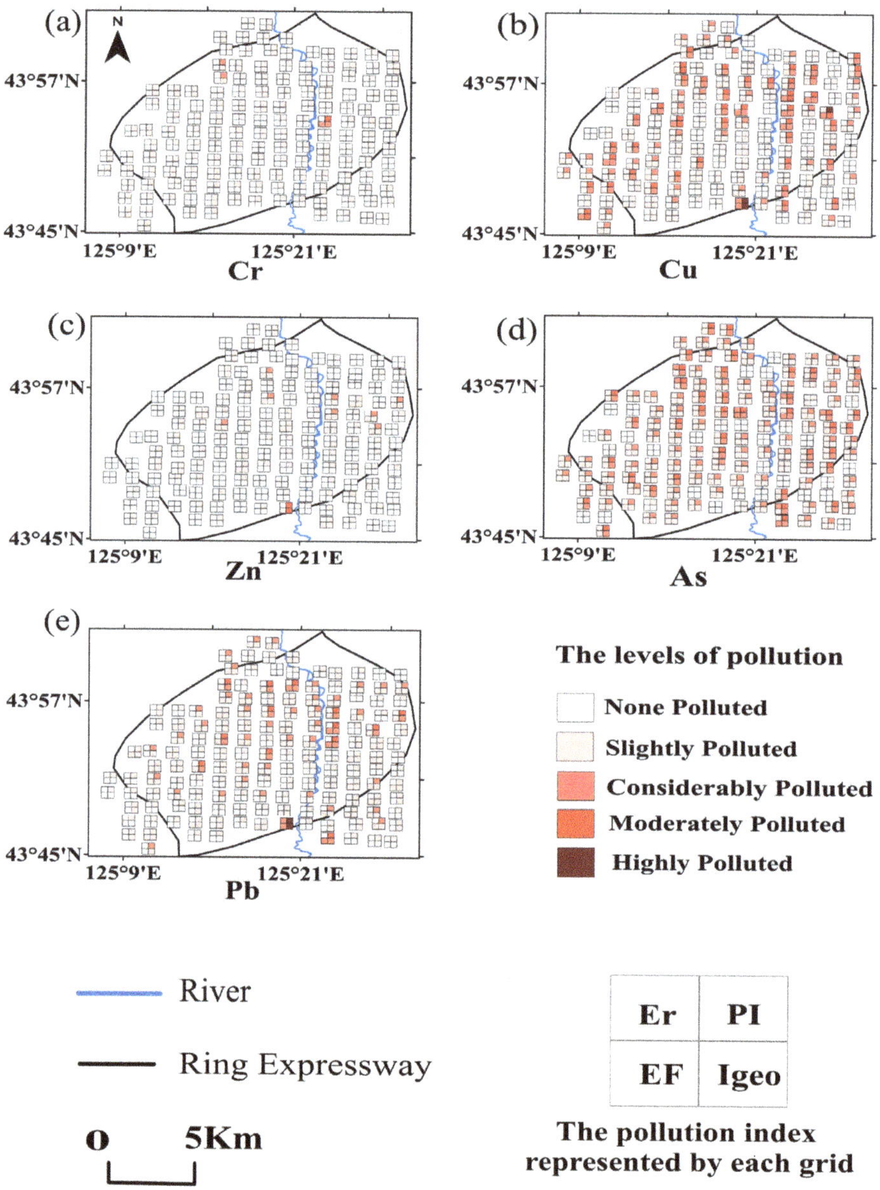

Figure 5. Spatial distribution of point pollution in Changchun City based on four pollution assessment methods. The subplot represents the degree of heavy metal contamination at each sampling site in the main urban area of Changchun, (**a**) the pollution index level of element Cr; (**b**) the pollution index level of element Cu; (**c**) the pollution index level of element Zn; (**d**): the pollution index level of element As; (**e**): the pollution index level of element Pb.

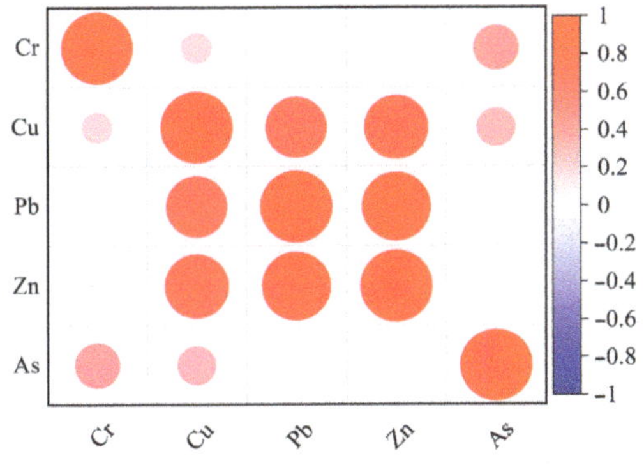

Figure 6. Correlation analysis of heavy metals in soil in the main urban area of Changchun City.

Table 7. Heavy metal element principal component analysis matrix and contribution of pollution sources.

Elements	Composition Matrix		Rotated Composition Matrix		Pollution Contribution		
	Component 1	Component 2	Component 1	Component 2	Source 1	Source 2	Unknown Source (s)
Cr	0.192	0.816	0.015	0.838	0.16%	69.37%	30.47%
Cu	0.916	0.081	0.878	0.273	30.89%	37.50%	31.61%
Zn	0.941	−0.253	0.974	−0.049	50.64%	13.17%	36.19%
As	0.257	0.809	0.081	0.845	2.30%	83.20%	14.50%
Pb	0.923	−0.217	0.948	−0.017	67.65%	8.26%	24.09%

Source 1 primarily contributes to the levels of Pb and Zn, while Source 2 has a greater impact on As and Cr. Additionally, the unknown sources show some correlation with Cu, Pb, Zn, and Cr. Notably, Cu is significantly associated with all three sources of pollution. We scrutinized the descriptive records at the time of sampling, as well as the views of satellite maps and the site. The contour map indicates that high concentrations of Cu, Pb, and Zn are present in the lower reaches of the river (Figure 7). Our sampling records reveal multiple sewage pipes in these areas, as well as metal product manufacturing facilities that discharge wastewater, including from electroplating processes, into the watershed. This has resulted in soil pollution in the surrounding regions [22]. Notably, these contaminated sites overlap with several transportation routes, such as Line 2 of the underground system and railway stations. Previous studies have demonstrated that traffic contributes to the accumulation of Pb and Zn content in the soil [23]. Therefore, the primary source of pollution can be attributed to metal product manufacturing activities. Source 2 primarily contributes to the levels of As and Cr. In the spatial distribution map of As, this coincides with numerous residential areas. In the vicinity of the high-concentration zones, our records reveal the presence of several abandoned factories [5]. Changchun City experiences severe winters, often relying on coal or electric heating. The combustion of coal releases atmospheric soot that contains heavy metals such as As, Cr, Pb, etc., which accumulate in the soil through atmospheric deposition, resulting in a broad area of impact [27]. Therefore, Source 2 is predominantly a coal-fired heating source. Similarly, the unidentified Source 3 contributes to the levels of a comprehensive assessment suggesting that these elements are primarily associated with the chemical manufacturing industry [65]. The production of chemical raw materials or products generates complex chemical products, which not only produce wastewater but also result in significant emissions and residual waste [32]. Consequently, this unidentified source is mainly for chemical pollution.

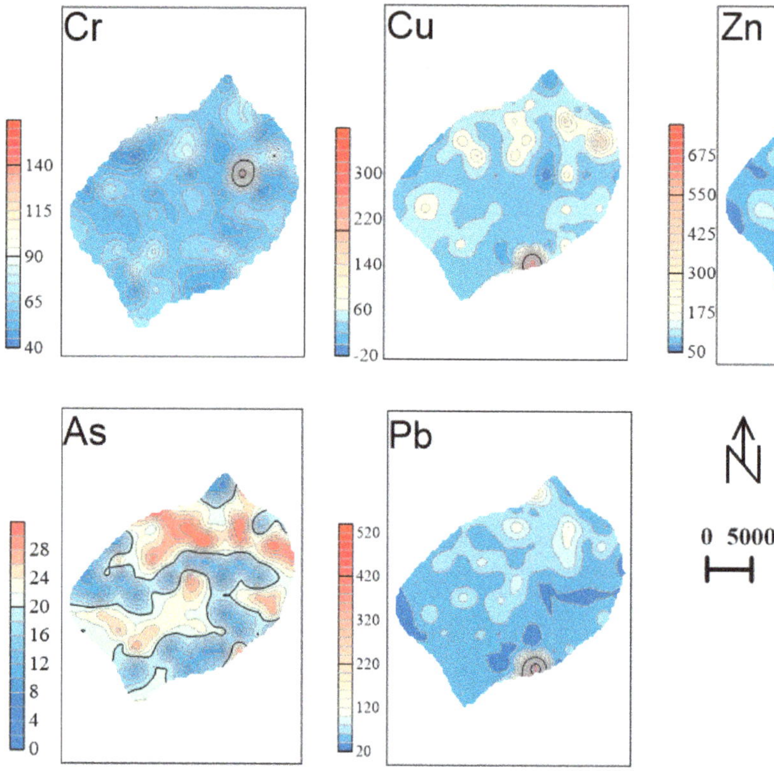

Figure 7. Spatial distribution of elements (Cu, Pb, Zn, As, Cr) in the main urban area of Changchun City.

3.3. Assessment of Leopold Matrix

The Leopold Matrix was evaluated mainly by several environmental experts associated with Changchun City (Table 8). In the Leopold Matrix, our list of environmental activities with the highest impact on the soil are the waste recycling activities of the automotive industry and the manufacturing of pharmaceutical and chemical products, with negative impact evaluation scores of twelve and nine, respectively. The chemical–pharmaceutical industry has the highest score (20.10) for each environmental factor, representing a significant negative impact. The average value of the impact on soil for each activity is 2.67 less than 3.0, which indicates that the activities are all still within the acceptable range. The impact value for all activities is 32.04 for soil, and the total impact values for water and air are 27.88 and 28.36, respectively. Activity factors such as automobile manufacturing and chemical–pharmaceutical industries currently have a relatively large negative impact on the environment, but all are still within the alert range and require treatment activities.

Table 8. Environmental impact analysis based on the Leopold Matrix.

	Physiological Environment			Social Environment		Economic Environment	
Impact Activities	Soil	Water	Air	Infrastructure	Health	Employment	Value of Land
Raw materials for metal production	−2/3	−2/3	−1/3	−2/2	−1/3	1/3	−1/2
Vehicle manufacturing and assembly	−2/3	−2/2	−1/3	−1/2	−2/3	4/3	−1/2

Table 8. *Cont.*

Impact Activities	Physiological Environment			Social Environment		Economic Environment	
	Soil	Water	Air	Infrastructure	Health	Employment	Value of Land
Recycling and disposal of abandoned cars	−4/3	−2/2	−2/3	−1/2	−2/3	2/3	−2/2
Input of chemical raw materials	−3/2	−3/3	−1/3	-	−3/3	1/3	-
Product manufacturing	−4/3	−3/3	−2/3	−1/2	−3/3	4/3	−1/2
Treatment of waste biological products	−3/3	−2/3	−1/3	-	−3/3	3/3	−2/2
Fossil fuel usage	−3/3	−1/2	−4/3	−1/1	−3/3	1/2	-
Waste gas and waste residue emissions	−2/3	−3/3	−4/3	-	−3/3	1/2	−1/2
Fertilizer and pesticide usage	−1/2	−3/3	−1/3	−2/2	−3/3	-	−3/3
Vehicle exhaust emission	−2/2	−1/3	−2/3	-	−1/3	-	−1/3
Sewage disposal	−2/3	−3/3	−1/2	−1/2	−3/3	−1/2	-
Solid waste emissions	−3/3	−2/3	−1/2	−2/2	−3/3	−1/2	−1/3

4. Conclusions

The conclusions we can draw are as follows:

1. The Changchun built-up area as a whole is slightly to moderately polluted, but it needs to be alerted to the contamination of elemental As, as well as Cu and Pb, with the main sources of pollution being metal-related industrial manufacturing, the manufacturing of chemical products, and coal-fired heating. The environmental impacts of activities in the urban areas of Changchun are all within manageable limits and soil remediation should be carried out for the corresponding response sites immediately.
2. The on-site test data obtained by pXRF can be considered as a reliable dataset after processing by the correction model. The order of correction for each element under this simple correction model is as follows: Cu > Pb > Cr > Zn > As. This exploratory correction method can be extended to the correction of other elements, which also provides a valuable reference for the correction of in situ measurements of other potential soil pollutants.
3. The pXRF is efficiently calibrated for real-time scanning of regional soil contamination and large-scale sustainable rapid assessment. Therefore, we advocate that calibrated pXRF data from proximal sensors can be used by government agencies or monitoring organizations as complementary information to enhance spatial monitoring of potentially contaminated sites at the local and regional levels to ensure the safety and health of populations in urban environments.

Author Contributions: All authors contributed to the study's conception and design. X.Z. (Xiaoxiao Zou): Conceptualization, Investigation, Methodology, Software, Writing—original draft. J.L.: Resources, Supervision, Funding acquisition, Formal analysis, Writing—review and editing. X.Z. (Xinyun Zhao): Writing—review and editing, Data curation. Q.W. and Z.G.: Materials, Writing—review and editing. Y.H. and Y.L.: Writing—review and editing. All authors have read and agreed to the published version of the manuscript.

Funding: The authors are grateful for the support from the State Key Research and Development Program (2023YFC2907105), the Special study on mineral resources planning in Changchun (JM-2020-11-13594) and the Jilin Province Provincial School Joint Construction Program Special Project (SXGJXX2017-2).

Institutional Review Board Statement: Not applicable.

Informed Consent Statement: Not applicable.

Data Availability Statement: The data presented in this study are available on request from the corresponding author.

Conflicts of Interest: The authors have no relevant financial or non-financial interests to disclose.

Appendix A. Pollution Index Classification Scale

Index	Numerical Range	Classes
Single pollution index (PI)	PI < 1	non-polluting
	1−2	lightly polluted
	2−3	moderately polluted
	PI > 3	heavily polluting
Geo-accumulation index (I-geo)	I−geo < 0	non-polluted
	0−1	Uncontaminated to moderately contaminated
	1−2	Moderately contaminated
	2−3	Moderately to strongly contaminated
	3−4	Strongly contaminated
	4−5	Strongly to extremely contaminated
	I-geo > 5	Extremely high contaminated
Contamination Factor (CF)	CF < 1	non-polluting
	1−3	lightly polluted
	3−6	moderately polluted
	CF > 6	heavily polluted
Enrichment factor (EF)	EF < 2	Minimal enrichment
	EF = 2−5	Moderate enrichment
	EF = 5−20	Significant enrichment
	EF = 20−40	Very high enrichment
	EF > 40	Extremely high enrichment
The pollution load index (PLI)	PLI < 1	Clearly
	PLI > 1	Polluted
Risk factor (RI)	RI < 20	Low ecological risk
	20−40	Moderate ecological risk
	40−80	Considerable ecological risk
	80−160	High ecological risk
	>160	Serious ecological risk
Degree of contamination ($C\text{-}_{deg}$)	$C-_{deg} < 8$	non-polluting
	8−16	lightly polluted
	16−32	moderately polluted
	$C-_{deg} > 32$	heavily polluted
Nemerow pollution Index (PI-Nemerow)	PIN < 0.7	Class I soils (unpolluted)
	0.7−1	Class II soils (Safety)
	1−2	Class III soils (Mild pollution)
	2 −3	Super tertiary soils (Moderated)
	PIN > 3	Severe pollution

References

1. Tang, S.; Wang, C.; Song, J.; Ihenetu, S.C.; Li, G. Advances in Studies on Heavy Metals in Urban Soil: A Bibliometric Analysis. *Sustainability* **2024**, *16*, 860. [CrossRef]
2. Zhao, M.; Zhao, Y.; Shen, T.; Li, S. Co-Kriging Interpolation of Mn and Zn Pollution Distribution and High-Score Mapping Based on in situ PXRF Data. *Res. Environ. Sci.* **2023**, *36*, 600–609. [CrossRef]
3. Tešić, M.; Stojanović, N.; Knežević, M.; Đunisijević-Bojović, D.; Petrović, J.; Pavlović, P. The Impact of the Degree of Urbanization on Spatial Distribution, Sources and Levels of Heavy Metals Pollution in Urban Soils—A Case Study of the City of Belgrade (Serbia). *Sustainability* **2022**, *14*, 13126. [CrossRef]
4. Chen, L.; Ren, B.; Deng, X.; Yin, W.; Xie, Q.; Cai, Z.; Zou, H. Black shale bedrock control of soil heavy metal typical high geological background in China Loushao Basin: Pollution characteristics, source and Influence assessment based on spatial analysis. *J. Hazard. Mater.* **2024**, *477*, 135072. [CrossRef] [PubMed]
5. Sun, Q.; Sun, B.; Wang, D.; Pu, Y.; Zhan, M.; Xu, X.; Wang, J.; Jiao, W. A review on the chemical speciation and influencing factors of heavy metals in Municipal Solid Waste landfill humus. *Waste Dispos. Sustain. Energy* **2024**, *6*, 209–218. [CrossRef]
6. Bao, W.; Wan, W.; Sun, Z.; Hong, M.; Li, H. Spatial Distribution and Migration of Heavy Metals in Dry and Windy Area Polluted by Their Production in the North China. *Processes* **2024**, *12*, 160. [CrossRef]

7. He, A.; Li, X.; Ai, Y.; Li, X.; Li, X.; Zhang, Y.; Gao, Y.; Liu, B.; Zhang, X.; Zhang, M.; et al. Potentially toxic metals and the risk to children's health in a coal mining city: An investigation of soil and dust levels, bioaccessibility and blood lead levels. *Environ. Int.* **2020**, *141*, 105788. [CrossRef]
8. Jiang, R.; Wang, M.; Xie, T.; Chen, W. Site-specific ecological effect assessment at community level for polymetallic contaminated soil. *J. Hazard. Mater.* **2023**, *445*, 130531. [CrossRef]
9. Shao, Y.; Chen, W.; Li, J.; Yan, B.; He, H.; Zhang, Y. Influencing Factors, Risk Assessment, and Source Identification of Heavy Metals in Purple Soil in the Eastern Region of Guang'an City, Sichuan Province, China. *Minerals* **2024**, *14*, 495. [CrossRef]
10. Adnan, M.; Xiao, B.; Ali, M.U.; Xiao, P.; Zhao, P.; Wang, H.; Bibi, S. Heavy metals pollution from smelting activities: A threat to soil and groundwater. *Ecotoxicol. Environ. Saf.* **2024**, *274*, 116189. [CrossRef]
11. Crocombe, R.A. Portable Spectroscopy. *Appl. Spectrosc.* **2018**, *72*, 1701–1751. [CrossRef] [PubMed]
12. da Costa, M.V.; Lima, G.J.d.O.; Guilherme, L.R.G.; Carneiro, M.A.C.; Ribeiro, B.T. Towards direct and eco-friendly analysis of plants using portable X-ray fluorescence spectrometry: A methodological approach. *Chemosphere* **2023**, *339*, 139613. [CrossRef] [PubMed]
13. Da Silva, A.; Triantafyllou, A.; Delmelle, N. Portable x-ray fluorescence calibrations: Workflow and guidelines for optimizing the analysis of geological samples. *Chem. Geol.* **2023**, *623*, 121395. [CrossRef]
14. Santos, N.F.; Guilherme, L.R.G.; Carneiro, M.A.C.; Guerra, M.B.B. A simple and reliable calibration method for direct analysis of ground-roasted coffee by portable XRF: An accurate analytical tool for total diet studies. *J. Anal. At. Spectrom.* **2024**, *39*, 1152–1159. [CrossRef]
15. Ravansari, R.; Wilson, S.C.; Tighe, M. Portable X-ray fluorescence for environmental assessment of soils: Not just a point and shoot method. *Environ. Int.* **2020**, *134*, 105250. [CrossRef]
16. Shrestha, G.; Calvelo-Pereira, R.; Roudier, P.; Martin, A.; Turnbull, R.; Kereszturi, G.; Jeyakumar, P.; Anderson, C. Quantification of multiple soil trace elements by combining portable X-ray fluorescence and reflectance spectroscopy. *Geoderma* **2022**, *409*, 115649. [CrossRef]
17. Wang, D.; Lu, J.; Wu, J.; Li, B.; Nyasha, N.K. Enrichment Characteristics of Hazardous Trace Elements in Feed Coal and Coal Ash in Huaibei Area under Leaching. *Toxics* **2023**, *11*, 308. [CrossRef]
18. Korbel, C.; Mezoued, N.; Demeusy, B.; Fabre, C.; Cauzid, J.; Filippova, I.V.; Filippov, L.O. Quantification of lithium using handheld instruments: Application of LIBS and XRF spectroscopy to assay the lithium content of mineral processing products. *J. Anal. At. Spectrom.* **2024**, *39*, 1838–1853. [CrossRef]
19. Adimalla, N.; Chen, J.; Qian, H. Spatial characteristics of heavy metal contamination and potential human health risk assessment of urban soils: A case study from an urban region of South India. *Ecotoxicol. Environ. Saf.* **2020**, *194*, 110406. [CrossRef]
20. Xia, F.; Hu, B.; Zhu, Y.; Ji, W.; Chen, S.; Xu, D.; Shi, Z. Improved Mapping of Potentially Toxic Elements in Soil via Integration of Multiple Data Sources and Various Geostatistical Methods. *Remote Sens.* **2020**, *12*, 3775. [CrossRef]
21. Shi, J.; Zhao, D.; Ren, F.; Huang, L. Spatiotemporal variation of soil heavy metals in China: The pollution status and risk assessment. *Sci. Total. Environ.* **2023**, *871*, 161768. [CrossRef] [PubMed]
22. Sun, Y.; Zhao, Y.; Hao, L.; Zhao, X.; Lu, J.; Wei, Q.; Shi, Y.; Ma, C. Evaluation and Source Identification of Heavy Metal Pollution in Black Soils, Central-Eastern Changchun, China. *Sustainability* **2023**, *15*, 7419. [CrossRef]
23. Wang, Z.; Zhang, Y.; Wang, L.; Li, X.; Zhou, X.; Li, X.; Yan, M.; Lu, Q.; Tang, Z.; Zhang, G.; et al. Characteristics and Risk Assessments of Mercury Pollution Levels at Domestic Garbage Collection Points Distributed within the Main Urban Areas of Changchun City. *Toxics* **2021**, *9*, 309. [CrossRef]
24. Zhang, P.; Dong, Y.; Guo, Y.; Wang, C.; Wang, G.; Ma, Z.; Zhou, W.; Zhang, D.; Ren, Z.; Wang, W. Urban forest soil is becoming alkaline under rapid urbanization: A case study of Changchun, northeast China. *Catena* **2023**, *224*, 106993. [CrossRef]
25. Liu, Y.; Zhang, J.; Li, C.; Zhou, G.; Fu, Z.; Liu, D. Influential intensity of urban agglomeration on evolution of eco-environmental pressure: A case study of Changchun, China. *Chin. Geogr. Sci.* **2017**, *27*, 638–647. [CrossRef]
26. Ma, Z.; He, X.G.; Tong, X.Z.; Duan, H.Y.; Wang, X.E.; Dong, D.M. The Study on Carbon Emission Influencing Factors of Industrial Energy Consumption of Changchun City. *Appl. Mech. Mater.* **2012**, *164*, 302–305. [CrossRef]
27. Wang, J.; Yang, J.; Chen, T. Source appointment of potentially toxic elements (PTEs) at an abandoned realgar mine: Combination of multivariate statistical analysis and three common receptor models. *Chemosphere* **2022**, *307*, 135923. [CrossRef]
28. Ministry of Ecology and Environment, PRC. Potentiometric Method for the Determination of Soil pH: HJ962-2018[S]. 2018. Available online: https://www.scirp.org/reference/referencespapers?referenceid=3795318 (accessed on 27 October 2024).
29. Ministry of Ecology and Environment, PRC. Determination of Organic Matter in Forest Soil and Calculation Carbon-Nitrogen Ratio:LY/T1237-1999[S]. 1999. Available online: https://www.cnki.com.cn/Article/CJFDTOTAL-NXNL903.006.htm (accessed on 27 October 2024).
30. Xie, S.; Yang, F.; Feng, H.; Wei, C.; Wu, F. Assessment of Potential Heavy Metal Contamination in the Peri-urban Agricultural Soils of 31 Provincial Capital Cities in China. *Environ. Manag.* **2019**, *64*, 366–380. [CrossRef]
31. Hao, L.; Tian, M.; Zhao, X.; Zhao, Y.; Lu, J.; Bai, R. Spatial distribution and sources of trace elements in surface soils, Changchun, China: Insights from stochastic models and geostatistical analyses. *Geoderma* **2016**, *273*, 54–63. [CrossRef]
32. Tardani, D.; Vera, F.; Álvarez-Amado, F.; Tolorza, V.; Lacassie, J.P.; Jullian, D.; Sepúlveda, C.; Sánchez-Alfaro, P.; Daniele, L.; Gutiérrez, L. Evaluating natural and anthropogenic inputs on the distribution of potentially toxic elements in urban soil of Valdivia, Chile. *Environ. Geochem. Health* **2023**, *45*, 7841–7859. [CrossRef]

33. Horf, M.; Gebbers, R.; Vogel, S.; Ostermann, M.; Piepel, M.-F.; Olfs, H.-W. Determination of Nutrients in Liquid Manures and Biogas Digestates by Portable Energy-Dispersive X-ray Fluorescence Spectrometry. *Sensors* **2021**, *21*, 3892. [CrossRef]
34. Piikki, K.; Söderström, M.; Eriksson, J.; John, J.M.; Muthee, P.I.; Wetterlind, J.; Lund, E. Performance Evaluation of Proximal Sensors for Soil Assessment in Smallholder Farms in Embu County, Kenya. *Sensors* **2016**, *16*, 1950. [CrossRef] [PubMed]
35. Guo, J.; Lu, J.; Chen, Z.; Zhao, X.; Wei, Q.; Fan, Y.; Lan, T. Geochemical study of the Ashele Copper-Zinc Deposit using portable X-ray fluorescence spectrometry based on matrix effect correction, Northwest China. *Appl. Geochem.* **2022**, *146*, 105461. [CrossRef]
36. Kazimoto, E.O.; Messo, C.; Magidanga, F.; Bundala, E. The use of portable X-ray spectrometer in monitoring anthropogenic toxic metals pollution in soils and sediments of urban environment of Dar es Salaam Tanzania. *J. Geochem. Explor.* **2018**, *186*, 100–113. [CrossRef]
37. Wang, C.; Lv, J.; Wang, X.; Khumalo, A.; Hong, A. A novel algorithm combined X-ray fluorescence and Neural Network (XRF-NN) for coal ash content prediction: Algorithm design and performance evaluation. *Physicochem. Probl. Miner. Process.* **2024**, *60*, 193187. [CrossRef]
38. Lu, J.; Guo, J.; Wei, Q.; Tang, X.; Lan, T.; Hou, Y.; Zhao, X. A Matrix Effect Correction Method for Portable X-ray Fluorescence Data. *Appl. Sci.* **2022**, *12*, 568. [CrossRef]
39. Tulla, P.S.; Kumar, P.; Vishwakarma, D.K.; Kumar, R.; Kuriqi, A.; Kushwaha, N.L.; Rajput, J.; Srivastava, A.; Pham, Q.B.; Panda, K.C.; et al. Daily suspended sediment yield estimation using soft-computing algorithms for hilly watersheds in a data-scarce situation: A case study of Bino watershed, Uttarakhand. *Theor. Appl. Clim.* **2024**, *155*, 4023–4047. [CrossRef]
40. He, S.; Zhang, Y.; Luo, L.; Song, Y. Establishment of Remote Sensing Inversion Model and Its Application in Pollution Source Identification: A Case Study of East Lake in Wuhan. *Remote Sens.* **2024**, *16*, 3402. [CrossRef]
41. da Silva, T.R.; de Almeida, E.; Tavares, T.R.; Melquiades, F.L.; Baesso, M.M.; de Camargo, R.F.; Gomes, M.H.F.; de Carvalho, H.W.P. In situ determination of soybean leaves nutritional status by portable X-ray fluorescence: An initial approach for data collection and predictive modelling. *Biosyst. Eng.* **2024**, *247*, 143–152. [CrossRef]
42. Bashir, O.; Bangroo, S.A.; Shafai, S.S.; Senesi, N.; Kader, S.; Alamri, S. Geostatistical modeling approach for studying total soil nitrogen and phosphorus under various land uses of North-Western Himalayas. *Ecol. Informatics* **2024**, *80*, 102520. [CrossRef]
43. Horta, A.; Azevedo, L.; Neves, J.; Soares, A.; Pozza, L. Integrating portable X-ray fluorescence (pXRF) measurement uncertainty for accurate soil contamination mapping. *Geoderma* **2021**, *382*, 114712. [CrossRef]
44. Zhang, J.; Liu, Z.; Tian, B.; Li, J.; Luo, J.; Wang, X.; Ai, S.; Wang, X. Assessment of soil heavy metal pollution in provinces of China based on different soil types: From normalization to soil quality criteria and ecological risk assessment. *J. Hazard. Mater.* **2023**, *441*, 129891. [CrossRef] [PubMed]
45. Zhang, Y.; Lu, X.; Deng, S.; Zhu, T.; Yu, B. Bibliometric and visual analysis of heavy metal health risk assessment: Development, hotspots and trends. *Arch. Environ. Prot.* **2024**, *50*, 56–71. [CrossRef]
46. Dange, S.; Arumugam, K.; Vijayaraghavalu, S.S. Geochemical Insights into Heavy Metal Contamination and Health Hazards in Palar River Basin: A Pathway to Sustainable Solutions. *Ecol. Indic.* **2024**, *166*, 112568. [CrossRef]
47. Guo, G.; Wang, Y.; Zhang, D.; Lei, M. Source-specific ecological and health risks of potentially toxic elements in agricultural soils in Southern Yunnan Province and associated uncertainty analysis. *J. Hazard. Mater.* **2021**, *417*, 126144. [CrossRef]
48. Ferreira, S.L.; da Silva, J.B.; dos Santos, I.F.; de Oliveira, O.M.; Cerda, V.; Queiroz, A.F. Use of pollution indices and ecological risk in the assessment of contamination from chemical elements in soils and sediments—Practical aspects. *Trends Environ. Anal. Chem.* **2022**, *35*, e00169. [CrossRef]
49. Yang, Y.; Lu, X.; Yu, B.; Zuo, L.; Wang, L.; Lei, K.; Fan, P.; Liang, T.; Rennert, T.; Rinklebe, J. Source-specific risk judgement and environmental impact of potentially toxic elements in fine road dust from an integrated industrial city, North China. *J. Hazard. Mater.* **2023**, *458*, 131982. [CrossRef]
50. Buat-Menard, P.; Chesselet, R. Variable influence of the atmospheric flux on the trace metal chemistry of oceanic suspended matter. *Earth Planet. Sci. Lett.* **1979**, *42*, 398–411. [CrossRef]
51. Ardila, P.A.R.; Álvarez-Alonso, R.; Árcega-Cabrera, F.; Valsero, J.J.D.; García, R.M.; Lamas-Cosío, E.; Oceguera-Vargas, I.; DelValls, A. Assessment and Review of Heavy Metals Pollution in Sediments of the Mediterranean Sea. *Appl. Sci.* **2024**, *14*, 1435. [CrossRef]
52. Hossain, M.M.; Mojumdar, S.; Islam, S.; Rahman, A.; Alahmadi, T.A.; Ansari, M.J.; Mistry, S.K. A comprehensive analysis of health risks from metal contamination in the Sundarbans mangrove forest ecosystem in Bangladesh. *Toxicol. Environ. Health Sci.* **2024**, 1–29. [CrossRef]
53. Alqattan, Z.A.; Artiola, J.F.; Walls, D.; Ramírez-Andreotta, M.D. Evaluating the portable X-ray fluorescence reliability for metal(loid)s detection and soil contamination status. *Environ. Monit. Assess.* **2024**, *196*, 765. [CrossRef] [PubMed]
54. Plak, A.; Telecka, M.; Charzyński, P.; Hanaka, A. Evaluation of hazardous element accumulation in urban soils of Cracow, Lublin and Torun (Poland): Pollution and ecological risk indices. *J. Soils Sediments* **2024**, *24*, 3286–3296. [CrossRef]
55. Gong, C.; Quan, L.; Chen, W.; Tian, G.; Zhang, W.; Xiao, F.; Zhang, Z. Ecological risk and spatial distribution, sources of heavy metals in typical purple soils, southwest China. *Sci. Rep.* **2024**, *14*, 11342. [CrossRef] [PubMed]
56. Arani, M.H.; Mohammadzadeh, M.; Kalantary, R.R.; Rad, S.H.; Moslemzadeh, M.; Jaafarzadeh, N. Environmental impact assessment of a steel industry development plan using combined method involving Leopold matrix and RIAM. *J. Environ. Health Sci. Eng.* **2021**, *19*, 1997–2011. [CrossRef]
57. Hnin, H.W.; Bonnet, S.; Gheewala, S.H. Environmental impact assessment of electricity production from municipal solid waste in Yangon, Myanmar. *Environ. Dev. Sustain.* **2024**, 1–36. [CrossRef]

58. Ayiwouo, M.N.; Sriram, S.; Ngounouno, F.Y.; Rajagopal, K.; Ngounouno, I. Assessment of the environmental impacts of gold mining activities at Gankombol (Adamawa-Cameroon) using Leopold matrix, Fecteau grid and remote sensing approach. *J. Afr. Earth Sci.* **2023**, *207*, 105050. [CrossRef]
59. Valizadeh, S.; Hakimian, H. Evaluation of waste management options using rapid impact assessment matrix and Iranian Leopold matrix in Birjand, Iran. *Int. J. Environ. Sci. Technol.* **2018**, *16*, 3337–3354. [CrossRef]
60. Xu, D.; Chen, S.; Xu, H.; Wang, N.; Zhou, Y.; Shi, Z. Data fusion for the measurement of potentially toxic elements in soil using portable spectrometers. *Environ. Pollut.* **2020**, *263*, 114649. [CrossRef]
61. Zhao, M.; Chen, Z.; Qian, C.; Zhao, Y.; Xu, Y.; Liu, Y. Correcting correlation quality of portable X-ray fluorescence to better map heavy metal contamination by spatial co-kriging interpolation. *Ecotoxicol. Environ. Saf.* **2024**, *271*, 115962. [CrossRef]
62. Shahab, A.; Hui, Z.; Rad, S.; Xiao, H.; Siddique, J.; Huang, L.L.; Ullah, H.; Rashid, A.; Taha, M.R.; Zada, N. A comprehensive review on pollution status and associated health risk assessment of human exposure to selected heavy metals in road dust across different cities of the world. *Environ. Geochem. Health* **2022**, *45*, 585–606. [CrossRef]
63. Ma, M.; Fang, L.; Zhao, N.; Ma, X. Detection of Cadmium and Lead Heavy Metals in Soil Samples by Portable Laser-Induced Breakdown Spectroscopy. *Chemosensors* **2024**, *12*, 40. [CrossRef]
64. Mugudamani, I.; Oke, S.A.; Gumede, T.P. Influence of Urban Informal Settlements on Trace Element Accumulation in Road Dust and Their Possible Health Implications in Ekurhuleni Metropolitan Municipality, South Africa. *Toxics* **2022**, *10*, 253. [CrossRef] [PubMed]
65. Ren, Z.; Christakos, G.; Lou, Z.; Xu, H.; Lv, X.; Fei, X. Contamination Assessment and Source Apportionment of Metals and Metalloids Pollution in Agricultural Soil: A Comparison of the APCA-MLR and APCA-GWR Models. *Sustainability* **2022**, *14*, 783. [CrossRef]
66. Li, Y.; Zhou, S.; Liu, K.; Wang, G.; Wang, J. Application of APCA-MLR receptor model for source apportionment of char and soot in sediments. *Sci. Total. Environ.* **2020**, *746*, 141165. [CrossRef]
67. Zhou, Y.; Du, S.; Liu, Y.; Yang, T.; Liu, Y.; Li, Y.; Zhang, L. Source identification and risk assessment of trace metals in surface sediment of China Sea by combining APCA-MLR receptor model and lead isotope analysis. *J. Hazard. Mater.* **2024**, *465*, 133310. [CrossRef]
68. Zhu, Y.; Liu, B.; Jin, G.; Wu, Z.; Wang, D. Identifying the Local Influencing Factors of Arsenic Concentration in Suburban Soil: A Multiscale Geographically Weighted Regression Approach. *Toxics* **2024**, *12*, 229. [CrossRef]

toxics

MDPI

Article

Lead Concentrations in Tissues of Pigeons (*Columba livia*) in the Urban Area of Comarca Lagunera, Mexico

Andrea Ocampo-Lopez [1], Cristo Omar Puente-Valenzuela [2], Homero Sánchez-Galván [2], Ana Alejandra Valenzuela-García [2], Josué Raymundo Estrada-Arellano [2], Ramón Alfredo Delgado-González [1], Jorge Alejandro Aguirre-Joya [3], Cristian Torres-León [3], Alejandra Ocampo-Lopez [3] and David Ramiro Aguillón-Gutiérrez [3,*]

1 Veterinay Diagnostic Unit, Laguna Unit, Antonio Narro Agrarian Autonomous University, Periférico Raúl López Sánchez, Col. Valle Verde s/n, Torreon 27054, Mexico; ocampoandrea2304@gmail.com (A.O.-L.); raldego@gmail.com (R.A.D.-G.)

2 Faculty of Biology, Juarez University of the State of Durango, Av. Universidad s/n, Fracc. Filadelfia, Gomez Palacio 35010, Mexico; puvacrom@ujed.mx (C.O.P.-V.); hosagafcb@gmail.com (H.S.-G.); ale.valenzuela@ujed.mx (A.A.V.-G.); j.estradarellano@gmail.com (J.R.E.-A.)

3 Research center and Ethnobiological Garden, Autonomous University of Coahuila, Dr. Francisco González 37, Viesca 27480, Mexico; jorge_aguirre@uadec.edu.mx (J.A.A.-J.); ctorresleon@uadec.edu.mx (C.T.-L.); jandisoc@gmail.com (A.O.-L.)

* Correspondence: david_aguillon@uadec.edu.mx

Abstract: The Comarca Lagunera is one of Mexico's most important productive areas. Its main economic activities are livestock, agriculture, and the processing industry. A wide variety of industries emit wastes that are considered highly toxic environmental pollutants, which have strong negative impacts on public health. The objective of this work was to determine the lead concentrations present in tissues of pigeons (*Columba livia*) belonging to the urban area of the Comarca Lagunera, Mexico. Specimens were collected from the localities that comprise the region and the tissue extracted; the organs were dried, calcined, and diluted in an acidic HCl solution. Lead concentrations were obtained by atomic absorption spectrometry using the graphite furnace technique. The results demonstrate the presence of lead in all the tissues analyzed, with maximum concentrations of 191.14 mg/kg and minimum concentrations of 0.86 mg/kg, the area with the highest average concentration being Torreón, Coahuila (p = 0.030). The organ with the highest concentration was the bone (p = 0.000). Evidence of lead poisoning is presented in *Columba livia* tissues in the Comarca Lagunera, thus demonstrating the presence of this contaminant and the ability of these pigeons to function as bioindicators of environmental contamination.

Keywords: heavy metals; contamination; pigeon; bioindicator; urban area

1. Introduction

The environment comprises the sum of biotic factors, which describe living organisms, from plants and animals to microorganisms, and abiotic factors, comprising those non-living entities vital for the development of biotic factors [1]. These factors are in states of equilibrium and exchange on a continuous basis, but can be altered by the introduction of harmful substances to living entities, leading to contamination of the environment and increased risk of disease [2]. Pollution is defined as the introduction of any harmful substance, whether in solid, liquid, or gaseous form, which at a certain concentration alters the environment into which it is introduced [1]. At present, environmental pollution is one of the problems with the greatest impact on the environment, not only because of the alterations caused in the environment but also because of the effect it has on public and individual health; contact with this type of pollutant can represent an important risk factor for the increase in morbidity and mortality in different populations of living beings [3].

Large cities and metropolitan areas face serious pollution problems due to overcrowded places, high flows of motorized vehicles, and unplanned industrial developments [4]. The effects of exposure to pollutants affect not only current but also future generations, bringing with them health problems such as cancers, reduced fertility, respiratory problems, and various inflammatory disorders, among others [5,6]. There are different types of pollution. In the case of air pollution, the primary sources come from industrial activities, burning fossil fuels, and various wastes [7]. Large industrial complexes constitute stationary sources of various pollutants such as fine dusts, sulfur and nitrogen dioxide, carbon monoxide, ozone, volatile compounds, and heavy metals [8].

The Comarca Lagunera, southwest of Coahuila and northeast of Durango, is one of Mexico's most important productive areas. Its main economic activities are livestock, agriculture, and the processing industry. Anthropogenic activities in the region have led to increased environmental pollution, bringing with them a growing need for monitoring, remediation, and prevention measures [9]. The Comarca Lagunera is home to the world's largest silver refining plant, the leading producer of gold and lead in Latin America. The metallurgical complex in the region is composed of lead smelting and lead–silver refining and has a capacity of 118,000 tonnes of bullion in the lead smelter [10].

Heavy metals have been present on Earth since its formation, but their concentrations have increased in terrestrial and aquatic environments in an emergent manner due to anthropogenic activities. They are defined as metals and metalloids with densities above 5 g/cm^3, with bioaccumulate potential once in food chains, and commonly with high toxicity to living organisms that come into contact with them [11]. Media such as soil and air and accumulation in living organisms can reflect heavy metal exposure to the population, with inhalation, ingestion, and direct skin contact being the usual routes of entry [12]. Heavy metals have a natural origin on our planet, but their concentrations have started to fluctuate and increase due to human activity. Many have harmful effects on the health of living beings, which have access to them through inhalation, ingestion, and contact. Once inside the organism, they cause various types of physiological damage. Among the most important metals with negative health impacts are lead, cadmium, arsenic, chromium, aluminum, iron, and mercury [13]. Lead is a toxic element that can accumulate in blood and bone and reach significant concentrations in the kidneys, liver, brain, and skin; exposure to it can cause nervous system, circulatory, gastrointestinal, and hormonal problems [14]. Among the most common anthropogenic activities through which lead can be found in the environment are mining, waste from the battery industry, and its use as an additive to substances such as gasoline, paints, and varnishes, as well as the raw material processing industry, including mining resources [15]. In nature, lead does not exist in pure form; it is part of the components of other minerals, the predominant example being galena, where silver is mostly found [16].

The main routes of entry of lead into the body are considered to be digestive, respiratory, and cutaneous [17], and it also has an important impact at the gestational level due to its ability to cross the placental barrier in mammals and its ability to mobilize during the laying season in birds together with the calcium that makes up the eggshell [18]. It plays an important toxic role by substituting for calcium in the processes in which it is involved, thus affecting normal neuronal functions such as the release of neurotransmitters, causing excitotoxicity and alteration of myelin synthesis [19]. This calcium substitution process also causes alterations in bone composition; it can alter bone homeostasis by competing with calcium for its place in apatite networks, and once this metal is absorbed, it is stored in mineralized tissues [20]. In the case of the liver, it is considered one of the major soft organ accumulators of lead and can have an acute and chronic presentation [21]; when exposure to this substance is chronic, it causes dyslipidemia, hormonal depression, and hypercholesterolemia [22].

The pigeon *C. livia*, also called rock pigeon, is a bird species with an average size ranging from 30.5 to 35.5 cm wing length and a weight of 180–355 g; it does not present sexual dimorphism, and its plumage varies between a light grey pattern and two large blackish stripes on the wings, white rump, and purple and green iridescence on the neck; it is possible to find some species where these color characteristics are not met. The shades can range from white to brown [23]. Pigeons of the species *Columba livia* have a high potential as environmental bioindicators; it is a species that is distributed worldwide and is even considered urban fauna, although it is possible to find them in feral and domesticated states, given that their populations are abundant in cities and rural areas. They are considered suitable bioindicators for measuring environmental health because they do not migrate easily from their nesting sites. From the conditions in which the concentration of Pb in these individuals is found, the presence of this and other pollutants in different animal species can be assumed, functioning as a reflection of the environmental conditions of the urban area of the Comarca Lagunera, Mexico, and as a method of monitoring possible risks to human health and the affectation of the diversity of local ecosystems.

Several studies, such as the one conducted by Delgado [24] in Mexico City, demonstrate the ability of *C. livia* to resist these pollutants and reflect the environmental condition. Nam [25] used domestic pigeons of the species *C. livia* from urban and industrial areas to measure the concentration of lead and cadmium in the city of Seoul, South Korea, using the bone, kidney, liver, and lung for analysis, reporting lead concentrations of 29.5 ± 21.1 µg/wet weight, 4.13 ± 1.31 µg/wet weight, 2.33 ± 0.78 µg/wet weight, and 1.72 ± 0.66 µg/wet weight, respectively, in adults. In 2010, in the city of Rabat-Salé, Morocco, a study was carried out to measure air pollution by heavy metals using feral pigeons as bioindicators, determining the concentration of these metals through samples of kidney, liver, lung, heart, and blood tissue, reporting the highest concentration of lead and cadmium in kidneys (0.56 ± 0.06 µg/kg and 3.07 ± 0.78 µg/kg, respectively) [26]. Begum and Sehrin [27] evaluated heavy metal levels in birds of this species and focused on human consumption, where concentrations of 0.067 ± 0.52 µg/g of lead were obtained. In 2021, Valladares [28] carried out an analysis of lead, cadmium, and arsenic content in the liver and bone of *C. livia* pigeons found in areas known to be previously contaminated with mining waste, reporting averages of 17.027 µg/g and 108.436 µg/g of lead, respectively.

Considering this background, the Comarca Lagunera being a region with high environmental pollution, and that there are no studies in this area of bioindicators of heavy metal levels using birds, the purpose of the present research is to determine the levels of lead in tissues of domestic pigeons.

2. Materials and Methods

2.1. Study Site

The Comarca Lagunera is located between the states of Durango and Coahuila, has an extension of more than 44,887 km^2, and most of it is in the Chihuahuan Desert [29]. It comprises 15 municipalities, 4 of which were selected for trapping (Figure 1). It is one of the ten most productive areas in the country and has more than 1.2 million inhabitants [30]. Its temperatures are warm, with characteristic semi-dry temperate climates in the higher elevations and dry temperate climates in the lower elevations; in the middle elevations, the climate is very dry and semi-warm [31], which corresponds to the classification BWhw″ (e′) [32].

2.2. Biological Model

The pigeon *C. livia* is an invasive bird native to Africa and Eurasia, belonging to the Columbidae family (Table 1), with an excellent adaptive system to human environments that allows it to have very high population growth [33]. Its close relationship with humans stems from the anthropogenic food source and nesting sites provided by urban constructions [34]. The pigeons have demonstrated a wide potential as bioindicators and have been one of the most widely used biological models in determining metals in tissues [35].

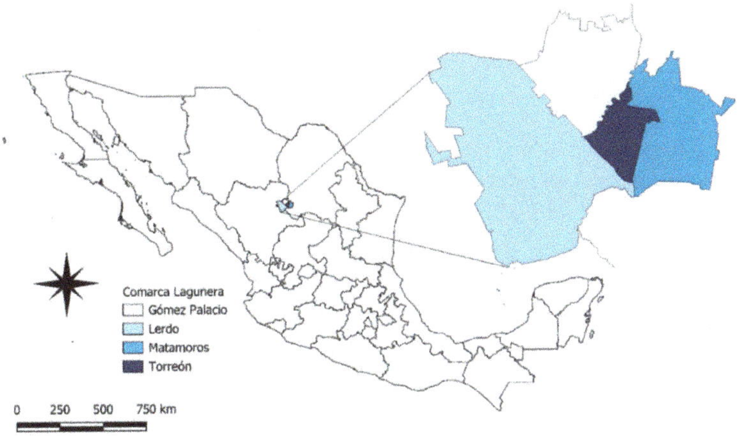

Figure 1. Sampling points in the Comarca Lagunera, Mexico.

Table 1. Taxonomic classification of *Columba livia* [36].

Kingdom	Animalia
Filum	Chordata
Class	Birds
Order	Columbiformes
Family	Columbidae
Subfamily	Columbinae
Genus	*Columba*
Specie	*Columba livia*

2.3. Fieldwork

Birds were taken from each of the municipalities (Gómez Palacio, Durango, $n = 25$; Lerdo, Durango, $n = 12$; Torreón, Coahuila, $n = 16$; Matamoros, Coahuila, $n = 12$); a specific area was chosen for sampling based on the criteria of the number of people and businesses in the area, its central location in each city sampled, and the fact that they are wooded areas and have the necessary urban infrastructure for the presence of *C. livia* pigeon populations. These points are public squares with similar climates and flora. In order to carry out the capture activities, a permit was obtained from the Ministry of the Environment and Natural Resources (SEMARNAT), obtaining the folio SPARN/DGVS/06267/22. The captures were carried out during daylight hours from 20 July to 12 August 2022. The collection sites and their coordinates are shown below:

- Parque Victoria, Lerdo, Durango (25°32′33″ N 103°31′23″ W).
- Plaza de Armas, Gómez Palacio, Durango (25°34′05″ N 103°30′00″ W).
- Alameda Zaragoza, Torreón, Coahuila (25°32′21″ N 103°26′43″ W).
- Plaza Mayor, Matamoros, Coahuila (25°31′41″ N 103°13′46″ W).

The specimens were captured using manual nets and carried out by a collection brigade of two people from 20 to 30 June 2022, from 9:00 to 13:00 h. The specimens were taken the same day of their capture to the Pathology Laboratory of the Diagnostic Unit of the Universidad Autónoma Agraria Antonio Narro (Torreón, Mexico), Laguna Unit, where they were processed.

2.4. Specimen Processing

Birds belonging to all groups collected were slaughtered by cervical dislocation in accordance with the provisions of the Mexican Official Standard for the Humane Slaughter of Domestic and Wild Animals (NOM-033-ZOO-2014). The primary necropsy procedure was performed using a dissection kit. Brain, lung, liver, kidney, and bone were extracted using plastic tools; organs that were not processed during this study were preserved in 5% formalin and frozen (0 to 4 °C) for further studies. During necropsy, identification of the reproductive organs of the specimens was carried out to determine sex.

2.5. Sample Processing and Analysis

Porcelain crucibles were brought to constant weight at a temperature of 200 °C for 2 h in an oven, after which they were left to stand in a desiccator for 2 h, and their weight was recorded. The organs chosen for the study were dried using an oven at 105 °C and handled in the crucibles at constant weight. The weight of each organ was recorded, and they were then incinerated at 500 °C for 12 h in the muffle. After this period, the resulting ashes were suspended in 25 mL of HCl solution with a 2N concentration following the methodology of Delgado [24]. The ashes suspended in 2N HCl were diluted using distilled water to concentrations favoring their reading in the spectra (Figure 2). Analysis of the concentrations was performed using graphite furnace atomic absorption spectrophotometry, the blanks were Tritón 100x and array modifier, and the 2N HCl was also used as a blank during the analysis to check the presence of contamination of samples. This study used a NIST 1577b bovine liver sample, with a recovery of >92%, a quantification detection limit of 10 ng/g, and a detection limit of 3 ng/g.

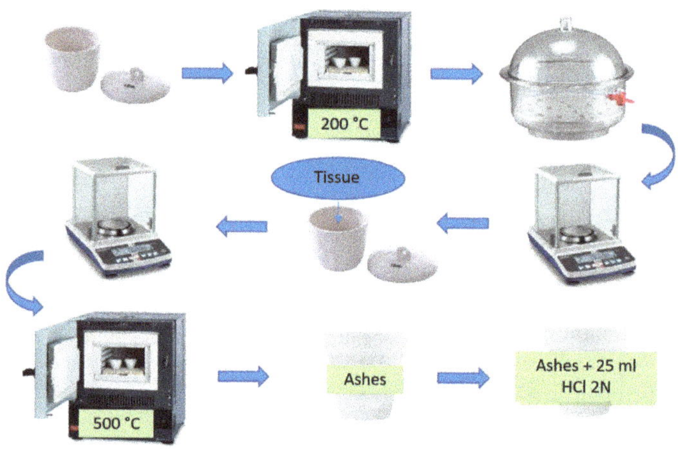

Figure 2. Sample processing and analysis.

2.6. Statistical Analysis

Statistical analysis was performed using Kruskal–Wallis and post hoc Mann–Whitney. Normality tests were performed using Shapiro–Wilk and Kolmogorov–Smirnov. The statistical program used was GraphP 10.1.1.

3. Results

Sex organ analysis determined that of the 64 pigeons collected, 16 (25%) were male, and 48 (75%) were female. The average weight of the birds was 264.5 g, with a minimum of 185 g and a maximum of 340 g. Figure 3 compares all the samples' concentrations, with bone being the organ with the highest concentrations of all and liver and lung when comparing soft organs; Figure 4 compared only soft tissues. Torreón, Coahuila, is the city with the highest values in bone, heart, lungs, and brain; Lerdo, Durango, is the city with the highest average lead concentration in liver; the average lead concentration in kidney was highest in the city of Matamoros, Coahuila, compared to the other cities. There is lead concentration in all tissues and cities sampled. The concentrations of metals are shown in Figure 5 by organ, city, and bird sex. The tissue and city with the highest lead concentration were bones ($p = 0.000$) and Torreón Coahuila ($p = 0.030$), respectively, using the Kruskal–Wallis test and a Mann–Whitney post hoc.

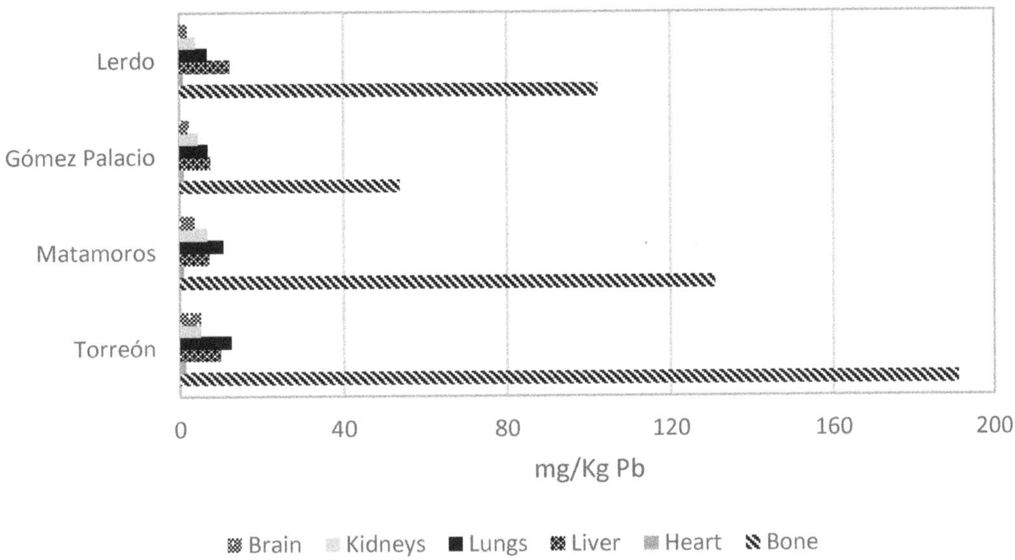

Figure 3. Comparison of lead concentration among all organs and cities. The concentration in bones is more elevated compared with the soft tissues.

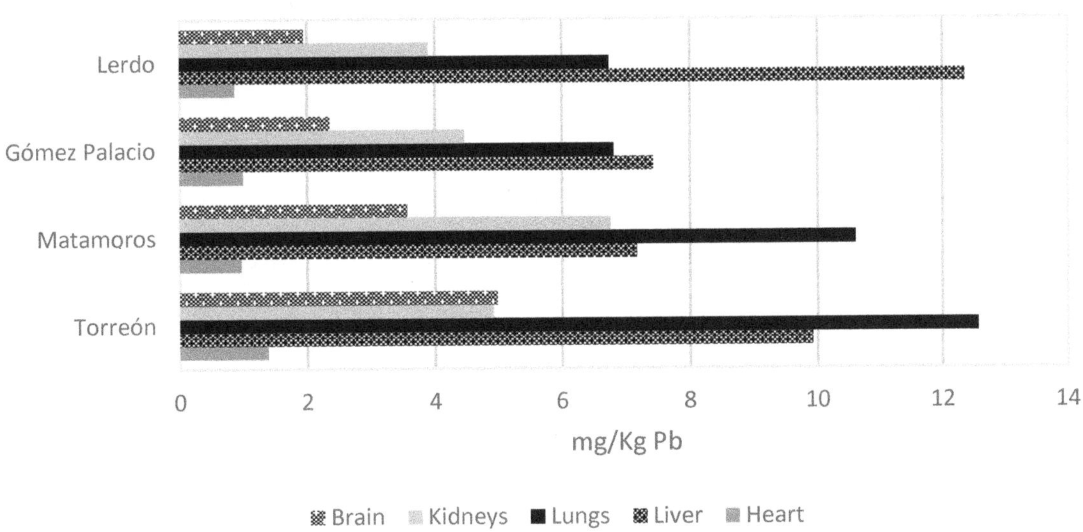

Figure 4. Comparison of lead concentration among soft tissues.

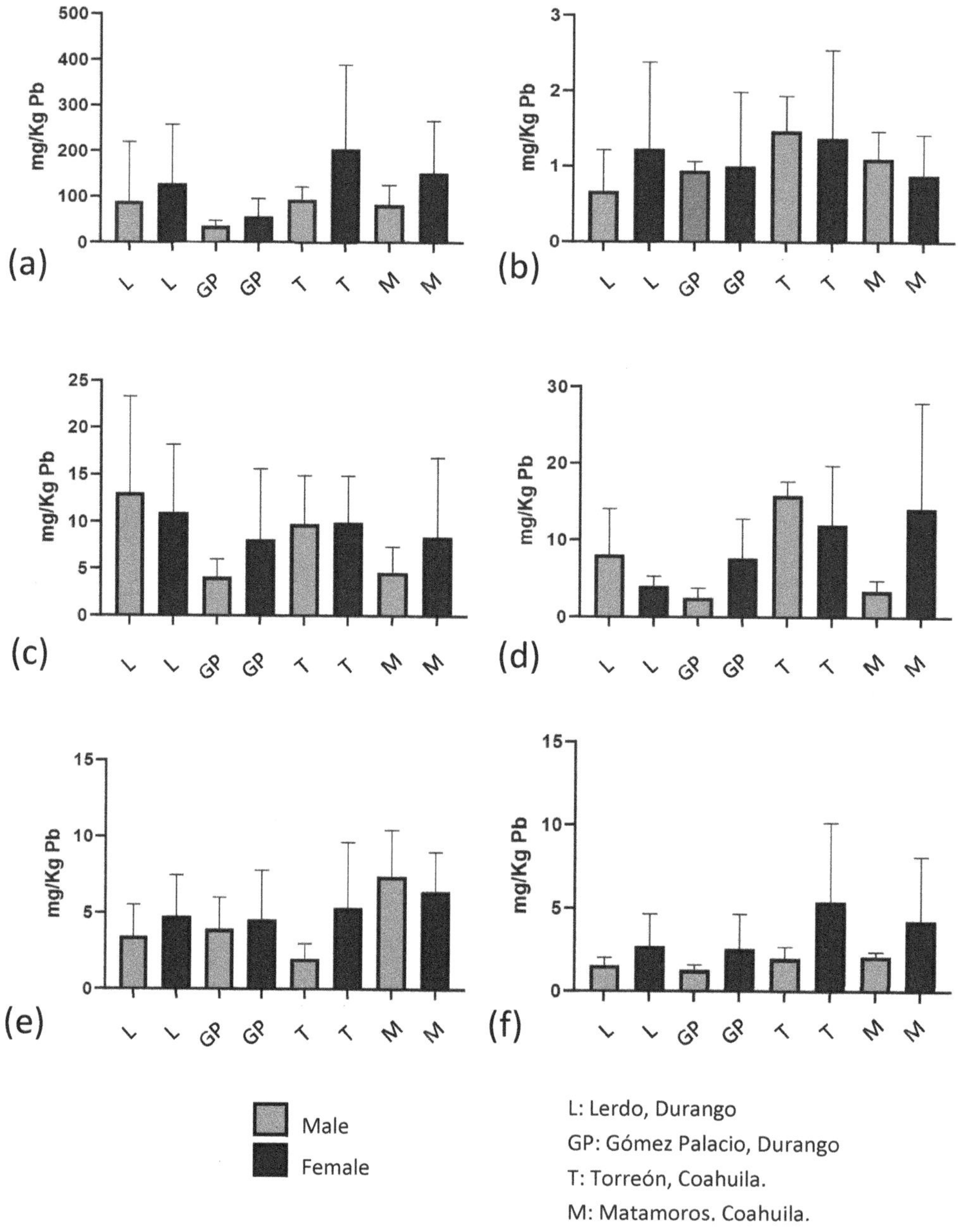

Figure 5. Lead concentrations in tissues of *C. livia*. (**a**) Bone, (**b**) Heart, (**c**) Liver, (**d**) Lung, (**e**) Kidney, (**f**) Brain.

4. Discussion

The presence of lead was found in the different tissues of *Columba livia* specimens collected in the region, which indicates that other species of urban fauna could probably be affected by heavy metals, and that *Columba livia* pigeons function as bioindicators of environmental pollution. This work coincides with that reported by Delgado [24] on the use of the *Columba* livia pigeon as a bioindicator in our country, being the first and only record in Mexico, and is very close to the results obtained by Valladares [28] in Chile and Guevara-Torres [35] in Peru. One of the previous works carried out in the region with other biological models is that of Domínguez-Zuñiga [37], who analyzed lead concentrations in fungi with urban growth in the region, finding that Torreon, Coahuila, was the city most affected by this pollutant, where lead concentrations were found in various species of fungi collected in different parts of the city.

The organ with the highest lead concentration was bone, where the maximum value was 191.04 mg/kg; these results are similar to those found by Valladares [28], where *Columba livia* pigeons were used to measure the concentrations of lead and other metals in an area contaminated by mining in the city of Arica, Chile; the two areas sampled share a history of contamination by mining or related activities, and it is an urban area inhabited not only by numerous people but also by a great diversity of urban species such as this pigeon. Physiologically, when lead exposures occur over long periods, there is a cumulative trend in the bones of the bird throughout its life [38]; this could be an indication of chronic exposure to this pollutant for individuals living in the sampled areas. Some authors explain this type of accumulation in bone in different animal models as the product of exposure in recurrent periods; however, in nature, the *Columba livia* pigeon does not achieve such a long life. According to Ishii [39], lead concentrations in the trabecular bones of birds increase more rapidly than in other bones due to the activities of lead in blood. Trabecular bones are tissues highly irrigated from blood vessels, and have an interaction within the exchange of calcium, iron, and magnesium, minerals that lead has the ability to substitute for in the bones for their chemical characteristics like divalent cation charge [40]. This could suggest this bone is the tissue with the greatest potential for lead accumulation in the face of prolonged exposure, which is consistent with what has been reported by Lee [41], where the bone and eggshell where the tissues with the higher concentration of lead.

The average concentrations of lead in the heart were higher than those reported by Elabidi [26] in Rabat-Sale, Morocco, suggesting that this contaminant is present in this way because this organ cannot filter substances. At present, there are not many studies that focus on the analysis of heart tissue of the pigeon *Columba livia*, but there are studies on other bird species, such as the one carried out by Millaku [42], where they analyzed hearts of house sparrows (*Passer domesticus*) and obtained concentrations higher than those found in this work. This study shares with ours the historical characteristics of contamination of the areas sampled by mining activities involving lead. However, it uses another species of bird that does not have an ethology like that of *C. livia* pigeons, which, unlike *Passer domesticus*, do not travel long distances, and their behavior is more gregarious, which may be the reason why the two biological models, despite being birds, present variations in their ability to accumulate metals in this tissue.

The average maximum lead concentration in the liver was 12.36 mg/kg; these results are higher than those found by Guevara-Torres [35], who analyzed specimens of *Columba livia* from the city of Lima, Peru, and reported an average of 0.624 mg/kg in the analysis of this tissue. This difference could be explained due to the disparity in the sources of contamination considered in the two studies. In their work, Guevara-Torres [35] analyzed a city polluted by vehicular influx and a variety of industries, unlike our study and others that compared that focus on the analysis of areas polluted by mining or related activities, and this is also the possible reason why higher results are presented than the analysis done by Romero [43], where specimens were affected systemically by lead bullets. In contrast to this, Valladares [28], who worked with *Columba livia* pigeons from an area contaminated by mining (City of Arica, Chile) and with whom the characteristics of the sources of these pollutants are shared, found similar results with average lead concentrations of 13.007 and 21.084 mg/kg.

For the lungs case, the lead concentrations that were found are higher than those reported by Delgado [24] in their study, carried out in Mexico City. The average lead concentrations were 3.65 mg/kg, while the average values obtained in our work were

12.57 mg/kg, which could suggest the inhalation route as the main entry access of this pollutant in the *Columba livia* pigeons inhabiting the region.

Regarding the average concentrations of lead in the kidney, Lee [41] reported in their study that the kidney is the second organ where lead accumulates the most (the first being bone), with concentrations of 2.116 mg/kg, a presentation three times lower than that obtained in our work with averages of 6.76 mg/kg. These studies are similar in the choice of rural areas as sampling points; the difference between the concentrations could be due to the disparity in the sources of contamination. Delgado [24] found similar results to those in this work, with maximum average concentrations of 7.61 mg/kg in *Columba livia* pigeons sampled in Mexico City, Mexico.

The average concentration of lead in the brain of *Columba livia* was 4.98 mg/kg; this result is lower than that obtained by Delgado [24], where the average concentration of lead in this tissue in Mexico City was 7.5 mg/kg. Both works share the development methodology, but there are differences in the sources of contamination sampled. Mexico City has a much higher population influx than the Comarca Lagunera, even in those years, which could be the reason for this difference. At present, there are not many studies that analyze lead concentrations in the brains of *Columba livia*. Aloupi [44] analyzed the livers and brains of nine different species of waterfowl from the Evros Delta National Park in Greece, but none reached the levels found in this study.

Among the four cities analyzed within the Comarca Lagunera, it was found that there was a significant difference in the city of Torreón, Coahuila, in relation to the rest of the cities; no significant statistical difference was found among the sampled localities. However, the maximum values of this pollutant were found in three of the six organs belonging to the individuals sampled in this city. Torreón, Coahuila, has one of the most important sources of lead emissions in northern Mexico (mining and metallurgical industry). This is the possible reason why in this and other studies, it is considered to be the city with the highest concentrations of this pollutant in the Comarca Lagunera; also, in this region, there is an import background of lead toxicosis in children, where the reported concentration have decreased over time [45].

In the case of Lerdo, Durango, in our work, it was reported as the second city with the highest lead concentrations in our model. However, Domínguez-Zuñiga [37] reported it as the third; Gómez Palacio, Durango, was, for them, the second city with the highest impact of lead contamination. This disparity can be explained by the difference in the chosen biological models and by a geophysical issue of the region. The possible existence of an unknown secondary source of contamination near the city of Lerdo, Durango, which could affect lead levels in these organisms, is considered.

The city of Matamoros, Coahuila, showed lower lead concentrations than Torreón, Coahuila, and Lerdo, Durango, which had the highest average values for lead concentration in kidneys. Currently, there is no previous record of analysis of lead concentrations in the city of Matamoros, Coahuila, using biological models.

In Gómez Palacio, Durango, lead concentrations were detected in all tissues of all pigeons analyzed. Domínguez-Zuñiga [37] found concentrations of 100.49 mg/kg in mushrooms belonging to the species *Coprinellus truncorumn* found in this municipality. Santoyo-Martinez [46] reported maximum concentrations of lead in dusts sampled outdoors of 464 mg/kg, which could suggest the presence of this pollutant in the city's environment regularly. The organ with the highest average lead concentration in this municipality was bone, with a value of 53.74 mg/kg, although this concentration does not reach the bones of other *Columba livia* pigeons analyzed in this study. Likely, exposure to lead in this area is not as chronic as in the rest of the cities but recurrent.

5. Conclusions

Lead was found in all *C. livia* tissues analyzed and at all sampling sites. The organ with the highest lead concentration was bone; a significant difference was found in this tissue compared to the others. The city with the highest lead concentration in individuals was Torreón, Coahuila, which is significantly different from other cities. In the case of the municipality of Matamoros, these results represent the first report of lead contamination in animal models. It also proves the ability of *C. livia* to function as bioindicators of environmental contamination.

Author Contributions: Conceptualization, A.O.-L. (Andrea Ocampo-Lopez), D.R.A.-G. and R.A.D.-G.; Methodology, A.O.-L. (Andrea Ocampo-Lopez), R.A.D.-G., H.S.-G., D.R.A.-G. and C.O.P.-V.; Validation, C.O.P.-V., J.R.E.-A. and D.R.A.-G.; Formal Analysis, C.O.P.-V. and A.O.-L. (Alejandra Ocampo-Lopez); Investigation, J.A.A.-J. and C.T.-L.; Resources, A.O.-L. (Andrea Ocampo-Lopez) and J.R.E.-A.; Data Curation, A.O.-L. (Alejandra Ocampo-Lopez); Writing—Original Draft Preparation, A.A.V.-G.; J.A.A.-J. and C.T.-L.; Writing—Review and Editing, A.A.V.-G., J.A.A.-J. and C.T.-L.; Visualization, A.A.V.-G. and H.S.-G.; Supervision, D.R.A.-G., R.A.D.-G. and C.O.P.-V.; Project Administration, A.O.-L. (Andrea Ocampo-Lopez); Funding Acquisition, A.O.-L. (Andrea Ocampo-Lopez) and J.R.E.-A. All authors have read and agreed to the published version of the manuscript.

Funding: This research was funded by the Mexican Council for Humanities, Science, and Technology CONAHCYT (Proyecto RENAJEB-2023-17).

Institutional Review Board Statement: This work was carried out in accordance with NOM-033-SAG/ZOO-2014: Methods for killing domestic and wild animals. Permit for SEMARNAT folio SPARN/DGVS/06267/22.

Informed Consent Statement: Not applicable.

Data Availability Statement: The raw data supporting the conclusions of this article will be made available by the authors on request.

Acknowledgments: Thank you very much to Miroslava Rihan Lemus, Paola Cruz Hernández, Felipe de Jesús Galindo Hernández, Patricia Soberón Nakasima Cerda, Fernando Hernández Terán, Ashanti Díaz Ocampo, and Unidad de Diagnóstico UAAAN UL.

Conflicts of Interest: The authors declare no conflicts of interest.

References

1. Manisalidis, I.; Stavropoulou, E.; Stavropoulos, A.; Bezirtzoglou, E. Environmental and Health Impacts of Air Pollution: A Review. *Front. Public Health* **2020**, *8*, 14. [CrossRef] [PubMed]
2. Li, A.M. Ecological determinants of health: Food and environment on human health. *Environ. Sci. Pollut. Res. Int.* **2017**, *24*, 9002–9015. [CrossRef] [PubMed]
3. Santos, U.P.; Arbex, M.A.; Braga, A.L.F.; Mizutani, R.F.; Cançado, J.E.D.; Terra-Filho, M.; Chatkin, J.M. Environmental air pollution: Respiratory effects. *J. Bras. Pneumol.* **2021**, *8*, e20200267. [CrossRef] [PubMed]
4. Meo, S.A.; Adnan-Abukhalaf, A.; Sami, W.; Hoang, T.D. Effect of environmental pollution PM2.5, carbon monoxide, and ozone on the incidence and mortality due to SARS-CoV-2 infection in London, United Kingdom. *J. King Saud Univ. Sci.* **2021**, *33*, 101373. [PubMed]
5. Schraufnagel, D.E.; Balmes, J.R.; Cowl, C.T.; De Matteis, S.; Jung, S.H.; Mortimer, K.; Perez-Padilla, R.; Rice, M.B.; Riojas-Rodriguez, H.; Sood, A.; et al. Air Pollution and Noncommunicable Diseases: A Review by the Forum of International Respiratory Societies' Environmental Committee, Part 2: Air Pollution and Organ Systems. *Chest* **2018**, *155*, 417–426. [CrossRef]
6. Tiffon, C. The Impact of Nutrition and Environmental Epigenetics on Human Health and Disease. *Int. J. Mol. Sci.* **2018**, *1*, 3425. [CrossRef]
7. Academy of Science of South Africa; Brazilian Academy of Sciences; German National Academy of Sciences Leopoldina; U.S. National Academy of Medicine; U.S. National Academy of Sciences. Air Pollution and Health—A Science-Policy Initiative. *Ann. Glob. Health.* **2019**, *16*, 140.
8. Eom, S.-Y.; Choi, J.; Bae, S.; Lim, J.-A.; Kim, G.-B.; Yu, S.-D.; Kwon, H.-J. Health effects of environmental pollution in population living near industrial complex areas in Korea. *Environ. Health. Toxicol.* **2018**, *33*, e2018004. [CrossRef]
9. Morán-Martínez, J. *La Contaminación Ambiental y Ocupacional por Plomo y sus Efectos en la Salud Reproductiva Masculina, Evidencia de Daño al ADN/Occupational and Environmental Contamination by Lead and Its Effects on Male Reproductive Health, Evidence of DNA Damage*; Royal Institution of Chartered Surveyors: London, UK, 2014; Volume 1, pp. 1–36.
10. Téllez-Ramírez, I.; Azamar-Alonso, A. Los niños de plomo: Justicia ambiental y conflictividad minera en la ciudad de Torreón, México. *Utopía Y Prax. Latinoam.* **2023**, *28*, e8027784.
11. Witkowska, D.; Słowik, J.; Chilicka, K. Heavy Metals and Human Health: Possible Exposure Pathways and the Competition for Protein Binding Sites. *Molecules* **2021**, *7*, 6060. [CrossRef]

12. Filipoiu, D.C.; Bungau, S.G.; Endres, L.; Negru, P.A.; Bungau, A.F.; Pasca, B.; Radu, A.-F.; Tarce, A.G.; Bogdan, M.A.; Behl, T.; et al. Characterization of the Toxicological Impact of Heavy Metals on Human Health in Conjunction with Modern Analytical Methods. *Toxics* **2022**, *10*, 716. [CrossRef] [PubMed]
13. Jaishankar, M.; Tseten, T.; Anbalagan, N.; Mathew, B.B.; Beeregowda, K.N. Toxicity, mechanism and health effects of some heavy metals. *Interdiscip. Toxicol.* **2014**, *7*, 60–72. [CrossRef] [PubMed]
14. Charkiewicz, A.; Backstrand, J. Lead Toxicity and Pollution in Poland. *Int. J. Environ. Res. Public Health* **2020**, *17*, 4385. [CrossRef] [PubMed]
15. Raj, K.; Das, A. Lead pollution: Impact on environment and human health and approach for a sustainable solution. *J. Environ. Chem. Ecotoxicol.* **2023**, *5*, 79–85. [CrossRef]
16. Williams, R.J.; Tannenbaum, L.V.; Williams, S.M.; Holladay, S.D.; Tuckfield, R.C.; Sharma, A.; Humphrey, D.J.; Gogal, R.M., Jr. Ingestion of a Single 2.3 mm Lead Pellet by Laying Roller Pigeon Hens Reduces Egg Size and Adversely Affects F1 Generation Hatchlings. *Arch. Environ. Contam. Toxicol.* **2017**, *73*, 513–521. [CrossRef]
17. Sani, A.; Musa, A. Lead: A concise review of its toxicity, mechanism and health effect. *GSC Biol. Pharm. Sci.* **2021**, *15*, 055–062. [CrossRef]
18. Rokanuzzaman, B.M.; Salma, U.; Akter-Bristy, N.; Kundu, S.; Alam, S.S.; Khalil, I. Assessment of Heavy Metals and Trace Elements in Eggs and Eggshells of *Gallus gallus domesticus*, *Coturnix coturnix* and *Anas platyrhynchos* from Bangladesh. *Saudi J. Biomed. Res.* **2022**, *7*, 137–142. [CrossRef]
19. Takeuchi, H.; Taki, Y.; Nouchi, R.; Yokoyama, R.; Kotozaki, Y.; Nakagawa, S.; Sekiguchi, A.; Iizuka, K.; Hanawa, S.; Araki, T.; et al. Lead exposure is associated with functional and microstructural changes in the healthy human brain. *Commun. Biol.* **2021**, *26*, 912. [CrossRef]
20. Álvarez-Lloret, P.; Benavides-Reyes, C.; Lee, C.M.; Martínez, M.P.; Conti, M.I.; Rodríguez-Navarro, A.B.; González-López, S.; Perez-Huerta, A.; Terrizzi, A.R. Chronic Lead Exposure Alters Mineral Properties in Alveolar Bone. *Minerals* **2021**, *11*, 642. [CrossRef]
21. Teerasarntipan, T.; Chaiteerakij, R.; Prueksapanich, P.; Werawatganon, D. Changes in inflammatory cytokines, antioxidants and liver stiffness after chelation therapy in individuals with chronic lead poisoning. *BMC Gastroenterol.* **2020**, *8*, 263. [CrossRef]
22. Firoozichahak, A.; Rahimnejad, S.; Rahmani, A.; Parvizimehr, A.; Aghaei, A.; Rahimpoor, R. Effect of occupational exposure to lead on serum levels of lipid profile and liver enzymes: An occupational cohort study. *Toxicol. Rep.* **2022**, *26*, 269–275. [CrossRef] [PubMed]
23. De la Ossa, J.; De la Ossa-Lacayo, A.; Monroy-Pineda, M. Abundance of domestic dove (*Columba livia* domestica Gmelin, 1789) in Santiago de Tolú, Sucre, Colombia. *Rev. MVZ Córdoba* **2017**, *22*, 5718–5727. [CrossRef]
24. Delgado, R.; Fortoul, T.; Rosiles, R. Concentraciones de plomo, cadmio y cromo y su relación con algunas modificaciones morfológicas en tejidos de palomas *Columba livia* de la ciudad de México e Ixtlahuaca, Estado de México. *Vet. Méx.* **1994**, *25*, 109–115.
25. Nam, D.H.; Doo-Pyo, L.; Tae-Hoe, K. Comparison of Heavy Metal Concentration and Reproduction of Feral Pigeons (*Columba livia*) between Urban and Industrial Complex Areas from Korea. *Korean J. Ecol.* **2002**, *25*, 389–393. [CrossRef]
26. Elabidi, A.; Fekhaqui, M.; Ghouli, A.; Piervittori, R.; Yahyaqui, A. Use of pigeons as bioindicators of air pollution from heavy metals at Rabat-Salé (Morocco). *Avocetta* **2010**, *34*, 29–34.
27. Begum, A.; Sehrin, S. Levels of Heavy Metals in Different Tissues of Pigeon (*Columba livia*) of Bangladesh for Safety Assessment for Human Consumption. *Bangladesh Pharm. J.* **2013**, *16*, 81–87. [CrossRef]
28. Valladares-Faundez, P.; Cáceres, G.; Valdés, J. Content of lead, cadmium and arsenic in biological tissues of the feral pigeons (*Columba livia*) present in an urban area previously contaminated with minig residues. *Rev. Int. Contam. Ambient.* **2021**, *36*, 485–490.
29. Enriquez-Robledo, A.; Alvarado, H.G.; Pérez, J. Hidroarsenicismo en la Comarca Lagunera y Políticas Públicas 1 Hydroarsenism in the Lagunera Region and Public Policies. *Rev. Enfoques* **2021**, *19*, 21–43.
30. Barrios, J. *Más allá del Agua: Desarrollo Urbano de la Comarca Lagunera, México, Después de la Regulación del río Nazas. XI Seminario Internacional de Investigación en Urbanismo, Barcelona-Santiago de Chile*; Universitat Politècnica de Catalunya: Barcelona, Spain, 2019.
31. CONAGUA. Actualización de la Disponibilidad Media Anual de Agua en el Acuífero Principal-Región Lagunera (0523), Estado de Coahuila. Available online: https://sigagis.conagua.gob.mx/gas1/Edos_Acuiferos_18/coahuila/DR_0523.pdf (accessed on 22 July 2024).
32. García, E. *Modificaciones al Sistema de Clasificación Climática de Köppen (Para Adaptarlo a las Condiciones de la República Mexicana)*; Instituto de Geografía: Mexico City, Mexico, 1964; UNAM.
33. Rodríguez-De La Barrera, A.E.; Causil-Vargas, L.A.; Causil-Vargas, O. Determinación de la diversidad genética de la paloma doméstica *Columba livia* (Columbidae) a partir de genes polimórficos asociados con el color del plumaje en San Antero, Córdoba, Colombia. *Rev. Acad. Colomb. Cienc. Ex. Fis. Nat.* **2019**, *43*, 78–83. [CrossRef]
34. Soh, M.C.K.; Pang, R.Y.T.; Ng, B.X.K.; Lee, B.P.Y.-H.; Loo, A.H.B.; Er, K.B.H. Restricted human activities shift the foraging strategies of feral pigeons (*Columba livia*) and three other commensal bird species. *Biol. Conserv.* **2021**, *253*, 108927. [CrossRef]
35. Guevara-Torres, D.R.; Williams, M.; Palacios, G. Use of feral pigeons (*Columba livia*) as biomonitors of trace metal pollution in Lima, Peru. *Ecol. Apl.* **2022**, *21*, b103–b111. [CrossRef]

36. Integrated Taxonomic Information System. *Columba livia*. Available online: https://www.itis.gov/servlet/SingleRpt/SingleRpt?search_topic=TSN&search_value=177071#null (accessed on 12 March 2024).
37. Domínguez-Zúñiga, L.I.; Puente-Valenzuela, C.O.; Estrada-Arellano, J.R.; Aguirre-Acosta, E.; Aguillón-Gutiérrez, D.R. Concentración de metales pesados en hongos de la zona metropolitana de la Comarca Lagunera, México. *Sci. Fungorum* **2021**, *52*, e1389. [CrossRef]
38. Pain, D.; Mateo, R.; Green, R. Effects of lead from ammunition on birds and other wildlife: A review and update. *Ambio* **2019**, *48*, 935–953. [CrossRef] [PubMed]
39. Ishii, C.; Nakayama, S.M.M.; Kataba, A.; Ikenaka, Y.; Saito, K.; Watanabe, Y.; Ishizuka, M. Characterization and imaging of lead distribution in bones of lead-exposed birds by ICP-MS and LA-ICP-MS. *Chemosphere* **2018**, *212*, 994–1001. [CrossRef] [PubMed]
40. Rodríguez, J.; Mandalunis, P.M. A Review of Metal Exposure and Its Effects on Bone Health. *J. Toxicol.* **2018**, *2018*, 4854152. [CrossRef]
41. Lee, J.; Lee, J.; Lee, S.H.; Kim, M.; Lee, E.; Han, A.; Shim, K. The Characteristics of Heavy Metal Accumulations in Feral Pigeon (*Columba livia*) Feathers for Environmental Monitoring. *J. Environ. Impact Assess.* **2014**, *23*, 492–504. [CrossRef]
42. Millaku, L.; Imeri, R.; Trebicka, A. House sparrow (*Passer domesticus*) as bioindicator of heavy metals pollution. *Eur. J. Exp. Biol.* **2014**, *4*, 77–80.
43. Romero, D.; de Jose, A.; Theureau-Raigon, M.D.; Torregrosa, J.B. Lead in terrestrial game birds from Spain. *Environ. Sci. Pollut. Res.* **2019**, *27*, 1585–1597. [CrossRef]
44. Aloupi, M.; Karagianni, A.; Kazantzidis, S.; Akriotis, T. Heavy Metals in Liver and Brain of Waterfowl from the Evros Delta, Greece. *Arch. Environ. Contam. Toxicol.* **2017**, *72*, 215–234. [CrossRef]
45. Soto-Jímenez, M.F.; Flegal, A.R. Childhood lead poisoning from the smelter in Torreon, Mexico. *Environ. Res.* **2011**, *111*, 590–596. [CrossRef]
46. Santoyo-Martínez, M.; Aguilera, A.; Gallegos, Á.; Puente, C.; Goguitchaichvili, A.; Bautista, F. Pollution Levels and Potential Health Risks of Potentially Toxic Elements in Indoor and Outdoor Dust during the COVID-19 Era in Gómez Palacio City, Mexico. *Land* **2023**, *12*, 29. [CrossRef]

toxics

MDPI

Article

Trace Element Speciation and Nutrient Distribution in *Boerhavia elegans*: Evaluation and Toxic Metal Concentration Across Plant Tissues

Tahreer M. Al-Raddadi [1,2,3], Lateefa A. Al-Khateeb [1], Mohammad W. Sadaka [4] and Saleh O. Bahaffi [1,*]

1 Chemistry Department, Faculty of Science, King Abdulaziz University, Jeddah 21589, Saudi Arabia; tmradade@uqu.edu.sa (T.M.A.-R.); laalkhatib@kau.edu.sa (L.A.A.-K.)
2 Chemistry Department, Al-Qunfudah University College, Umm Al-Qura University, Makkah 24382, Saudi Arabia
3 King Fahd Medical Research Center, King Abdulaziz University, Jeddah 21589, Saudi Arabia
4 College of Health Technology, Cihan University-Erbil, Kurdistan Region, Erbil 44001, Iraq; waleed.sadaka@gmail.com
* Correspondence: sbahaffi@kau.edu.sa

Abstract: This study investigated the elemental composition of *Boerhavia elegans*, addressing the gap in comprehensive trace element profiling of this medicinal plant. The research aimed to determine the distribution of macronutrients, micronutrients, and beneficial and potentially toxic elements across different plant parts (seeds, leaves, stems, and roots). Using ICP-OES analysis, two digestion methods were employed to capture both complex and labile elements. The study revealed distinct elemental distribution patterns, with iron and nickel concentrating in stems, manganese and zinc in leaves, and copper in roots. Magnesium emerged as the most abundant macronutrient, particularly in leaves. Importantly, all detected toxic elements (arsenic, chromium, lead, and cadmium) were below WHO safety limits. These findings provide crucial insights into the nutritional and safety profile of *B. elegans*, potentially informing its use in traditional medicine and highlighting its potential as a source of essential elements.

Keywords: *Boerhavia elegans*; chemical speciation; essential and toxic elements; medicinal plants; plant tissues; PCA

1. Introduction

In the last century, there has been an increasing global concern over soil trace element contamination, which has coincided with an escalation of economic and societal activity. Anthropogenic sources are widely recognized as the primary driver of elevated soil trace element levels. Trace elements, including Cu, Ni, and Zn, are crucial for agricultural productivity and environmental health. They play a vital role in maintaining the quality of crops, food, and consumer health. These elements are required by plants in specific, small quantities. Their presence in the soil directly impacts the nutrient balance in animal feed (fodder). However, exceeding these recommended trace element levels can be detrimental, causing toxicity in plants and soil organisms. Therefore, maintaining optimal trace element availability in soil is essential for sustainable agriculture [1]. Edible and medicinal plants are substances with biological activity that have been used for millennia to cure many human ailments due to their low adverse effects [2,3]. However, some medicinal plants and their mixtures may pose health risks owing to toxic elements. The contamination may result from environmental pollution [4]. Creating, regulating, and using herbal or traditional

https://doi.org/10.3390/toxics13010014

medicines in many parts of the world can be challenging. Regulatory status, quality control, safety and efficacy assessment, safety monitoring, and a need for more awareness about conventional, complementary, and alternative medicines among public drug regulatory bodies are common issues in many nations [5]. Numerous mineral elements present in plants are necessary for human nutrition [6]. The literature extensively documents that they possess a range of metabolic functions, from cell defense to primary and secondary metabolism [7]. In addition to being necessary for forming active compounds in medicinal plants, trace metals also cause toxicity in those plants [8].

Trace elements are essential in plant metabolism, forming chemical constituents in plants and acting as cofactors for enzymes [9]. The contents of major and trace elements in plants are governed by the geochemical features of the soil where the plant grows and by the ability of plants to accumulate elements selectively. Moreover, plants can accumulate elements from the surrounding aerial or aquatic environment, enabling some plants to be used as biomonitors [10–12]. The concentration of toxic heavy metals in plants can be influenced by the geochemical parameters of the soil, air, and water pollution, as well as the ability of plants to accumulate certain elements selectively [13]. Metals may also be related to the geographical origin, harvesting, or gathering of these plant materials. Plants serve as an essential medium for trace elements to transit from the soil to humans [14]. Consequently, quality control measures for trace element content in these medicinal plants are necessary.

"Heavy metals" are metals with a density greater than 5 g/cm^3. This group contains more than 60 metals. The primary sources of heavy metals are mineral fertilizers, some base rocks, sewage sludge, pesticides, wastewater, urban wastes, motor vehicle exhaust gases, and mining [13]. Because of their nature, these elements are typically found in the earth's crust as oxide, carbonate, sulfides, and silicates or in stable compounds. All living creatures require some heavy metals in their surroundings to continue their daily activities. Elements that plants need to survive are called "plant nutrients". Analyzing plant tissues reveals almost all the elements found in nature. Although plants are selective about taking up nutrient ions, as the concentration of nutrient forms of elements increases in the growth medium, some non-essential heavy metals can passively enter the plant body and enter the food chain [15]. Some metals, such as iron (Fe), copper (Cu), manganese (Mn), cadmium (Cd), molybdenum (Mo), silicon (Si), and boron (B), are essential heavy metals required for plant metabolism. Cobalt (Co), copper (Cu), iron (Fe), manganese (Mn), Molybdenum (Mo), zinc (Zn), selenium (Se), and iodine (I) are also essential heavy metals needed for animals. However, some metals can be toxic at high concentrations [13]. Various heavy metals (HMs), such as arsenic (As), cadmium (Cd), chromium (Cr), lead (Pb), and mercury (Hg), cause severe toxicity in plants when they enter agricultural soil ecosystems through human activities or natural processes [16].

Medicinal plants are an ancient practice that dates to historical times. They are widely utilized across the globe and are prevalent in many countries. Over 70% of the world's population uses medicinal plants to treat various ailments. However, medicinal plants can also pose a risk of exposure to toxic elements, depending on their origin and characteristics [17]. Plants are crucial in transferring trace elements from the soil to humans [14]. Analyzing the metal ion compositions of medicinal plants is essential for understanding their medicinal, nutritional, and toxic properties [18]. With increasing awareness about elemental composition, elemental analysis is becoming more significant. Consequently, there is a growing interest in developing rapid, simple, and sensitive methods that can be used for routine analysis. Sample preparation methods are typically critical in chemical analysis. Acid digestion, the most commonly employed sample treatment for determining elemental concentration in biological materials like plants, necessitates high energy input and concentrated mineral acids [19]. Quantifying trace elements in medicinal plants is important for several reasons. It aids in the treatment of various diseases by understanding the pharmacological effects of these plants. It also helps determine the appropriate dosage of herbal medicines derived from medicinal plants. Furthermore, it enables the evaluation of the effect of medicinal plants on human health [8].

The plant genus *Boerhavia* has increasingly interested physiochemists as past studies have demonstrated its wide range of pharmacological and biological effects, signifying potential therapeutic applications. Research has found that a methanolic extract from *B. diffusa* leaves exhibits antibacterial properties and noteworthy inhibition against Staphylococcus aureus, a known pathogenic bacterium [20]. According to traditional medicine systems, *B. elegans* has been used to treat various other health conditions, including painful menstruation, urinary tract infections or other issues, intestinal infections, inflammation, jaundice, and overall weakness or fatigue [21]. *Boerhavia elegans* plant, as part of the plant kingdom, requires various mineral elements for growth, development, and reproduction. These mineral elements are mobilized from the soil matrix and absorbed by the roots as mineral ions.

To the best of our knowledge, no study has been reported regarding trace metal concentration and speciation of selected *Boerhavia elegans* tissues. Therefore, the current research is focused on I- determining the total trace metal concentration in various tissues and II- speciation of labile and complex fractions of selected elements. The concentrations of trace elements in the root, stem, leaf, and seed of the *Boerhavia elegans* plant were determined using the inductively coupled plasma optical emission spectroscopy (ICP-OES) technique.

Several techniques have been used for the quantity of trace elements in food samples, such as the ICP-OES, fame atomic absorption spectrometry (FAAS), and graphite furnace atomic absorption spectrometry (GFAAS) [22–25]. Among these techniques, ICP-OES was reported as the most sensitive method, time-saving, and specificity [26–28].

2. Materials and Methods

2.1. Sampling

Seeds, stems, leaves, and roots of *Boerhavia elegans* were analyzed in this study. The plant seeds of *B. elegans* were purchased from a local market in Jeddah, Saudi Arabia, as they are used in preparing porridge, a famous food in the Arabian Peninsula. The roots, stems, and leaves were collected from the same field plots in Hadramout, Alborgat, South Yemen. Samples were cleaned with tap water and deionized water, then air-dried in an oven (JS Research Inc., Gongju, Republic of Korea) at 50 °C. The dried samples were ground with an electrical grinder and sieved through a 100-mesh standard sieve. All the samples were initially stored in closed plastic bags until analysis.

2.2. Reagents and Chemicals

A multi-elemental standard solution of 1000 mg/L containing all analyzed elements (Cd, Co, Cr, Cu, Fe, Mg, Mn, Ni, Pb, Al, As, Se, Zn, K, Na, B, Ca, and P) supplied by Merck (Darmstadt, Germany) was used for calibration. Analytica grade HNO_3 65% and H_2O_2 30% from Merck (Darmstadt, Germany) analytical grade were used for sample digestion. Ultrapure water delivered by a Milli Q system (Millipore, Merck Millipore, France) was used to dilute and prepare a more diluted and amore-diluted working solution.

2.3. Instrumental Analysis

A PerkinElmer Optima 7000 DV ICP-Optical Emission Spectrometer (Shelton, CT, USA), equipped with Win Lab 32 for ICP Version 4.0 software, was utilized to measure the levels of Cd, Co, Cr, Cu, Fe, Mg, Mn, Ni, Pb, Al, As, Se, Zn, K, Na, B, Ca, and P. The

ICP-OES was set with a wet plasma aerosol type and an axial view, using a nebulizer startup instant condition that included a flow rate (Ar) of 15 L/min, an auxiliary flow (Ar) of 0.2 L/min, a nebulizer flow (Ar) of 0.8 L/min, a sample uptake rate of 1.5 mL/min, and a sample flush time of 5 s. Calibration was performed using a standard mixture with concentrations of 1, 5, 10, and 15 μg/mL for each element, ensuring high accuracy and precision (%RSD).

2.4. Recommended Procedures

2.4.1. Wet Digestion

Boerhavia elegans seeds, leaves, stems, and roots were washed with tap water and deionized water and dried at 100 °C. for 24 h until they achieved a constant weight to remove moisture. The plant parts were ground to fine particles using an electrical grinder. An accurate mass (10.0 ± 0.001 g) (except for seeds and leaves, 5.0 g was taken) of each part of the plant sample was digested in a 100 mL beaker containing 50 mL HNO_3 (69%). The beaker was covered with a watch glass and left overnight at room temperature, followed by the addition of 12.5 mL of H_2O_2 30% (Sigma-Aldrich, Gillingham, UK), and heated on a hot plate to 110–120 °C until the light brown-colored fume disappeared. The volume became nearly dry, and 5 mL of deionized water was added. The digest solutions were filtered through Whatman No. 42 filter paper and diluted to 50 mL using HNO_3 (0.1 M). The concentration of elements was finally measured.

2.4.2. Analysis of Labile (Water-Soluble) Trace Elements in *B. elegans* Tissues

Parts of different plant tissues (seeds, leaves, stems, and roots) were dried at 100 °C until they achieved a constant weight to remove moisture and then homogenized using an electric grinder. An accurate mass (10.0 ± 0.001 g) of dried *B. elegans* tissues was transferred to deionized water (20 mL 18.2 $M\Omega \cdot cm^{-1}$) and left overnight in a shaker at room temperature. The beaker was covered with a watch glass and heated on a hot plate to 90–120 °C for 30 min. Sample solutions were vacuum-filtered through the Whatman No. 2 filter paper. The filtrate and washing solutions were transferred to a 50 mL volumetric flask, completed to the mark using deionized water and stored in Low-Density Polyethylene (LDPE) bottles. An appropriate volume of the *B. elegans* infusion was filtered through a 0.45 μm membrane filter, and the labile water-soluble elements were analyzed versus blank solution. The ICP-OES signal intensity (I_1) for *B. elegans* is a measure of labile concentration (C_L) of the element in the extract in boiling water, which was determined with the help of a consistent calibration plot employing the equation:

$$\text{Average } C_L \ \mu g \ g^{-1} = (C - B) \times \frac{V}{m} \quad (1)$$

where C is the element concentration (μg mL^{-1}) of *B. elegans* tissue, B is the blank result, V is the sample volume (mL), and m is the mass (g) of the *B. elegans* tissue sample.

2.5. Statistical Method

For statistical analysis, XLSTAT Software 26.4.0 was used to determine PCA techniques.

3. Results and Discussion

3.1. Trace Metals in Plant

Trace metal concentrations in the seeds, leaves, stems, and roots of *B. elegans* in Table 1. The results indicate variable metal levels in different types and parts.

Table 1. Mean concentration mg/L of trace metals in the tissue of *B. elegans* using ICP-OES.

Tissue	Seeds		Leaves		Stems		Roots	
Elements	HNO_3	H_2O	HNO_3	H_2O	HNO_3	H_2O	HNO_3	H_2O
Cd	0.009 ± 0.0003	0 ± 0.0012	0.054 ± 0.0019	0 ± 0.0007	0.021 ± 0.0049	0 ± 0.0014	0.018 ± 0.0002	0.061 ± 0.0025
Co	0.124 ± 0.0009	0.051 ± 0.0011	0.097 ± 0.0004	0.08 ± 0.0001	0.056 ± 0.0009	0.061 ± 0.0009	0.118 ± 0.0004	0.059 ± 0.0008
Cr	0.793 ± 0.0025	0 ± 0.0003	0.565 ± 0.003	0 ± 0.0006	1.004 ± 0.001	0 ± 0.0001	0.738 ± 0.0014	0 ± 0.0003
Cu	1.13 ± 0.0052	4.958 ± 0.0354	0.91 ± 0.0066	0.207 ± 0.0019	0.954 ± 0.0189	18.21 ± 0.053	1.692 ± 0.0126	0.292 ± 0.0008
Fe	8.021 ± 0.0653	0.081 ± 0.0015	25.37 ± 0.15	0 ± 0.0004	30.66 ± 0.083	0.655 ± 0.0045	18.54 ± 0.13	0 ± 0.0004
Mg	367.8 ± 0.75	14.5 ± 0.121	914.7 ± 3.89	259.9 ± 1.32	683.4 ± 2.85	102 ± 0.33	392.2 ± 2.91	59.84 ± 0.351
Mn	3.034 ± 0.0091	0.194 ± 0.0003	7.343 ± 0.0314	1.264 ± 0.0062	4.033 ± 0.0449	1.004 ± 0.0141	1.887 ± 0.0119	0.157 ± 0.0007
Ni	0.5 ± 0.0019	0.323 ± 0.002	0.377 ± 0.0026	0 ± 0.0012	0.551 ± 0.0023	0.752 ± 0.0037	0.423 ± 0.0027	0 ± 0.0007
Pb	0.033 ± 0.0061	0.282 ± 0.0009	0.165 ± 0.0053	0 ± 0.0078	0 ± 0.0162	1.333 ± 0.0061	0.113 ± 0.0078	0.04 ± 0.0048
Al	24.49 ± 2.154	0 ± 0.0018	59.71 ± 1.255	0 ± 0.002	67.53 ± 1.062	0 ± 0.0059	40.67 ± 0.426	0 ± 0.0079
As	1.068 ± 0.1653	0.109 ± 0.0264	0.728 ± 0.0407	0.177 ± 0.0252	0.742 ± 0.0164	0.099 ± 0.0138	1.374 ± 0.0185	0.221 ± 0.0306
Se	1.386 ± 0.1493	0.126 ± 0.0134	0.501 ± 0.0063	0.147 ± 0.0149	0.499 ± 0.0072	0.119 ± 0.0128	1.253 ± 0.0375	0.258 ± 0.0191
Zn	4.619 ± 0.1817	34.52 ± 0.194	7.117 ± 0.0995	1.279 ± 0.0059	3.108 ± 0.0453	64.03 ± 0.37	2.754 ± 0.0035	0.643 ± 0.0039
Na	0 ± 0.0128	9.737 ± 0.2996	0.234 ± 0.1354	30.3 ± 5.058	9.728 ± 1.3931	50.68 ± 1.328	0 ± 0.0385	29.2 ± 0.61
K	2.649 ± 0.0068	103.2 ± 0.37	9.937 ± 0.6758	67.76 ± 0.064	152.9 ± 6.95	76.53 ± 3.66	173 ± 5.24	66.46 ± 0.357
P	0 ± 0.0031	9.151 ± 0.2234	0 ± 0.026	4.922 ± 0.3786	1.012 ± 0.8824	0.31 ± 0.0252	0 ± 0.0032	15.61 ± 0.279
B	0 ± 0.0012	0.714 ± 156	0 ± 0.0012	2.669 ± 2051	0 ± 0.0041	0.173 ± 0.0184	0 ± 0.0003	0.244 ± 0.004
Ca	1.061 ± 0.0501	20.99 ± 0.149	30.17 ± 0.0501	255.4 ± 3.61	102.8 ± 7.08	8.533 ± 0.0554	6.939 ± 0.8491	1.38 ± 0.0587

Mean ± Standard Deviation.

In *Boerhavia elegans*, ten essential elements, including macronutrients (potassium, calcium, magnesium, phosphorus) and micronutrients (iron, manganese, zinc, copper, boron, nickel), along with aluminum (not essential for plants), beneficial elements (sodium, selenium, cobalt), and potentially toxic elements (cadmium, lead, chromium, arsenic) were analyzed, with the latter posing risks to plant physiology.

Determining the distribution and speciation of metals in plants is essential for understanding the mechanisms underlying their toxicity to plants. In this study, the total concentrations of micro- and macro-elements in *B. elegans* were found to exceed the critical deficiency levels that are known to limit plant growth and survival. Table 2 presents the statistical analysis of metal concentrations in the seeds, leaves, stems, and roots of *B. elegans*.

Table 2. Variance of element concentrations across plant tissues.

Groups	Count	Sum	Average	Variance
seed	18	2083.585	115.7547	185,790.4
leaves	18	5289.89	293.8828	1,146,789
stems	18	5294.99	294.1661	653,386
roots	18	3208.595	178.2553	239,611.8

The data reveal distinct differences among the four groups. Seeds exhibited the lowest average metal concentration (115.75) but had the highest variability, with a variance of 185,790, indicating significant differences within this group. Leaves and stems recorded the highest average concentrations, nearly identical at 293.88 and 294.16, respectively. However, leaves displayed the highest variability (1,146,789), suggesting substantial internal differences, while stems showed lower variability (653,386), reflecting greater consistency. Roots had a moderate average concentration of 178.25 and the lowest variance (239,611), indicating the most uniform data among the groups.

Overall, while leaves and stems have comparable average concentrations, leaves exhibit significantly higher variability. Roots are the most consistent group, and seeds, despite their low average concentration, demonstrate considerable variability. These findings highlight the tissue-specific distribution and variability of metal concentrations in *B. elegans*.

3.2. Toxic Elements

The presence of hazardous elements such as arsenic (As), cadmium (Cd), and lead (Pb) in contaminated soils poses a risk to crop plants and, ultimately, human health. The World Health Organization (WHO) has set limits for the concentration of these heavy metals in wild plants, specifying maximum levels of 5 mg/kg for arsenic, 10 mg/kg for lead, 0.3 mg/kg for cadmium, and 2 mg/kg for chromium [29].

This study investigated the distribution of cadmium, lead, arsenic, and chromium in different parts of the *B. elegans* plant. The results indicate that seeds contain the lowest levels of cadmium, while leaves exhibit the highest concentration, as illustrated in Figure 1 and Table 3. Lead concentrations are highest in leaves and lowest in seeds. Arsenic concentrations, on the other hand, are highest in roots and lowest in leaves. The concentration of arsenic in seeds ranges from 5.34 µg/g to 3.71 µg/mL. Chromium concentration shows a more balanced distribution across all plant tissues, with the highest concentration observed in seeds. Toxic metals enter root cells and are moved to other plant tissues via membrane transporters that typically handle essential or beneficial nutrients. This occurs because of the physicochemical similarities between toxic metals and nutrients, especially those within the same periodic table group, and the versatility of transporter proteins [30].

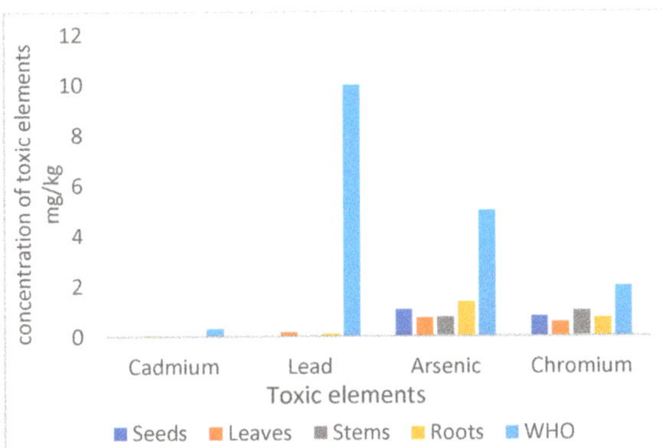

Figure 1. The concentration of cadmium, lead, and arsenic in *B. elegans* mg/kg.

Table 3. Total concentration µg/mL, labile µg/mL, and complex fraction µg/mL in *B. elegans* using wet and labile digestion toxic element.

Tissue	Seed			Leaves			Stem			Root		
Elements	Total Concentration	Labile	Complex Fraction	Total Concentration	Labile	Complex Fraction	Total Concentration	Labile	Complex Fraction	Total Concentration	Labile	Complex Fraction
cd	0.045	0	ND	0.27	0	ND	0.105	0	ND	0.09	1.22	ND
pb	0.165	2.82	ND	0.825	0	ND	0	13.33	ND	0.565	0.8	ND
As	5.34	1.09	4.25	3.64	3.54	0.1	3.71	0.99	2.72	6.87	4.42	2.45
Cr	3.965	0	ND	2.825	ND	ND	5.02	ND	6.01	3.69	0	ND

ND = not detect.

Cadmium toxicity induces the overproduction of reactive oxygen species (ROS), resulting in damage to plant membranes and the breakdown of cellular biomolecules and organelles. This toxicity also impairs the uptake of crucial nutrients like Fe and Zn, leading to leaf chlorosis. Cadmium also obstructs the transportation and absorption of vital minerals such as K, Mg, Ca, P, and Mn. Cd is primarily absorbed and transported within plants via transporters designed for divalent cations such as Fe (II), Mn, and Zn. In Arabidopsis, both the Fe (II) transporter IRT1 and the Mn transporter NRAMP1 can transport cadmium, significantly contributing to its uptake in the plant [31].

Lead (Pb) toxicity poses a significant threat to plants, animals, and humans due to its harmful effects. The rise in industrial activities involving Pb and the use of Pb-containing products like agrochemicals, oil, paint, and mining operations have led to environmental contamination, allowing Pb to infiltrate the food chain. As one of the most dangerous heavy metals, lead's presence in the food chain represents a serious health risk for both plants and humans [32]. Inorganic lead (Pb) is commonly found in dust, soil, old paint, and various consumer products, whereas organic lead, such as tetraethyl Pb, is mainly present in leaded gasoline. Both forms are toxic, but organic Pb complexes are particularly harmful to biological systems, posing a much greater threat than their inorganic counterparts [32,33]. Plants absorb free Pb ions either through capillary action or from the atmospheric air via cellular respiration. Once lead is absorbed into the soil from the external atmosphere, it enters the plant system. Through their well-developed root systems, plants absorb nutrients from the soil, including divalent-free Pb cations in contaminated soil. These absorbed lead ions are subsequently transported through the xylem vessels [34].

Arsenic (As) contamination in the environment poses a substantial global concern impacting environmental, agricultural, and health sectors because of its highly toxic and carcinogenic nature. Even at minimal concentrations, exposure of plants to As can trigger a wide range of morphological, physiological, and biochemical alterations [35]. Arsenic is considered non-essential for plants and other organisms [36]. Plant uptake of arsenic depends on its total concentration and, importantly, on its speciation in soil. In plants, arsenic primarily enters in the form of inorganic species, As (III) or As (V), via transporter proteins regulated by the concentration gradient between the growth medium and plant cells [37]. Xylem tissues facilitate the translocation of arsenic from roots to shoots and its distribution across various plant tissues. Arsenate competes with phosphate (Pi) for uptake through Pi transporters [38]. Once inside plant cells, As (V) is enzymatically reduced to As (III) by arsenate reductase (ACR2). Detoxification of As (III) involves its complexation with thiol-rich peptides, leading to its sequestration in vacuoles of root cells, which limits As (III) efflux and enables its long-distance transport to other plant tissues [39].

While chromium trioxide (Cr (III)) is essential for animal and human health in small amounts, plants don't require it. Chromium (Cr) contamination has become a major environmental concern due to its high levels in various agricultural and industrial activities [16]. Significant amounts of Cr are mined or produced annually, which contributes to its presence in soil and water. Natural sources of Cr include rocks, volcanic dust, gases, soil, animals, and plants, and Cr is often associated with primary rock-derived phases and well-crystallized iron oxides. Additionally, natural leaching from rocks and topsoil leads to substantial levels of Cr in water bodies [40].

Chromium (Cr) toxicity in plants adversely affects physiological, biochemical, and molecular traits, resulting in stunted growth and reduced yield. High Cr accumulation impacts seed germination, root and shoot growth rates, and overall biomass and yield. It inhibits photosynthesis, disrupts the cell cycle, and affects water and mineral balance, enzyme activity, nitrogen assimilation, the antioxidant system, and other metabolic processes. Cr accumulation also triggers the generation of reactive oxygen species (ROS), causing oxidative damage and cellular component damage, leading to cell death and altered morphology.

Because chromium (Cr) easily reacts with oxygen in the air, it's a very active metal. Cr can exist in multiple chemical states, ranging from zero to positive six. In nature, the most common forms are chromium trioxide (Cr (III)) and chromium hexoxide (Cr (VI)). However, Cr (VI) is more dangerous because it dissolves and moves around more easily in water [41]. Plants can absorb both types of chromium [42]. Cr (VI) actively gets into plant cells through specific channels, while Cr (III) enters more passively through exchange sites on cell walls [43]. Additionally, certain acids released by plant roots help dissolve Cr, making it easier for plants to take it up [41].

3.3. Macronutrient Elements

Levels of the macro (P, Mg, K, and Ca) elements in plant samples are given in Table 4, Figure 2.

Table 4. Total concentration μg/mL, labile μg/mL, and complex fraction μg/mL in *B. elegans* using wet and labile digestion of macronutrient element.

Tissue	Seed			Leaves			Stem			Root		
Elements	Total Concentration	Labile	Complex Fraction	Total Concentration	Labile	Complex Fraction	Total Concentration	Labile	Complex Fraction	Total Concentration	Labile	Complex Fraction
Mg	1839	145	1694	4573.5	5198	ND	3417	1020	2397	1961	1196.8	764.2
Ca	5.305	209.9	ND	150.85	5108	ND	514	85.33	428.67	34.695	27.6	7.095
K	13.245	1032	ND	49.685	1355.2	ND	764.5	765.3	ND	865	1329.2	ND
P	0	91.51	ND	0	98.44	ND	5.06	3.1	1.96	0	312.2	ND

ND = not detect.

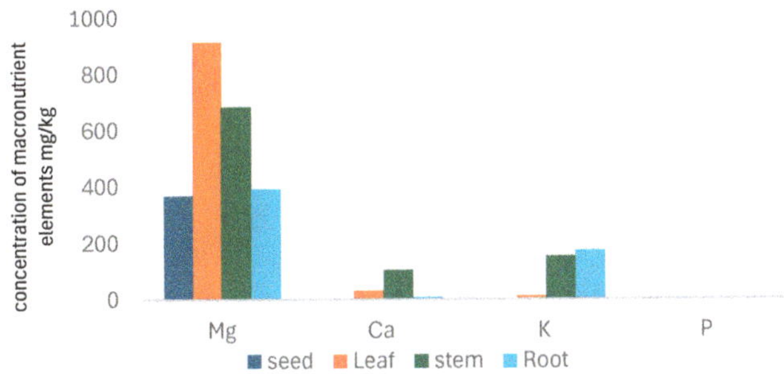

Figure 2. The concentration of macronutrient elements mg/kg in *B. elegans*.

The concentration of magnesium (Mg) in plants was observed to range between 1839.0 and 5198.0 μg/mL. Among different plant parts, leaves exhibited the highest total Mg concentration and readily available Mg (labile Mg), followed by seeds, stems, and roots, emphasizing the critical role of Mg in leaves, where it can exceed 70% of the total dry weight. Mg forms soluble complexes with organic anions like malate, citrate, and oxalate, facilitating its mobility and distribution within plants. Unlike calcium (Ca), Mg does not substitute for Ca in the diffusible fraction when Ca levels are low. Interestingly, cereal grains and seeds tend to maintain stable Mg levels even under varying Mg availability [44].

Plants regulate magnesium (Mg) distribution to support vital functions, transporting it through roots, vacuoles, and xylem. Mg is absorbed via specialized transporters and stored in vacuoles (20–120 mM concentrations), with its mobility facilitated by complexation with organic anions. Mg moves through apoplastic and symplastic pathways in roots, depending on transpiration and plant needs, before reaching shoots for photosynthesis and other processes. Most Mg remains in the symplastic pathway in shoots, accumulating in leaf cells. Research continues to explore Mg transport mechanisms to enhance plant health and crop yields [45–47].

The study showed variations in phosphorus (P) concentration among plant parts. Stems had the highest total P concentration (5.06 μg/mL), while roots contained the highest labile P (312.2 μg/mL), followed by leaves (98.44 μg/mL) and seeds (91.51 μg/mL). In stems, most P was stored in a complex form (1.96 μg/g), differing from other tissues. Phosphorus is challenging for plants to absorb due to its poor solubility and reliance on diffusion, a slow process supplying over 90% of a plant's P needs. Soil factors, including moisture, temperature, and clay mineral types, significantly affect P mobility, with increased moisture enhancing diffusion and colder temperatures slowing the process. Soil composition also impacts how tightly P binds to particles, influencing its availability.

Plants mainly absorb P in two small forms: HPO_4^{2-} (hydrogen phosphate) and H_2PO^{4-} (dihydrogen phosphate), allowing them to pass through cell membranes. Larger phosphorus molecules cannot directly enter the cell membrane but can enter the space outside the cell membrane (apoplast), where they might transform to become more soluble. Once phosphorus is in a soluble form, it can be absorbed by plant cells. Some phosphorus is used by root cells directly, while most are transported to the inner part of the root (stele) through a specialized process called symplastic transport.

Calcium (Ca) concentrations varied among plant parts, with stems showing the highest levels (514 to 5305 μg/g) and seeds the lowest. Ca distribution in plants is tightly regulated through the xylem, which transports water and nutrients. This transport depends on the plant's capacity and transpiration rate, both influenced by soil calcium availability and water loss through leaves. Ca moves in the xylem as free Ca^{2+} ions or complexes with organic acids like malate and citrate, with labile results indicating that most Ca is transferred as free ions. Ca can be transported within the xylem in two forms: as free Ca^{2+} ions or as complexes with organic acids like malate and citrate. The labile results indicate that calcium ions are transferred as free ions [48,49].

Potassium K is an essential nutrient that is absorbed by plants in more significant amounts than any other nutrient except N [50]. K is different from nitrogen in incorporation into the structures of organic compounds; instead, potassium remains in ionic form ($K+$) in solution in the cell and acts as an activator of many cellular enzymes [6]. Therefore, it has many functions in plant nutrition and growth that influence crop yield and quality. *B. elegans* has the second highest total concentration and labile of K in roots 915.0, 1329, leaves 784.0, 1355.2, seeds 639.0, 1032, and stems 285.75, 765.3 μg/g, respectively. K and Mg are critical in photosynthesis, carbohydrate partitioning, and protein synthesis.

3.4. Micronutrient Elements

The microelements analysis was done on four tissue plants obtained from *B. elegans*; the results of the analysis are presented in Table 5 and Figure 3.

Table 5. Total concentration μg/mL, labile μg/mL, and complex fraction μg/mL in *B. elegans* using wet and labile digestion of micronutrient elements.

Tissue	Seed			Leaves			Stem			Root		
Elements	Total Concentration	Labile	Complex Fraction	Total Concentration	Labile	Complex Fraction	Total Concentration	Labile	Complex Fraction	Total Concentration	Labile	Complex Fraction
Cu	5.65	49.58	ND	4.55	4.14	0.41	4.77	182.1	ND	8.46	5.84	2.62
Fe	40.105	0.81	39.295	126.85	0	126.85	153.3	6.55	146.75	92.7	0	92.7
Mn	15.17	1.94	13.23	36.715	25.28	11.435	20.165	10.04	10.125	9.435	3.14	6.295
Zn	23.095	345.2	ND	35.585	25.58	10.005	15.54	640.3	ND	13.77	12.86	0.91
Ni	2.5	3.23	ND	1.885	0	1.885	2.755	7.52	ND	2.115	0	2.115
B	0	7.14	ND	0	53.38	ND	0	1.73	ND	0	4.88	ND

ND = not detect.

Copper accumulation was predominantly observed in roots, with lower levels detected in leaves and no significant presence in stems or seeds. Roots had the highest total Cu concentration (8.46 μg/mL), while seeds exhibited the highest labile Cu concentration (49.58 μg/mL). No complex Cu fractions were detected in seeds or stems. Stems showed a total Cu concentration (4.77 μg/mL) similar to that of leaves but had a significantly higher labile Cu concentration (182.1 μg/mL) and no detectable complex Cu fraction.

These findings suggest that Cu distribution within the plant is tissue-specific, with seeds potentially storing Cu, leaves containing a mix of labile and complex Cu, and stems primarily accumulating labile Cu.

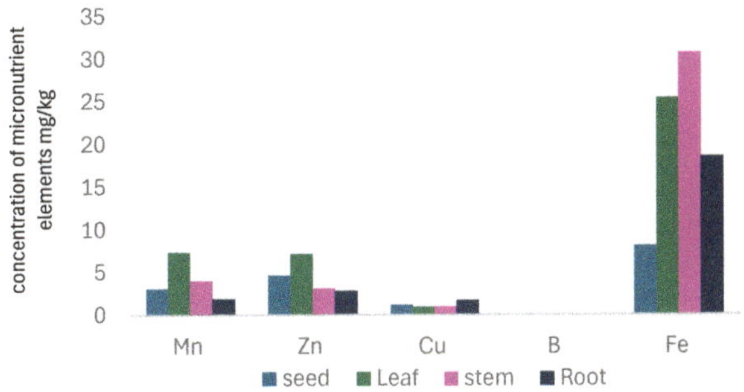

Figure 3. The concentration of micronutrient elements mg/kg in *B. elegans*.

Copper (Cu) distribution in plants shows most Cu concentrated in the rhizodermis and root cap, with lower levels in the inner cortex and stele, consistent with Cu's affinity for cell walls via polygalacturonic acid-binding. In roots and leaves, Cu predominantly exists as sulfur-coordinated Cu (I) species linked to glutathione/cysteine-rich proteins. Isotopic variations indicate reductive uptake and redox cycling during translocation. Cu uptake is facilitated by transporters like the COPT and CTR families, while chaperone proteins (e.g., HMA and CCS families) direct Cu to specific cellular compartments [19,51,52].

In the soil, copper (Cu) exists in various forms with an average concentration ranging from 6 to 80 mg kg^{-1}. Maintaining copper homeostasis in plants requires complex regulatory processes, with multiple transporters playing key roles. Among these, the copper transporter protein (COPT) family has been identified in plants based on sequence similarities with the Ctr protein or through functional complementation in yeast. Members of the COPT family are characterized by three predicted transmembrane segments, with most featuring an N-terminal region rich in methionine and histidine residues, indicative of potential metal-binding domains. Copper ions are uniquely capable of binding small molecules like O_2 as ligands, making them essential cofactors for various oxidases, including mitochondrial cytochrome oxidase. These transporters often operate within a network, forming complexes to enhance their functionality. Their activity is influenced by factors such as copper availability and the presence of other metals, enabling plants to regulate copper levels effectively and prevent either deficiency or toxicity [53,54].

Iron (Fe) is essential for chlorophyll synthesis and photosynthesis, with younger leaves requiring more Fe than older ones. Fe distribution in plants varies significantly across tissues. Seeds had the lowest labile Fe concentration (0.81 μg/mL) but most Fe in a complex form (39.295 μg/g). Roots had the highest total Fe concentration (153.3 μg/mL) but a low labile fraction (6.55 μg/mL), while leaves showed intermediate total Fe levels (126.85 μg/mL), entirely in complex form. This distribution suggests seeds store complex Fe, while roots and leaves primarily accumulate Fe in complex forms.

Nitrogen form also influences Fe accumulation. Ammonium (NH_4^+) increases medium acidity, enhancing Fe solubility and promoting higher Fe concentrations in young leaves, as seen in petunia plants with greener foliage. Conversely, nitrate (NO_3^-) supply causes more Fe accumulation in roots and iron deficiency (chlorosis) in young leaves, as observed in corn plants. These patterns highlight the interplay between nitrogen forms and Fe distribution in plants.

Manganese is a vital micronutrient in plant health, playing a crucial role in light-induced water oxidation during photosynthesis and ATP synthesis. The data reveals manganese (Mn) concentration variations across different plant parts. Seeds contained the lowest labile

Mn concentration (1.94 μg/mL), with most Mn existing in a complex form (13.23 μg/mL). Leaves exhibited a higher total Mn concentration (36.715 μg/g) than seeds, with a significant portion (25.28 μg/mL) found in the labile fraction. Stems also showed a moderate total Mn concentration (20.165 μg/mL) with a nearly equal distribution between labile (10.04 μg/mL) and complex Mn (10.125 μg/mL). Interestingly, roots had the highest total Mn concentration (9.435 μg/mL) but a lower labile Mn fraction (3.14 μg/mL) than leaves. Plants primarily absorb Mn^{2+} cations directly from soil solutions, which are transported to aerial plant parts in this form. Mn is further transported within plant tissues through xylem complexed with specific proteins or non-protein amino acids. The efficiency of Mn absorption depends on its transfer across the soil–root interface and the total amount of bioavailable Mn in the soil.

The interaction between manganese (Mn) and iron (Fe) in plant absorption varies across species. Fe reduces Mn toxicity by limiting its excessive uptake and accumulation rather than by increasing Fe^{2+} absorption or its corrective action within the plant. Research indicates that high Mn concentrations in barley seeds can promote plant growth and boost grain yield in Mn-deficient soils.

This study highlights Nickel (Ni) distribution patterns in plants, with seeds showing the highest labile Ni concentration (3.23 μg/mL) but the lowest total concentration (2.5 μg/mL). Leaves contain only complex Ni (1.885 μg/g), while stems have higher labile Ni (7.52 μg/mL) and total Ni (2.755 μg/mL). Roots show total Ni concentration (2.115 μg/mL), all in complex form. Ni uptake occurs primarily through roots, influenced by soil properties and the rhizosphere. Competing cations (e.g., Fe^{2+}, Cu^{2+}) can hinder Ni absorption. Foliar Ni concentrations in non-contaminated soil typically range from 0.05 to 10 μg/mL dry weight.

The study highlights variations in Zinc (Zn) distribution across plant tissues. Seeds have the highest labile Zn content, while leaves show a higher total Zn concentration but less labile Zn. Stems have the lowest total Zn content but a notable labile Zn fraction, with no complex forms detected. Roots have the second-highest total Zn content, entirely in the labile form. These patterns suggest seeds use labile Zn for germination, leaves balance Zn forms for functions and stems' high labile Zn levels merit further investigation.

Zn is primarily absorbed as Zn^{2+} ions through roots, with some grasses also utilizing Zn–phytosiderophore (Zn–PS) pathways. Once absorbed, Zn is transported to shoot tissues via symplast, apoplast, and phloem pathways. Healthy leaves typically contain 15–100 mg Zn per kg dry weight, while Zn-deficient plants exhibit much lower levels. Zn tolerance does not directly correlate with low leaf Zn levels, as Zn distribution within leaves depends on the plant's overall Zn status. Understanding these processes is essential for maintaining optimal Zn levels without surpassing safe thresholds.

3.5. Beneficial Elements

Table 6 and Figure 4 illustrate the findings on the total content, labile fractions, and complex fractions of beneficial elements.

Table 6. Total concentration μg/mL, labile μg/mL, and complex fraction μg/mL in *B. elegans* using wet and labile digestion of beneficial elements.

Tissue	Seed			Leaves			Stem			Root		
Elements	Total Concentration	Labile	Complex Fraction	Total Concentration	Labile	Complex Fraction	Total Concentration	Labile	Complex Fraction	Total Concentration	Labile	Complex Fraction
Na	0	97.37	ND	1.17	606	ND	48.64	506.8	ND	0	584	ND
Co	0.62	0.51	0.11	0.485	1.6	ND	0.28	0.61	ND	0.59	1.18	ND
Se	6.93	1.26	5.67	2.505	2.94	ND	2.495	1.19	1.305	6.265	5.16	1.105
Al	122.45	0	122.45	298.55	0	298.55	337.65	0	337.65	203.35	0	203.35
Na	0	97.37	ND	1.17	606	ND	48.64	506.8	ND	0	584	ND

ND = Not detected.

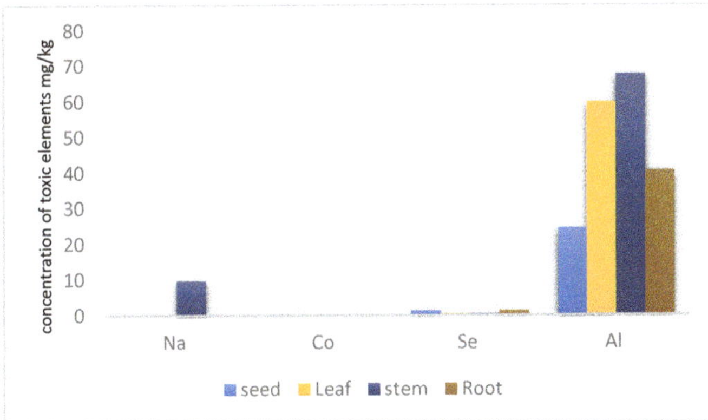

Figure 4. The concentration of the beneficial element mg/kg in *B. elegans*.

Selenium (Se) is an essential micronutrient absorbed by plants primarily as inorganic selenates, which are converted into organic forms like selenocysteine (SeCys) and selenomethionine (SeMet). Its availability varies with soil pH, existing as selenate in alkaline soils and selenite in acidic soils, each differing in mobility and plant absorption.

The study shows that seeds have the highest total Se concentration (6.93 µg/mL), mostly in complex form. Leaves and stems have lower Se concentrations (2.505 µg/mL and 2.495 µg/mL, respectively), with Se in leaves entirely labile and stems evenly split between labile and complex forms. Roots contain the lowest total Se concentration (6.265 µg/mL) but a higher labile fraction. The mobility of Se is confirmed by the presence of selenite and selenomethionine in leaves, suggesting stems act primarily as a conduit between roots and leaves [55,56]

Sodium (Na) distribution varies among plant parts. Roots exhibited the highest labile Na concentration (584 µg/mL), and seeds had 97.37 µg/mL, both with no complex Na fraction identified, indicating a readily available pool of Na in these tissues. Leaves showed a very low total Na concentration (1.17 µg/mL), all in the labile form. Stems had a higher total Na concentration among above-ground parts (48.64 µg/mL), with all Na in the labile fraction as well.

While both sodium and potassium (K) are abundant alkali metals and monovalent cations, they serve different roles in plants. Potassium is an essential nutrient crucial for charge balance, enzyme activation, and osmotic regulation due to its larger ionic diameter and smaller hydration shell, which make it more suitable for these functions [57]. Sodium can be beneficial or toxic depending on the plant species and environmental conditions. In saline environments, plants must balance Na^+ and K^+ uptake to prevent Na toxicity and ensure sufficient K^+ levels. High K^+ concentrations can suppress Na^+ uptake in some halophytes [46]. Understanding the interplay between Na and K is vital for maintaining plant health, especially in high-salinity areas [57,58].

The search results indicate that Cobalt is absorbed from the soil by plant roots and then translocated and distributed to other plant parts like stems and leaves. Cobalt concentrations can vary significantly across different plant species, ranging from 0.28 to 0.62 µg/mL. No complex Co fraction was detected in leaves, stems, and roots. In contrast, Co fraction was detected as complex in seed concentration 0.11 µg/mL.

Cobalt is an essential component of cobalamin, which is needed for the activities of several enzymes and coenzymes and is responsible for forming leghemoglobin, which is involved in nitrogen fixation in nodules of leguminous plants. Cobalt has both beneficial and harmful effects on plants. A relatively lower concentration of cobalt helped in better modulation and, consequently, better growth and yield. However, a higher level of cobalt reduced the bacterial population in the rhizosphere, and as a result, nodulation was hampered, leading to lower crop growth and yield [59].

3.6. Principal Component Analysis (PCA)

The PCA biplot provides a detailed and comprehensive visualization of the relationships between the concentrations of various elements (active variables) and different plant tissues (active observations). The elements analyzed include Pb, Co, Zn, Cd, Mn, Fe, Al, Mg, Ca, Na, P, Ni, Cr, K, Cu, As, and Se. These elements are represented as red vectors, with their direction and length indicating their contributions and influence on the two principal components, F1 (49.23%) and F2 (38.58%), which together explain 87.81% of the total variance in the dataset. The plant tissues—leaf, stem, root, and seed—are distributed across the plot based on their elemental composition.

Chemometric analysis was used in this study to analyze the eighteen elements in different tissues of *B. elegans* by reducing the dimensionality of the data and summarizing the relationships in graphs. This method provided an evaluation of the main components affecting the samples. As illustrated in Figure 5, leaves were closely associated with elements such as Zn, Cd, and Mn, reflecting higher concentrations of these elements in the leaf tissue. In contrast, stems aligned with elements like Ca, Na, and P and were positioned in the negative F1 and F2 quadrants. The root and seed were clustered together near elements such as Cu, As, and Se, indicating similarities in their elemental profiles, particularly for potentially toxic elements.

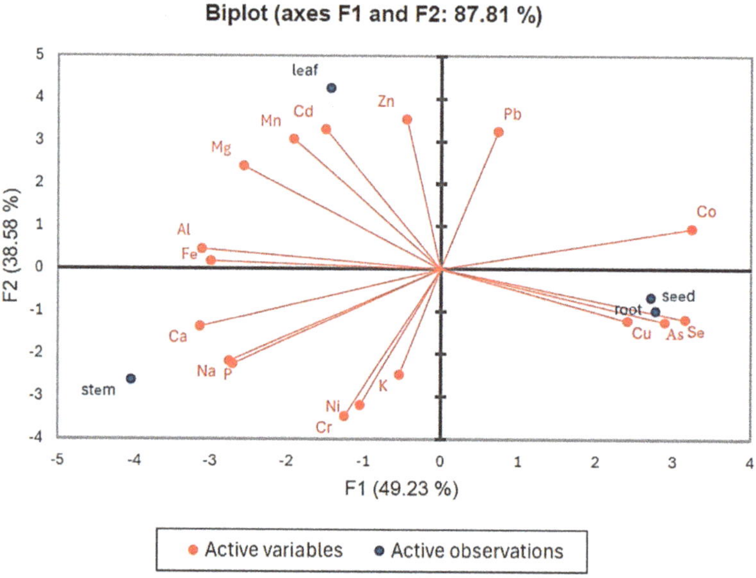

Figure 5. Biplot of principal component analysis.

The scatter of data points in the PCA biplot indicates variation in the mineral element concentrations among leaf and stem samples. This variability may be attributable to biological factors, such as genetics, or environmental factors, such as soil composition. Roots and seeds, on the other hand, exhibit greater similarity in their mineral compositions compared to leaves and stems, likely due to their shared function as storage organs. Roots serve as reservoirs for essential nutrients and water, while seeds provide nourishment for the developing embryo [60]. This common role in resource storage likely explains their similar mineral profiles, contrasting with the distinct elemental distributions observed in leaves and stems. This analysis effectively highlights the nutrient allocation and potential toxicity levels within the plant, providing valuable insights into its elemental composition.

4. Conclusions and Future Perspectives

The study identified distinct distribution patterns of macronutrients, micronutrients, beneficial elements, and toxic metals in various parts of *B. elegans* (seeds, leaves, stems, and roots). The seeds contained the highest magnesium, phosphorus, selenium, and labile zinc levels. The leaves showed high calcium, manganese, iron, and total zinc concentrations. The roots accumulated significant amounts of potassium, copper, and arsenic, while the stems had elevated sodium and labile nickel levels. Toxic elements like cadmium and lead were primarily found in the leaves, whereas arsenic was concentrated in the roots. This compartmentalization underscores the plant's capacity to regulate and distribute elements according to each tissue's specific roles and requirements.

Future research on *Boerhavia elegans* should focus on several key areas to enhance understanding of its physiology and potential applications. First, investigating the molecular mechanisms behind the observed element distribution patterns in *B. elegans* is essential. Additionally, exploring the plant's potential for phytoremediation, particularly in the context of arsenic-contaminated soils, could reveal new environmental applications. Assessing the nutritional value and safety of *B. elegans* seeds for human consumption is another important area, given their rich concentration of essential nutrients. Moreover, studies should examine the plant's adaptability to diverse environmental conditions and its potential for cultivation in marginal lands. Investigating the effects of different cultivation practices on the elemental composition of *B. elegans* tissues is also crucial. Finally, comparing the elemental distribution patterns of *B. elegans* with those of other species in the Brassica genus may uncover unique traits or shared characteristics. These future research directions will provide valuable insights into *B. elegans* and its potential role in agriculture, environmental management, and addressing global food security challenges.

Author Contributions: T.M.A.-R.: investigation, and writing—original draft, formal analysis, Investigation; S.O.B.: project supervision, Conceptualization, writing review, and editing; L.A.A.-K.: project co-super-visor, Formal analysis, methodology; M.W.S.: Formal analysis, Visualization. All authors have read and agreed to the published version of the manuscript.

Funding: This research received no external funding.

Institutional Review Board Statement: Not applicable.

Informed Consent Statement: Not applicable.

Data Availability Statement: The data presented in this study are available on request from the corresponding author.

Acknowledgments: We thank Mohammed Soror El-Shahawi for the revision of the manuscript. We thank Khlood J. Edrees for her technical assistance.

Conflicts of Interest: The authors declare no conflict of interest.

References

1. Haque, F.U.; Faridullah, F.; Irshad, M.; Bacha, A.U.R.; Ullah, Z.; Fawad, M.; Hafeez, F.; Iqbal, A.; Nazir, R.; Alrefaei, A.F.; et al. Distribution and Speciation of Trace Elements in Soils of Four Land-Use Systems. *Land* **2023**, *12*, 1894. [CrossRef]
2. Ertaş, A.; Boğa, M.; Haşimi, N.; Yılmaz, M.A. Fatty acid and essential oil compositions of Trifolium angustifolium var. Angustifolium with antioxidant, anticholinesterase and antimicrobial activities. *Iran. J. Pharm. Res.* **2015**, *14*, 233–241.
3. Yener, I. Trace Element Analysis in Some Plants Species by Inductively Coupled Plasma Optical Emission Spectrometry (ICP-OES). *J. Inst. Sci. Technol.* **2019**, *9*, 1492–1502. [CrossRef]

4. Başgel, S.; Erdemoğlu, S.B. Determination of mineral and trace elements in some medicinal herbs and their infusions consumed in Turkey. *Sci. Total Environ.* **2006**, *359*, 82–89. [CrossRef] [PubMed]
5. Brantner, A.; Grein, E. Antibacterial activity of plant extracts used externally in traditional medicine. *J. Ethnopharmacol.* **1994**, *44*, 35–40. [CrossRef]
6. Merusomayajula, K.V.; Tirukkovalluri, S.R.; Kommula, R.S.; Chakkirala, S.V.; Vundavilli, J.K.; Kottapalli, P.K.S.R. Development and validation of a simple and rapid ICP-OES method for quantification of elemental impurities in voriconazole drug substance. *Future J. Pharm. Sci.* **2021**, *7*, 45. [CrossRef]
7. Barker, A.V.; Pilbeam, D.J. *Handbook of Plant Nutrition*; Taylor & Francis Group: Abingdon, UK, 2016; pp. 1–773. Available online: https://www.kufunda.net/publicdocs/Barker-2007-Handbook_of_Plant_Nutrition.pdf (accessed on 13 December 2024).
8. Abugassa, I.O.; Bashir, A.T.; Doubali, K.; Etwir, R.H.; Abu-Enawel, M.; Abugassa, S.O. Characterization of trace elements in medicinal herbs by instrumental neutron activation analysis. *J. Radioanal. Nucl. Chem.* **2008**, *278*, 559–563. [CrossRef]
9. Maiga, A.; Diallo, D.; Bye, R.; Paulsen, B.S. Determination of some toxic and essential metal ions in medicinal and edible plants from Mali. *J. Agric. Food Chem.* **2005**, *53*, 2316–2321. [CrossRef]
10. Xie, G.; Ye, M.; Wang, Y.; Ni, Y.; Su, M.; Huang, H.; Qiu, M.; Zhao, A.; Zheng, X.; Chen, T.; et al. Characterization of pu-erh tea using chemical and metabolic profiling approaches. *J. Agric. Food Chem.* **2009**, *57*, 3046–3054. [CrossRef] [PubMed]
11. AL-Oud, S.S. Heavy Metal Contents in Tea and Herb Leaves. *Pak. J. Biol. Sci.* **2003**, *6*, 208–212. [CrossRef]
12. Altıntıg, E.; Altundağ, H.; Tuzen, M. Determination of Multi Element Levels in Leaves and Herbal. *Bull. Chem. Soc. Ethiop.* **2014**, *28*, 9–16. [CrossRef]
13. Izol, E.; Çiçek, İ.; Behçet, L.; Kaya, E.; Tarhan, A. Trace Element Analysis of Some Medicinal and Aromatic Plant Species by ICP-MS. *Türk Doğa Fen Dergisi* **2023**, *12*, 21–29. [CrossRef]
14. Bin, C.; Xiaoru, W.; Lee, F.S.C. Pyrolysis coupled with atomic absorption spectrometry for the determination of mercury in Chinese medicinal materials. *Anal. Chim. Acta* **2001**, *447*, 161–169. [CrossRef]
15. Leśniewicz, A.; Jaworska, K.; Zyrnicki, W. Macro- and micro-nutrients and their bioavailability in polish herbal medicaments. *Food Chem.* **2006**, *99*, 670–679. [CrossRef]
16. Ali, S.; Mir, R.A.; Tyagi, A.; Manzar, N.; Kashyap, A.S.; Mushtaq, M.; Raina, A.; Park, S.; Sharma, S.; Mir, Z.A.; et al. Chromium Toxicity in Plants: Signaling, Mitigation, and Future Perspectives. *Plants* **2023**, *12*, 1502. [CrossRef] [PubMed]
17. Brima, E.I. Toxic elements in different medicinal plants and the impact on human health. *Int. J. Environ. Res. Public Health* **2017**, *14*, 1209. [CrossRef] [PubMed]
18. Tokalioğlu, Ş. Determination of trace elements in commonly consumed medicinal herbs by ICP-MS and multivariate analysis. *Food Chem.* **2012**, *134*, 2504–2508. [CrossRef]
19. Marguí, E.; Dalipi, R.; Sangiorgi, E.; Štefan, M.B.; Sladonja, K.; Rogga, V.; Jablan, J. Determination of essential elements (Mn, Fe, Cu and Zn) in herbal teas by TXRF, FAAS and ICP-OES. *X-Ray Spectrom.* **2022**, *51*, 204–213. [CrossRef]
20. Satish, S.; Girish, H. V Antibacterial Activity of Important Medicinal Plants on Human Pathogenic Bacteria-a Comparative Analysis. *World Appl. Sci. J.* **2008**, *5*, 267–271.
21. Sadeghi, Z.; Valizadeh, J.; Azyzian Shermeh, O.; Akaberi, M. Antioxidant activity and total phenolic content of Boerhavia elegans (choisy) grown in Baluchestan, Iran. *Avicenna J. Phytomed.* **2015**, *5*, 1–9.
22. Mehrnia, M.A.; Bashti, A. Evaluation of toxic element contents in infant foods commercially available in Iran. *Bull. Environ. Pharmacol. Life Sci.* **2014**, *3*, 249–253.
23. El Aal, S.A.; Bayoumi, M.A.; A., E.I. Heavy metals residues and trace elements in milk powder marketed in Dakahlia Governorate. *Int. Food Res. J.* **2013**, *20*, 1807–1812.
24. Viñas, P.; Pardo-Martı, M.; Hernández-Córdoba, M. Rapid determination of selenium, lead and cadmium in baby food samples using electrothermal atomic absorption spectrometry and slurry atomization. *Anal. Chim. Acta* **2000**, *412*, 121–130. [CrossRef]
25. Kiani, A.; Arabameri, M.; Moazzen, M.; Shariatifar, N.; Aeenehvand, S.; Khaniki, G.J.; Abdel-Wahhab, M.; Shahsavari, S. Probabilistic Health Risk Assessment of Trace Elements in Baby Food and Milk Powder Using ICP-OES Method. *Biol. Trace Elem. Res.* **2022**, *200*, 2486–2497. [CrossRef] [PubMed]
26. Ikem, A.; Nwankwoala, A.; Odueyungbo, S.; Nyavor, K.; Egiebor, N. Levels of 26 elements in infant formula from USA, UK, and Nigeria by microwave digestion and ICP-OES. *Food Chem.* **2002**, *77*, 439–447. [CrossRef]
27. Fathabad, A.E.; Shariatifar, N.; Moazzen, M.; Nazmara, S.; Fakhri, Y.; Alimohammadi, M.; Azari, A.; Khaneghah, A.M. Determination of heavy metal content of processed fruit products from Tehran's market using ICP-OES: A risk assessment study. *Food Chem. Toxicol.* **2018**, *115*, 436–446. [CrossRef] [PubMed]
28. Al Khalifa, A.S.; Ahmad, D. Determination of key elements by ICP-OES in commercially available infant formulae and baby foods in Saudi Arabia. *Afr. J. Food Sci.* **2010**, *4*, 464–468.
29. WHO. *WHO Guidelines for Assessing Quality of Herbal Medicines with Reference to Contaminants and Residues*; WHO: Geneva, Switzerland, 2007.

30. Tang, Z.; Wang, H.Q.; Chen, J.; Chang, J.D.; Zhao, F.J. Molecular mechanisms underlying the toxicity and detoxification of trace metals and metalloids in plants. *J. Integr. Plant Biol.* **2023**, *65*, 570–593. [CrossRef]
31. Zhao, F.J.; Tang, Z.; Song, J.J.; Huang, X.Y.; Wang, P. Toxic metals and metalloids: Uptake, transport, detoxification, phytoremediation, and crop improvement for safer food. *Mol. Plant* **2022**, *15*, 27–44. [CrossRef] [PubMed]
32. Kumar, A.; Kumar, A.; Cabral-Pinto, M.; Chaturvedi, A.K.; Shabnam, A.A.; Subrahmanyam, G.; Mondal, R.; Gupta, D.K.; Malyan, S.K.; Kumar, S.S.; et al. Lead toxicity: Health hazards, influence on food Chain, and sustainable remediation approaches. *Int. J. Environ. Res. Public Health* **2020**, *17*, 2179. [CrossRef] [PubMed]
33. Wani, A.L.; Ara, A.; Usmani, J.A. Lead toxicity: A review. *Interdiscip. Toxicol.* **2015**, *8*, 55–64. [CrossRef] [PubMed]
34. Collin, M.S.; Venkatraman, S.K.; Vijayakumar, N.; Kanimozhi, V.; Arbaaz, S.M.; Stacey, R.G.S.; Anusha, J.; Choudhary, R.; Lvov, V.; Tovar, G.I.; et al. Bioaccumulation of lead (Pb) and its effects on human: A review. *J. Hazard. Mater. Adv.* **2022**, *7*, 100064. [CrossRef]
35. Abbas, G.; Murtaza, B.; Bibi, I.; Shahid, M.; Niazi, N.K.; Khan, M.I.; Amjad, M.; Hussain, M.; Natasha. Arsenic uptake, toxicity, detoxification, and speciation in plants: Physiological, biochemical, and molecular aspects. *Int. J. Environ. Res. Public Health* **2018**, *15*, 59. [CrossRef]
36. Khalid, S.; Shahid, M.; Niazi, N.K.; Rafiq, M.; Bakhat, H.F.; Imran, M.; Abbas, T.; Bibi, I.; Dumat, C. Arsenic behaviour in soil-plant system: Biogeochemical reactions and chemical speciation influences. *Enhancing Cleanup Environ. Pollut.* **2017**, *2*, 97–140. [CrossRef]
37. Neidhardt, H.; Kramar, U.; Tang, X.; Guo, H.; Norra, S. Arsenic accumulation in the roots of Helianthus annuus and Zea mays by irrigation with arsenic-rich groundwater: Insights from synchrotron X-ray fluorescence imaging. *Geochemistry* **2015**, *75*, 261–270. [CrossRef]
38. Khan Niazi, N.; Bibi, I.; Fatimah, A.; Shahid, M.; Tariq Javed, M.; Wang, H.; Sik Ok, Y.; Bashir, S.; Murtaza, B.; Ahmad Saqib, Z.; et al. Phosphate-assisted phytoremediation of arsenic by Brassica napus and Brassica juncea: Morphological and physiological response. *Int. J. Phytoremediat.* **2017**, *19*, 670–678. [CrossRef] [PubMed]
39. Ghosh, P.; Rathinasabapathi, B.; Ma, L.Q. Phosphorus solubilization and plant growth enhancement by arsenic-resistant bacteria. *Chemosphere* **2015**, *134*, 1–6. [CrossRef]
40. Quantin, C.; Ettler, V.; Garnier, J.; Šebek, O. Sources and extractibility of chromium and nickel in soil profiles developed on Czech serpentinites. *C. R.-Geosci.* **2008**, *340*, 872–882. [CrossRef]
41. Srivastava, D.; Tiwari, M.; Dutta, P.; Singh, P.; Chawda, K.; Kumari, M.; Chakrabarty, D. Chromium stress in plants: Toxicity, tolerance and phytoremediation. *Sustainability* **2021**, *13*, 4629. [CrossRef]
42. Shahid, M.; Shamshad, S.; Rafiq, M.; Khalid, S.; Bibi, I.; Niazi, N.K.; Dumat, C.; Rashid, M.I. Chromium speciation, bioavailability, uptake, toxicity and detoxification in soil-plant system: A review. *Chemosphere* **2017**, *178*, 513–533. [CrossRef] [PubMed]
43. Xu, Z.-R.; Cai, M.-L.; Chen, S.-H.; Huang, X.-Y.; Zhao, F.-J.; Wang, P. High-affinity sulfate transporter Sultr1; 2 is a major transporter for Cr (VI) uptake in plants. *Environ. Sci. Technol.* **2021**, *55*, 1576–1584. [CrossRef] [PubMed]
44. Kirkby, E.A.; Mengel, K. The role of magnesium in plant nutrition. *Z. Pflanzenernährung Bodenkd.* **1976**, *139*, 209–222. [CrossRef]
45. Gavrilescu, M. Water, soil, and plants interactions in a threatened environment. *Water* **2021**, *13*, 2746. [CrossRef]
46. Barker, A.V.; Pilbeam, D.J. *Handbook of Plant Nutrition*; CRC Press: Boca Raton, FL, USA, 2015; pp. 15–16. [CrossRef]
47. Gardner, R.C. Genes for magnesium transport. *Curr. Opin. Plant Biol.* **2003**, *6*, 263–267. [CrossRef]
48. González-Fontes, A.; Navarro-Gochicoa, M.T.; Ceacero, C.J.; Herrera-Rodríguez, M.B.; Camacho-Cristóbal, J.J.; Rexach, J. Understanding calcium transport and signaling, and its use efficiency in vascular plants. In *Plant Macronutrient Use Efficiency: Molecular and Genomic Perspectives in Crop Plants*; Elsevier Inc.: Amsterdam, The Netherlands, 2017; pp. 165–180. [CrossRef]
49. Gilliham, M.; Dayod, M.; Hocking, B.J.; Xu, B.; Conn, S.J.; Kaiser, B.N.; Leigh, R.A.; Tyerman, S.D. Calcium delivery and storage in plant leaves: Exploring the link with water flow. *J. Exp. Bot.* **2011**, *62*, 2233–2250. [CrossRef]
50. Njira, K.; Nabwami, J. A review of effects of nutrient elements on crop quality. *Afr. J. Food Agric. Nutr. Dev.* **2015**, *15*, 9777–9793. [CrossRef]
51. Kopittke, P.M.; Menzies, N.W.; de Jonge, M.D.; Mckenna, B.A.; Donner, E.; Webb, R.I.; Paterson, D.J.; Howard, D.L.; Ryan, C.G.; Glover, C.J.; et al. In situ distribution and speciation of toxic copper, nickel, and zinc in hydrated roots of cowpea. *Plant Physiol.* **2011**, *156*, 663–673. [CrossRef]
52. Ryan, B.M.; Kirby, J.K.; Degryse, F.; Harris, H.; Mclaughlin, M.J.; Scheiderich, K. Copper speciation and isotopic fractionation in plants: Uptake and translocation mechanisms. *New Phytol.* **2013**, *199*, 367–378. [CrossRef]
53. Yuan, M.; Li, X.; Xiao, J.; Wang, S. Molecular and functional analyses of COPT/Ctr-type copper transporter-like gene family in rice. *BMC Plant Biol.* **2011**, *11*, 69. [CrossRef] [PubMed]
54. Chen, G.; Li, J.; Han, H.; Du, R.; Wang, X. Physiological and Molecular Mechanisms of Plant Responses to Copper Stress. *Int. J. Mol. Sci.* **2022**, *23*, 12950. [CrossRef] [PubMed]
55. Hossain, A.; Skalicky, M.; Brestic, M.; Maitra, S.; Sarkar, S.; Ahmad, Z.; Vemuri, H.; Garai, S.; Mondal, M.; Bhatt, R.; et al. Selenium biofortification: Roles, mechanisms, responsesand prospects. *Molecules* **2021**, *26*, 881. [CrossRef] [PubMed]

56. Khanam, A.; Platel, K. Bioaccessibility of selenium, selenomethionine and selenocysteine from foods and influence of heat processing on the same. *Food Chem.* **2016**, *194*, 1293–1299. [CrossRef]
57. Kronzucker, H.J.; Coskun, D.; Schulze, L.M.; Wong, J.R.; Britto, D.T. Sodium as nutrient and toxicant. *Plant Soil* **2013**, *369*, 1–23. [CrossRef]
58. Maathuis, F.J.M. Sodium in plants: Perception, signalling, and regulation of sodium fluxes. *J. Exp. Bot.* **2014**, *65*, 849–858. [CrossRef] [PubMed]
59. Minz, A.; Sinha, A.K.; Kumar, R.; Kumar, B.; Deep, K.P.; Kumar, S.B. A review on importance of cobalt in crop growth and production. *Int. J. Curr. Microbiol. Appl. Sci.* **2018**, *7*, 2978–2984.
60. Dadlani, M.; Yadava, D.K. *Seed Science and Technology: Biology, Production, Quality*; Springer Nature: Berlin/Heidelberg, Germany, 2023; pp. 1–430. [CrossRef]

toxics

MDPI

Article

Consequences of Volcanic Ash on Antioxidants, Nutrient Composition, Heavy Metal Accumulation, and Secondary Metabolites in Key Crops of Cotopaxi Province, Ecuador

Raluca A. Mihai [1,*], Katherine Elizabeth Rodríguez Valencia [2], Nina G. Sivizaca Flores [2], Vivanco Gonzaga Ramiro Fernando [2], Cubi Isuaste Nelson Santiago [2] and Rodica D. Catana [3]

1 Army Scientific and Technological Research Center—CICTE, Department of Life Science and Agriculture, Universidad de Las Fuerzas Armadas—ESPE, Av. General Ruminahui s/n y, Sangolqui 171103, Ecuador
2 Department of Life Science and Agriculture, Universidad de Las Fuerzas Armadas—ESPE, Av. General Ruminahui s/n y, Sangolqui 171103, Ecuador; kerodriguez1@espe.edu.ec (K.E.R.V.); ngsivizaca@espe.edu.ec (N.G.S.F.); rfvivanco2@espe.edu.ec (V.G.R.F.); nscubi@espe.edu.ec (C.I.N.S.)
3 Institute of Biology Bucharest of Romanian Academy, 296 Splaiul Independentei, 060031 Bucharest, Romania; rodica.catana@ibiol.ro
* Correspondence: rmihai@espe.edu.ec

Abstract: This study investigates the consequences of volcanic ash on the antioxidant properties, nutrient composition, heavy metal levels, and secondary metabolites in *Phaseolus vulgaris* L. (common bean) and *Zea mays* L. (yellow corn), two crucial crops in Ecuador. The objective is to determine how volcanic ash exposure affects these crops, focusing on antioxidant properties and potential heavy metal accumulation. Field experiments were conducted in Cotopaxi Province, where both crops were cultivated under varying volcanic ash conditions. Secondary metabolites, particularly total phenols and flavonoids, were quantified using spectrophotometric methods, while heavy metal content was assessed via atomic absorption spectroscopy. Results showed a notable increase in the synthesis of secondary metabolites, especially phenols and flavonoids, in crops exposed to volcanic ash, enhancing their antioxidant capacity. Importantly, no significant heavy metal accumulation was detected, indicating that the benefits of volcanic ash application can be harnessed without associated toxicity risks. This research highlights the potential of volcanic ash to boost beneficial metabolites in yellow corn and common bean, advocating for careful agricultural practices in volcanic regions to optimize health benefits while mitigating toxicity risks.

Keywords: crops; heavy metals; *Phaseolus vulgaris* L.; *Zea mays* L.; volcanic ash; secondary metabolites

1. Introduction

Throughout Ecuador's history, agriculture and volcanic eruptions have been closely intertwined, with the agricultural sector consistently grappling with the impact and accumulation of volcanic ash on crops [1]. Cotopaxi Province (Ecuador) exemplifies this relationship, where agriculture faces significant challenges due to the continuous deposition of volcanic ash from the Cotopaxi volcano [2]. Volcanic eruptions produce ash that disperses over large areas, affecting soil and crops. Volcanic ash contains essential elements for plant biological processes, such as nickel (Ni), copper (Cu), iron (Fe), manganese (Mn), and zinc (Zn), which in the right amounts can benefit plant growth. However, it also contains non-essential, toxic elements such as arsenic (As), cadmium (Cd), lead (Pb), and mercury (Hg), which in excessive amounts can negatively impact physiological processes, becoming toxic [3]. This delicate balance between beneficial and harmful elements makes volcanic ash a crucial agricultural factor, especially in areas like Cotopaxi.

The elemental composition of volcanic ash and rocks varies considerably depending on factors such as the specific volcano's location and the type of eruption [4]. Volcanic ash is primarily composed of dense (lithic) fragments, slag, pumice, free fractured crystals, vitreous and aggregate particles, copper oxides, and iron minerals resulting from hydrothermal alteration by hot fluids [5]. This diversity in composition affects how the ash interacts with soil and crops. The long-term impact of volcanic ash on cultivable soils becomes evident in volcanic areas where soils exhibit surprisingly high fertility upon ash settling on the surface, resulting in soils with andic properties. However, agriculture practiced on the slopes of volcanoes and adjacent valleys exposes farmers to extreme conditions during volcanic eruptions [6]. The effects of ash exposure on crops depend on characteristics, such as layer thickness, grain size, and variations in vegetation structure. Although volcanic eruptions damage crops in the short term, their development is observed in volcanic regions in the long term due to the inherent fertility of volcanic soil [7]. This phenomenon highlights the duality of volcanic eruptions as both a natural disaster and a source of renewal for agriculture.

In Ecuador, *P. vulgaris* and *Z. mays* are the key crops, being vital economically and nutritionally, occupying first place in production and consumption [8]. Common beans play a crucial nutritional role, serving as a primary source of protein, underscoring their value in the Ecuadorian diet [9]. The culture of beans is particularly important due to the rich content of bioactive phytochemicals with beneficial biological properties [10].

Maize cultivation in Cotopaxi Province is of immense importance due to maize's pivotal role in food security, local economies, and human nutrition [1,11]. Maize is a key crop in Ecuador, significantly contributing to the agricultural economy and the cultural and dietary practices of the region. However, this crop is vulnerable to volcanic ash, which can profoundly alter its nutritional and biological properties. Investigating this impact is crucial for developing effective adaptation strategies that mitigate damage and enhance agricultural resilience in volcanic risk contexts [12,13]. Additionally, understanding the interaction between volcanic ash and maize cultivation is vital for ensuring the sustainability of agriculture in areas prone to volcanic eruptions.

Our research explores the relationship between *Phaseolus vulgaris* and *Zea mays* cultivation and volcanic ash. Both crops, pillars of the Ecuadorian diet, extend their influence far beyond borders. The impact of ash on the synthesis of secondary metabolites, which are key to human well-being and can influence potential health benefits as antioxidants, is significant. At the same time, there can be toxicity due to the absorption of heavy metals. This research is crucial for developing agricultural strategies that increase resilience and adaptation in volcanic areas. Understanding these effects will allow us to develop better agricultural practices that not only protect crops and soils but also ensure the health and well-being of the communities that depend on these crops.

2. Materials and Methods

2.1. Sample Collection

Mature fruits of *P. vulgaris* and *Z. mays* were collected from two zones: exposed (Cotopaxi Province, at 3080 m above sea level, with an ambient temperature ranging from 10 to 17 °C) and unexposed to volcanic ash (Alaques sector (0°50′37.6″ S; 78°35′15.5″ W), at an altitude of 2860 m above sea level, with temperature ranges from 15 to 20 °C).

2.2. Extraction of Bioactive Compounds

The extraction protocol for active compounds described by Claros [14] was employed to obtain the extracts of the two crops. Fresh and mature grains of *P. vulgaris* and *Z. mays* in the presence of volcanic ash were evaluated in two stages: Stage 1, where the soil was covered by a 3 cm layer of ash, and Stage 2, where the ash was no longer visible on the surface but was integrated into the soil due to rain. The grains were ground in a mortar to obtain flour, from which 1 g was weighed and used. Each gram of sample was macerated with 10 mL of 96% ethanol in 15 mL Falcon tubes. The mixture was manually stirred with a glass rod and then refrigerated at 5 °C for 72 h, allowing for the efficient extraction of active compounds. The assays were performed in triplicate, and the absorbance of the extracts was measured using a UV-Vis spectrophotometer.

2.3. Determination of Active Ingredients

In both cases, total phenolic compounds were determined using the Folin–Ciocalteu colorimetric method, following the protocol described by López-Froilán et al. [15]. For this, a small amount of water and 1.0 mL of a diluted sample (1:10) or standard solution was mixed with 1.0 mL of the Folin–Ciocalteu reagent. After 4 min, 4 mL of Na_2CO_3 (100 mg/L) was added, and the mixture was brought up to 25 mL with distilled water and incubated at room temperature, for 90 min, in the dark. All tests were performed in triplicate. A blank was prepared by replacing the sample with ethanol. The absorbance of the analyses was measured at a wavelength of 750 nm using a spectrophotometer. Results were calculated using a linear regression based on gallic acid (0–250 mg GAE/L) and expressed in milligrams of gallic acid per liter (mg GAE/L), yielding the equation $y = 0.0061x + 0.1393$ with a determination coefficient $R^2 = 0.9941$.

Flavonoid content in both crops was also determined using a colorimetric method based on the formation of complexes with aluminum chloride, as described by Pekal et al. [16]. For this, 1 mL of the crude extracts was mixed with 1.5 mL of a solvent, 100 µL of CH_3COONa (1 M), 100 µL of $AlCl_3$ (10% v/v), and 2.3 mL of distilled water. The samples were allowed to rest for 40 min at room temperature, after which absorbance at 435 nm was measured. The calibration curve was prepared using quercetin in a concentration range of 0 to 1.5 mg/L, yielding the equation $y = 0.0296x + 0.067$ with a correlation coefficient $R^2 = 0.9878$.

2.4. Evaluation of Antioxidant Capacity

The crops' antioxidant capacity, both exposed and unexposed to volcanic ash, was evaluated using three different methods: ferric-reducing antioxidant power (FRAP), α-diphenyl-α-picrylhydrazyl free radical scavenging (DPPH), and free-radical-scavenging activity (ABTS).

The FRAP assay was performed following the protocol described by Agudo [17], which measures the reduction of Fe^{3+} to Fe^{2+} based on the FRAP solution prepared by mixing 100 mL of an acetate buffer (300 mM, pH 3.6), 10 mL of $FeCl_3$ (20 mM), and distilled water (12 mL). For the analysis, 300 µL of the FRAP solution, 100 µL of the sample, and 300 µL of distilled water were added to each test tube, followed by a 4 min incubation in the dark, at ambient temperature. Absorbance was measured at 593 nm, and all tests were performed in triplicate, with a control prepared using ethanol instead of the sample.

The DPPH method was used to determine antioxidant activity based on the ability of antioxidants to reduce the DPPH radical, characterized by an intense purple color. Following the protocol described by Ramírez [18], 2.9 mL of the DPPH reagent and 0.1 mL of the sample were mixed, and incubated for 30 min in the dark, and absorbance was measured at 517 nm. The control was prepared similarly by replacing the sample with ethanol.

Finally, antioxidant activity was evaluated using the ABTS method, following Mendoza [19], where the $ABTS^{\bullet+}$ radical was generated and diluted to an absorbance of 0.76 ± 0.1 at 754 nm. To carry out the reaction, 2 mL of the $ABTS^{\bullet+}$ solution was mixed with 20 µL of the sample and kept for 7 min in the dark, after which the absorbance was measured at 754 nm.

2.5. Inductively Coupled Plasma Optical Emission Spectrometry (ICP-OES) Analysis

Inductively Coupled Plasma Optical Emission Spectrometry (ICP-OES) is a powerful analytical technique widely used for the detection of trace elements in various types of samples. The core principle of ICP-OES involves the excitation of atoms and ions in a high-temperature plasma, which is generated by coupling an inductively coupled plasma with an optical emission spectrometer. The high-temperature plasma (generated by passing an inert gas—argon—through a radiofrequency field) serves as an energetic source, atomizing and exciting the elements present in the sample, which are emitting light at characteristic wavelengths detected by a photomultiplier tube or a charge-coupled device (CCD) detector. Dry samples of *P. vulgaris* and *Z. mays* were meticulously prepared by weighing 0.5 g of finely ground plant material, which was then placed into high-resistance Teflon tubes to ensure integrity throughout the process. A carefully prepared reagent mixture, consisting of concentrated nitric acid (HNO_3), hydrogen peroxide (H_2O_2), and ultrapure water, was added to each tube. The samples were subjected to a controlled digestion process utilizing a microwave digestion system (specific model and parameters referenced), tailored specifically for plant materials [20]. This method ensures the complete breakdown of the plant matrix, allowing for the accurate determination of elemental content. Post-digestion, the resultant solutions were filtered to remove any particulate matter and diluted with ultrapure water to a final volume of 10 mL in volumetric flasks. This step is critical for minimizing any potential matrix effects during the subsequent analysis. For the elemental analysis, an inductively coupled plasma optical emission spectrometer (ICP-OES), specifically the Thermo Fisher Scientific 7400 Duo model, was employed. The calibration of the instrument was achieved using a multielement ICP Mix 33 standard, with a concentration range spanning from 0.01 to 7.5 mg/L. To enhance the accuracy and precision of the measurements, yttrium (Y) was incorporated as an internal standard at a concentration of 5 ppm. This internal standard corrects any instrumental drifts and potential matrix effects, ensuring the reliability of the analytical results [21]. The analysis was conducted during the second stage of the ash presence, ensuring that all measurements were taken under consistent conditions for comparability.

2.6. Determination of Bioactive Compounds by LC-MS

To identify bioactive compounds in crops affected by volcanic ash in Cotopaxi Province, a high-performance liquid chromatography–mass spectrometry (HPLC-MS) approach was employed, adapting the methodology proposed by Tohma et al. [22]. The study was conducted during the ash presence, Stage 2. Ethanolic extracts were prepared from lyophilized crop samples (1 g) using 20 mL of 80% ethanol, and incubated at 30 °C for 2 h [23]. The extracts were centrifuged at 5000 rpm for 10 min at 4 °C, followed by filtration and ethanol removal via rotary evaporation at 30 °C. The resulting samples were stored in sealed plastic tubes at −20 °C until the analysis.

The HPLC-MS analysis utilized a Vanquish HPLC system (Thermo Fisher Scientific, Waltham, MA, USA) paired with an Ion Trap mass spectrometer. Chromatographic separation was achieved on an Accucore Vanquish column (150 × 2.1 mm) maintained at 35 °C with a flow rate of 0.5 mL/min [24]. A 10 µL injection volume of 0.1% formic acid served as the mobile phase. Compound identification was based on mass spectra and retention times, comparing the detected peaks with ions from standard solutions and reference databases such as PubChem, ChEBI, Metlin, and HPLC datasets. Data processing and metabolite identification were conducted using MZmine 2.53 software, supplemented with information from the scientific literature [25].

2.7. Statistical Analysis

The statistical analysis was performed using RStudio software (R version 4.3.2), applying a two-factor ANOVA to evaluate significant differences between groups, with a significance level set at $p < 0.05$. All experiments were conducted in triplicate, and the results were expressed as the mean ± standard deviation (SD). The correlation between secondary metabolites and antioxidant capacity was determined using a Pearson Correlation Coefficient. For ICP data visualization, representative graphs were created, focusing exclusively on metals with concentrations greater than 0.001, allowing for a clearer comparison of trace element distributions across various samples.

3. Results

3.1. Active Ingredient Determination

The total phenolic content (TPC) displayed significant differences between bean samples exposed to volcanic ash and those not exposed. Samples grown in the presence of ash exhibited higher concentrations of phenolic compounds, with an average of 0.97 ± 0.04 mg GAE/g FW, compared to 0.59 ± 0.01 mg GAE/g FW in samples without ash exposure. The same situation was observed in the case of yellow corn samples exposed to ash with a slightly higher average compared with those without ash exposure (Table 1). The total flavonoid content (TFC) showed noticeable differences between yellow corn samples exposed to volcanic ash and those not exposed (Table 1).

Figure 1 illustrates the concentration of bioactive compounds, specifically total phenolic content (TPC) and total flavonoid content (TFC), in *P. vulgaris* and *Z. mays* under different volcanic ash exposure conditions. The results show a significant increase in both compounds in crops exposed to ash, particularly in *P. vulgaris*, highlighting the positive effect of abiotic stress on the accumulation of secondary metabolites. These differences align with the literature, which associates environmental stress exposure with enhanced biosynthesis of antioxidant compounds.

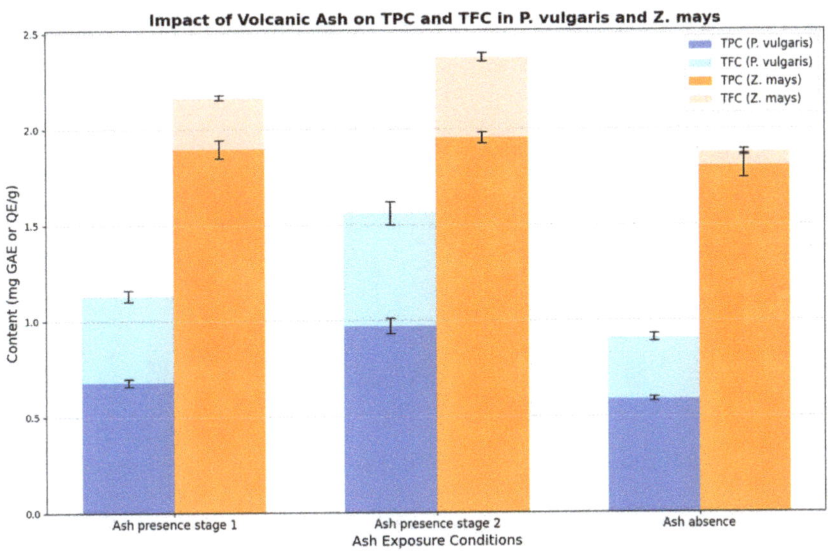

Figure 1. The concentration of total phenolic content (TPC) and total flavonoid content (TFC) in *P. vulgaris* and *Z. mays* under different volcanic ash exposure conditions. The data show significant differences in compound concentrations between the ash-exposed and non-exposed samples. Bars represent the mean ± standard deviation for triplicate measurements.

Table 1. The active ingredients in *P. vulgaris* and *Z. mays* samples exposed and not exposed to volcanic ash.

	P. vulgaris		*Z. mays*	
	TPC (mg GAE/g FW)	TFC (mg QE/g DW)	TPC (mg GAE/g FW)	TFC (mg QE/g DW)
Ash presence, Stage 1	0.68 ± 0.02 [b]	0.45 ± 0.03 [a]	1.89468 ± 0.0477 [a]	0.26956 ± 0.0127 [b]
Ash presence, Stage 2	0.97 ± 0.04 [a]	0.59 ± 0.06 [a]	1.95246 ± 0.0289 [a]	0.41915 ± 0.0216 [a]
Ash absence	0.59 ± 0.01 [b]	0.32 ± 0.02 [a]	1.80375 ± 0.0623 [a]	0.07193 ± 0.0162 [b]

Different letters denote significant differences.

3.2. Antioxidant Activity Determination

All three methods used for antioxidant activity determinations showed significant differences between bean samples exposed and those not exposed to volcanic ash. *P. vulgaris* samples grown in the presence of ash exhibited higher FRAP values (3.74 ± 0.34 µmol Fe^{2+}/g FW), DPPH values (9.10 ± 0.08 µmol TROLOX/g FW), and $ABTS^{*+}$ (8.28 ± 0.26 µmol TROLOX/g F) compared to samples grown in the zone without ash exposure (Table 2).

In the case of maize, antioxidant activity showed similar results for samples exposed to ash and those not exposed. Slight differences were observed between FRAP, DPPH, and ABTS radical scavenging values for both samples grown in the zone exposed to volcanic ash and unexposed (Table 2, Figure 2).

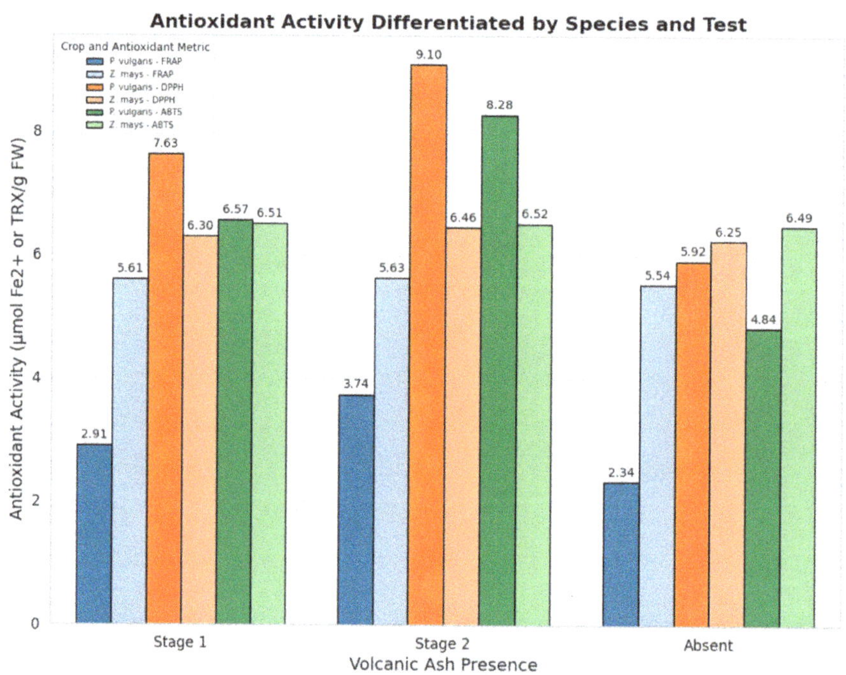

Figure 2. Antioxidant activity determined by FRAP, DPPH, and ABTS assays in *P. vulgaris* and *Z. mays* samples under the presence and absence of volcanic ash. The bar plot represents the mean values ± standard deviation, highlighting significant differences between conditions. The line graph illustrates trends across methods, emphasizing species-specific responses to volcanic ash exposure. Data represent triplicate measurements for each condition.

Table 2. Antioxidant activity in *P. vulgaris* and *Z. mays* samples exposed and not exposed to volcanic ash.

	P. vulgaris			*Z. mays*		
	FRAP (μmol Fe^{2+}/g FW)	DPPH (μmol TRX/g FW)	ABTS (μmol TRX/g FW)	FRAP (μmol Fe^{2+}/g FW)	DPPH (μmol TRX/g FW)	ABTS (μmol TRX/g FW)
Ash presence, Stage 1	2.91 ± 0.25 [b]	7.63 ± 0.10 [a]	6.57 ± 0.12 [b]	5.61 ± 0.37 [a]	6.30 ± 0.09 [a]	6.51 ± 0.28 [a]
Ash presence, Stage 2	3.74 ± 0.34 [a]	9.10 ± 0.08 [a]	8.28 ± 0.26 [a]	5.63 ± 0.49 [a]	6.46 ± 0.62 [a]	6.52 ± 0.81 [a]
Ash absence	2.34 ± 0.11 [b]	5.92 ± 0.24 [b]	4.84 ± 0.10 [b]	5.54 ± 0.45 [a]	6.25 ± 0.06 [a]	6.49 ± 0.31 [a]

Different letters denote significant differences.

The results reveal a marked increase in antioxidant capacity in *P. vulgaris* samples exposed to ash, emphasizing the influence of abiotic stress on enhancing antioxidant potential. In contrast, *Z. mays* exhibited minimal differences between the two conditions, suggesting species-specific responses to volcanic ash exposure.

The correlation between total phenolic content (TPC) and antioxidant capacity, measured by FRAP, DPPH, and ABTS methods, showed significant and strong relationships between the various variables in *P. vulgaris* plants in the presence and absence of ash. A very high correlation was observed between DPPH and TPC (r = 1.000 ***), as well as between ABTS and TPC (r = 0.991 ***), indicating that these antioxidant capacities are closely related in bean samples. Additionally, FRAP also showed a strong correlation with DPPH (r = 0.950 **) and with ABTS (r = 0.951 **), suggesting that phenolic compounds play a key role in the antioxidant capacity of bean plants affected by volcanic ash (Figure 3).

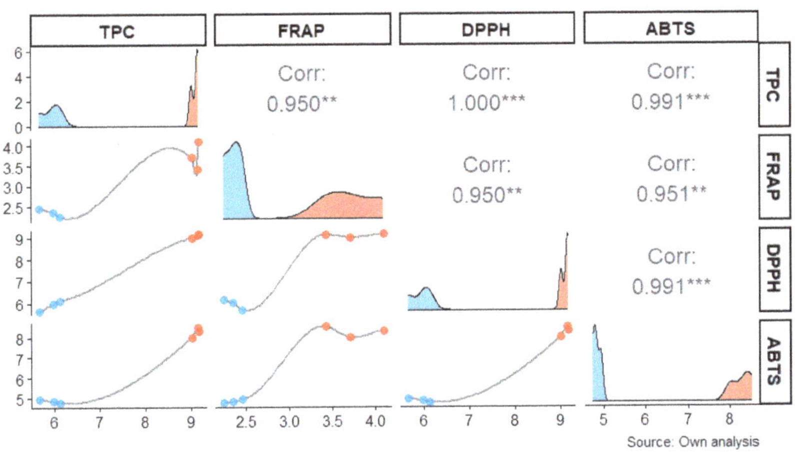

Figure 3. Correlation matrix between total phenolic content (TPC) and antioxidant capacity measured by FRAP, DPPH, and ABTS, in bean samples under the presence and absence of volcanic ash. Correlation coefficients (r) are indicated in each matrix cell, with statistical significance denoted by asterisks (** medium correlation, *** strong correlation). Different colors indicate data for each sample (presence and absence of volcanic ash), and the lines represent the fitted correlation trend for the respective assays. The data used include triplicate measurements for each condition.

3.3. Inductively Coupled Plasma

The ICP analysis of the bean samples exposed to volcanic ash revealed various metallic elements, only a few being present in significant concentrations. Potassium (K) had the highest concentration (14,370.87 mg/kg), followed by magnesium (Mg) with 1910.48 mg/kg and calcium (Ca) with 1005.83 mg/kg. Sodium (Na) also showed a significant concentration

(119.11 mg/kg). In comparison, elements such as iron (Fe), boron (B), and aluminum (Al) had much lower concentrations, being 68.30 mg/kg, 32.30 mg/kg, and 6.22 mg/kg, respectively. Other elements, such as zinc (Zn), manganese (Mn), and copper (Cu), were present in lower concentrations, while some, like bismuth (Bi), cobalt (Co), and thallium (Tl), were detected at levels near the detection limit, around 0.01 mg/kg (Figure 4).

In the case of yellow corn, the ICP analysis revealed that K stood out with the highest concentration (3762.22 mg/kg), followed by Mg (1224.56 mg/kg), Ca (307.47 mg/kg), and Na (120.01 mg/kg). Elements like Fe (12.55 mg/kg), B (11.13 mg/kg), and Al (2.76 mg/kg) showed much lower concentrations. Other elements present in low concentrations included Zn, Mn, and Cu, while elements such as Bi, Co, and Tl were detected at levels near the detection limit, around 0.01 mg/kg. The visual analysis underscores the significant differences in the concentration of metallic elements in maize, highlighting that only a few elements are present in appreciable amounts (Figure 5).

3.4. LC-MS Determination

The influence of volcanic ash deposition on the bioactive and metabolic composition of crops in Cotopaxi Province, Ecuador, was elucidated through the high-resolution HPLC-MS analysis. The study identified a diverse spectrum of compounds, underscoring the intricate biochemical adaptations of plants to environmental stressors. Key metabolites detected in the positive ion mode included daphnetin ($C_9H_6O_4$), a coumarin with potent antioxidant activity (RT: 4.424 min), and argininosuccinate ($C_{10}H_{18}N_4O_6$), a pivotal intermediate in the urea cycle (RT: 4.898 min). Arginine ($C_6H_{14}N_4O_2$), an amino acid essential for protein synthesis and nitrogen metabolism (RT: 4.98 min), further highlighted the relevance of nitrogen-based metabolites. Similarly, the identification of pumiloside ($C_{21}H_{24}O_{11}$), a glycosidic compound (RT: 5.447 min), and glycerophosphorylcholine ($C_8H_{20}NO_6P$), associated with phospholipid metabolism (RT: 5.861 min), underscores the structural and functional diversity of secondary metabolites.

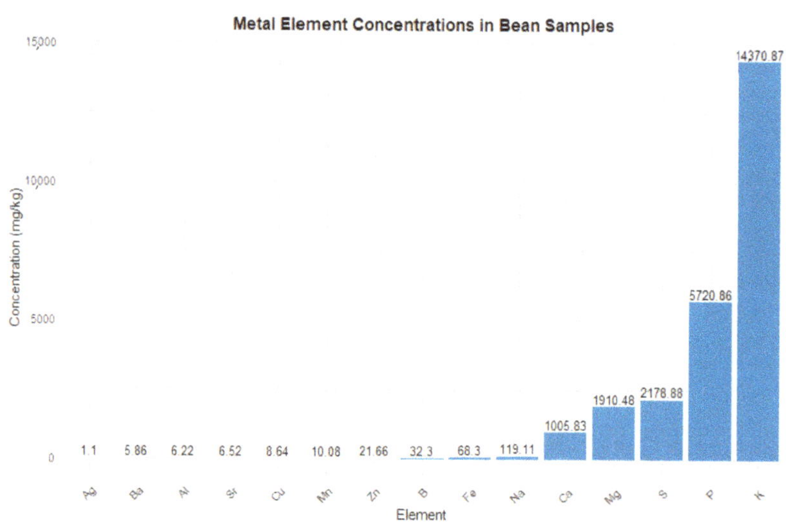

Figure 4. Concentrations of metallic elements in bean samples (ICP) represented in a bar graph. Legend: The graph illustrates the concentrations of various metallic elements in mg/kg found in bean samples, with potassium (K), magnesium (Mg), and calcium (Ca) being the most abundant. The *x*-axis displays the different elements, while the *y*-axis shows their respective concentrations. Metallic elements with concentrations below 0.01 mg/kg were excluded from this graph.

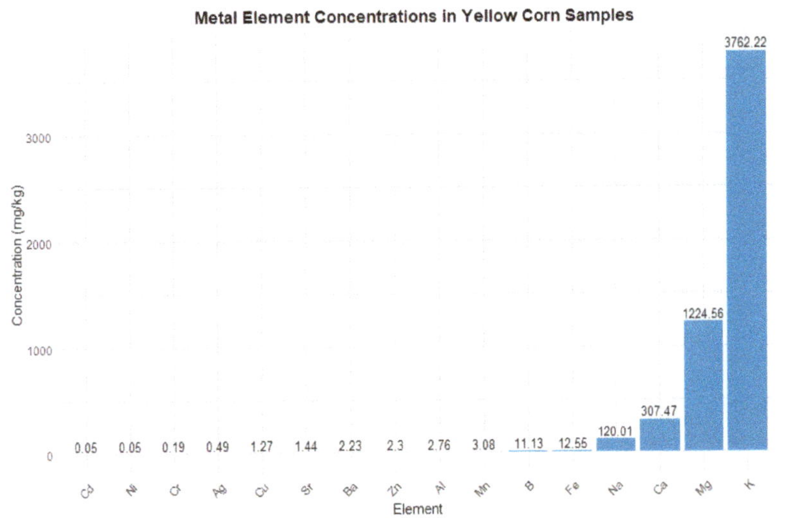

Figure 5. Concentrations of metallic elements in yellow corn (ICP) represented in a bar graph. Legend: The graph shows the concentrations of various metallic elements in mg/kg in yellow corn, highlighting potassium (K), magnesium (Mg), and calcium (Ca) as the most abundant elements. The *x*-axis indicates the different elements, while the *y*-axis represents the concentration.

Phenolic and flavonoid compounds were prominently represented, with notable examples such as caffeic acid ($C_9H_8O_4$), a hydroxycinnamic acid with dual antioxidant and anti-inflammatory properties (RT: 21.053 min), and pelargonidin-3-O-glucoside ($C_{21}H_{21}O_{10}$), an anthocyanin linked to stress mitigation and potential health benefits. Complex flavonoid conjugates, including vicenin ($C_{27}H_{30}O_{15}$) and isoschaftoside ($C_{26}H_{28}O_{14}$), were detected, highlighting the resilience of plants through the biosynthesis of bioactive molecules. In the negative ion mode, primary metabolites such as glucose-6-phosphate ($C_6H_{13}O_9P$) and sugars like sucrose ($C_{12}H_{22}O_{11}$) and sorbitol ($C_6H_{14}O_6$) revealed the pivotal role of carbon flux redistribution under stress conditions (Table 3).

Table 3. Bioactive and Metabolic Compounds Identified in Key Crops from Cotopaxi Province Using HPLC-MS.

	HPLC—MS POSITIVE IONS			
ID	**Proposed Compound Identity**	**Molecular Formula**	**Retention Time**	**Molecular Ion**
19	Daphnetin	$C_9H_6O_4$	4.424	M + H
49	Argininosuccinate	$C_{10}H_{18}N_4O_6$	4.898	M+
59	Arginine	$C_6H_{14}N_4O_2$	4.98	M + H
175	Pumiloside	$C_{21}H_{24}O_{11}$	5.447	M + H
218	Glycerophosphorylcholine	$C_8H_{20}NO_6P$	5.861	M+
284	Biotin	$C_{10}H_{16}N_2O_3S$	7.111	M + H
290	L-Tyrosine	$C_9H_{11}NO_3$	7.644	M + H
362	5′-Methylthioadenosine	$C_{11}H_{15}N_5O_3S$	15.649	M + H
407	gamma-Glutamylleucine	$C_{11}H_{20}N_2O_4$	16.912	M+
416	(2e)-3-(3,4-Dihydroxyphenyl)-N-[2-(4-Hydroxyphenyl)ethyl]acrylamide	$C_{17}H_{17}NO_4$	19.959	M+
548	Glucose_6-phosphate	$C_6H_{13}O_9P$	23.502	[M + Na]+
564	Sinigrin hydrate	$C_{10}H_{16}KNO_9S_2$	23.68	M + H
580	Isoshaftoside	$C_{26}H_{28}O_{14}$	24.811	[M + H]+
605	Vicenin	$C_{27}H_{30}O_{15}$	29.299	[M + Na]+
693	Abscisic acid	$C_{15}H_{20}O_4$	31.05	M + H
859	Ferulate	$C_{10}H_{10}O_4$	38.501	M + H
886	Pelargonidin-3-O-glucoside	$C_{21}H_{21}O_{10}$	19.959	M − H

Table 3. *Cont.*

ID	HPLC—MS NEGATIVE IONS Proposed Compound Identity	Molecular Formula	Retention Time	Molecular Ion
62	Glucose 1-phosphate	$C_6H_{13}O_9P$	5.175	M − H
70	Sorbitol	$C_6H_{14}O_6$	5.192	M − H
120	Sucrose	$C_{12}H_{22}O_{11}$	5.33	M − H
144	Glucose, fructose, mannose	$C_6H_{12}O_6$	5.3	M − H
459	alpha, alpha-Trehalose	$C_{12}H_{22}O_{11}$	19.878	M − H
503	Caffeic acid	$C_9H_8O_4$	21.053	M + H
563	Trihydroxyflavone-C-hexoside-C-pentoside	$C_{20}H_{20}O_{11}$	23.245	M − H
629	Orientin	$C_{21}H_{20}O_{11}$	24.594	M − H

4. Discussion

Cotopaxi's stunning landscapes and volcanic soil provide an interesting backdrop for agricultural endeavors. Culturing common beans in this province faces specific environmental influences due to its proximity to the Cotopaxi volcano, exposing crops to volcanic ash. This ash primarily comprises silicon, aluminum, sulfur, and iron, with alkali and alkaline earth metal oxides such as CaO, MgO, Na_2O, and K_2O in smaller quantities [26].

Volcanic ash has the potential to significantly alter the balance of agricultural ecosystems by impacting vegetation structure, affecting stem and leaf shape and density, which, in turn, influences the amount of ash retained in soil and plant leaves. Ash accumulation can cause direct damage, including leaf abrasion and foliage loss, as well as indirect damage by disrupting soil water and nutrient balance. These changes may affect the organic matter decomposition and nutrient availability in the soil. Despite these negative effects, volcanic ash can also have positive impacts by enhancing carbon and nitrogen availability in the soil, benefiting plant growth [6].

In small quantities, volcanic ash can act as a beneficial fertilizer spreading essential elements like copper (Cu), iron (Fe), manganese (Mn), and zinc (Zn), which are necessary for plant biological processes, but in larger amounts, it can pose significant challenges, such as obstructing sunlight, increasing plant weight, and acidifying leaves and fruits. Currently, volcanic ash soils are most widely used for growing high-value horticultural crops, leading to the growth of the local economy [27]. Under extreme conditions, critical ash exposure can markedly alter nitrogen cycles and water–oxygen exchange in the soil, affecting long-term fertility [28]. The recovery capacity of ash-affected agricultural ecosystems largely depends on the thickness of the deposited ash layer and the adaptability of plants and soil to these changing conditions [6]. Beyond its mineral bounty, volcanic ash contains non-essential toxic elements such as arsenic (As), cadmium (Cd), lead (Pb), and mercury (Hg) that can accumulate stealthily in the biosphere [3], their silent migration echoing through ecosystems. For common bean cultivation, this dual threat manifests absorption by the plants themselves and the looming specter of transfer up the food chain.

Our results showed an increase in secondary metabolite content, respectively, phenols and flavonoids in *P. vulgaris* and *Z. mays* exposed to volcanic ash in comparison with non-exposed cultivars. Similarly, Panico et al. [29] highlighted that strawberries cultivated in the volcanic soil of the Etna mountain region showed higher total phenolic content and antioxidant activity than those cultivated in soil without ash influence.

The findings regarding increased levels of phenolic compounds and flavonoids in plants exposed to volcanic ash, such as *P. vulgaris* and *Z. mays*, can be attributed to the plants' adaptive mechanisms to environmental stress. Volcanic ash is known to pose several abiotic stresses on soil and plant life, including nutrient imbalances, alterations in pH, and variations in water retention [30]. In response to these stresses, plants often enhance the synthesis of secondary metabolites like phenols and flavonoids, which play essential roles in defense against oxidative damage [31]. The increased quantity of phenolic compounds in

P. vulgaris and *Z. mays* may be an adaptive response. Phenols and flavonoids are recognized for their antioxidant properties, which help neutralize reactive oxygen species (ROS) generated during stress conditions [32]. Studies suggest that when plants are exposed to stressors such as volcanic ash, the increased production of antioxidant metabolites serves as a protective mechanism against cellular damage and enhances plant resilience [33]. This accumulation of secondary metabolites not only functions as a defense mechanism but also contributes to the nutritional value of the crops. Plants with higher levels of phenolic compounds and flavonoids can offer enhanced health benefits for consumers due to their bioactive properties [33]. Thus, cultivating *P. vulgaris* and *Z. mays* in volcanic ash could provide a valuable source of dietary antioxidants that promote human health. Additionally, it is known that volcanic ash improves soil fertility and alters the nutrient profile, leading to a potentially available source of minerals that may influence plant growth and development, including secondary metabolite production [34]. Minerals such as potassium and magnesium found in volcanic soils can stimulate metabolic pathways involved in the synthesis of phenols and flavonoids [35], underlining the significance of soil composition in the biochemical responses.

Finally, the genetic makeup of *P. vulgaris* and *Z. mays* may also play a significant role in their ability to synthesize secondary metabolites in response to environmental stress [36]. Biochemical pathways leading to phenolic and flavonoid production are often regulated by specific genes, and the phenotypic differences observed between exposed and non-exposed cultivars imply a strong genetic basis for the observed accumulation of these compounds. In conclusion, the elevated levels of phenols and flavonoids in *P. vulgaris* and *Z. mays* exposed to volcanic ash underline the complex interactions between environmental stresses, plant adaptive mechanisms, and the resulting implications for human health and nutrition.

The inductively coupled plasma (ICP) analysis on both species grown in Cotopaxi Province revealed a high potassium concentration. This abundance of potassium is essential, as this element plays a crucial role in regulating water balance and in key physiological processes such as photosynthesis [37]. However, relatively low concentrations of other essential elements such as copper (Cu) and zinc (Zn) were observed. This suggests that volcanic ash may influence the availability of these nutrients in the soil, thus affecting their uptake by plants [38]. In addition, the concentrations of other essential metals, such as calcium (Ca) and magnesium (Mg), were adequate in the analyzed samples, suggesting that volcanic ash may provide essential nutrients to the soil, potentially improving its fertility [39]. However, the presence of trace metals (copper and zinc) at relatively low concentrations indicates that continuous monitoring of these elements is essential, as they are crucial for plant health and crop yield [40].

Variations in mineral composition facilitate the identification of trends that could be related to soil conditions affected by volcanic ash [41]. Trace elements such as iron (Fe) and boron (B) showed adequate concentrations, indicating that *P. vulgaris* may possess accumulation mechanisms that allow it to adapt to these conditions [42].

The ICP analysis provides crucial information on the mineral composition of both species in soils affected by volcanic ash. The findings underline the importance of assessing nutrient availability in these contexts, which is critical for developing sustainable agricultural management strategies that ensure food security and soil health [43].

The deposition of volcanic ash on crops in the Cotopaxi Province has significantly influenced the composition of secondary metabolites, antioxidants, and essential nutrients. The HPLC-MS analysis identified key bioactive compounds such as daphnetin, argininosuccinate, and ferulic acid, which play critical roles in antioxidant defense and environmental stress modulation. Recent studies highlight that daphnetin, a coumarin with potent antioxidant properties, mitigates oxidative stress and inflammation under abiotic stress conditions and in experimental models of human diseases [35,44]. This suggests that the accumulation of this metabolite in exposed crops might be a crucial adaptive response to volcanic ash impact.

Additionally, the presence of biotin, glucose-6-phosphate, and sorbitol indicates alterations in carbohydrate metabolism and essential cofactors, likely associated with increased energy and antioxidant demands due to environmental stress. The identification of phenolic acids such as caffeic acid and flavonoids like vicenin and isoschaftoside further reinforces the potential of these crops to combat oxidative stress, aligning with previous findings on their ability to protect plants against adverse conditions [45,46]. These results not only highlight the adaptive effects of Cotopaxi crops but also underscore their potential for the development of nutraceutical products and sustainable agricultural applications in challenging environments.

Heavy metals in vegetables from Ecuador have been a subject of concern, particularly due to potential contamination from agricultural practices and environmental factors. In Ecuador, the non-essential metal contamination of crops has been the focus of domestic and international studies on agricultural products. Ecuador has specific regulations, excluding all non-essential metal contaminants (harmful in certain amounts). The prolonged consumption of foods with elevated levels of heavy metals can lead to various health problems, including neurological damage, kidney dysfunction, and an increased risk of certain cancers. The specific levels of heavy metals in corn from Ecuador can vary significantly depending on the region, farming practices, and environmental conditions. Studies have shown that corn grown in Ecuador can contain detectable levels of heavy metals such as cadmium, lead [47], and arsenium [48].

In the case of corn, our analysis recorded < 0.01 mg/kg lead, Pb, which is below the CODEX STAN 193-1995 (codex norms for canned sweet corn, 1 mg/kg). In the case of the other heavy metals identified in our paper, there is no information concerning corn or beans; however, all the values identified in our samples are lower than those registered in CODEX STAN 193-1995 for different cereals and legumes.

5. Conclusions

Our study on the impact of volcanic ash on *P. vulgaris* and *Z. mays* in Cotopaxi Province reveals significant increases in secondary metabolites, specifically phenols, and flavonoids, in response to the environmental stresses associated with volcanic soils. These compounds not only enhance the plants' defense against oxidative damage but also improve their nutritional value, offering potential health benefits for consumers.

The findings underscore the dual effects of volcanic ash on plants: while it contributes essential nutrients like potassium and magnesium that enhance soil fertility, it may also introduce toxic heavy metals such as arsenic and cadmium, necessitating careful monitoring of soil and crop health. In our case, these metals were not detectable. The genetic adaptability of these crops also plays a vital role in their response to environmental stress, suggesting that selective breeding could enhance resilience.

Overall, the cultivation of *P. vulgaris* and *Z. mays* in volcanic ash-affected soils presents both risks and opportunities; however, our findings demonstrate clear benefits. Understanding the implications of volcanic ash on nutrient dynamics and metabolite production can guide sustainable agricultural practices that maximize advantages while minimizing health risks, thereby supporting food production in volcanic regions. Future research should focus on long-term assessments to further explore these complex interactions.

Author Contributions: Conceptualization, R.A.M.; methodology, R.A.M.; formal analysis, K.E.R.V., N.G.S.F., V.G.R.F. and C.I.N.S.; investigation, K.E.R.V., N.G.S.F., V.G.R.F. and C.I.N.S.; resources, R.A.M.; writing—original draft preparation, R.A.M.; writing—review and editing, R.D.C.; supervision, R.A.M.; project administration, R.A.M.; funding acquisition, R.A.M. All authors have read and agreed to the published version of the manuscript.

Funding: This research was funded by Universidad de Las Fuerzas Armadas-ESPE, grant number CV-GNP-0066-2020. The APC was funded by Universidad de Las Fuerzas Armadas-ESPE.

Institutional Review Board Statement: Not applicable.

Informed Consent Statement: Not applicable.

Data Availability Statement: The original contributions presented in the study are included in the article; further inquiries can be directed to the corresponding author due to privacy.

Acknowledgments: The authors would like to express their sincere gratitude to the communities of Cotopaxi Province, particularly the members of Jatari Unancha College in Guasaganda, for their invaluable assistance during the plant collection process.

Conflicts of Interest: The authors declare no conflicts of interest. The funders had no role in the design of the study; in the collection, analyses, or interpretation of data; in the writing of the manuscript; or in the decision to publish the results.

References

1. Mihai, R.A.; Espinoza-Caiza, I.A.; Melo-Heras, E.J.; Cubi-Insuaste, N.S.; Pinto-Valdiviezo, E.A.; Catana, R.D. Does the mineral composition of volcanic ashes have a beneficial or detrimental impact on the soils and cultivated crops of Ecuador? *Toxics* **2023**, *11*, 846. [CrossRef] [PubMed]
2. León Cañar, D.J. Effect of entomopathogenic *Metarhizium anisopliae* fungus for the control of the cicada (*Dalbulus maidis*) in the cultivation of sweet corn (*Zea mays*) Milagro-Ecuador. Master's Program Plant Health Cohort. Ph.D. Thesis, Universidad Agraria del Ecuador, Guayaquil, Ecuador, 2023.
3. Carrera-Beltrán, L.; Gavilanes-Terán, I.; Idrovo-Novillo, J.; Valverde, V.H.; Rodríguez-Pinos, A.; Paredes, C.; Signes-Pastor, A.J.; Carbonell-Barrachina, Á.A. Environmental pollution by heavy metals within the area influenced by the Tungurahua volcano eruption–Ecuador. *Ecotoxicol. Environ. Saf.* **2024**, *270*, 115919. [CrossRef] [PubMed]
4. Jubera, F.J.; Vega, E.J.; Orihuela, R.R.; Montesdeoca, R.D.; Hermandez Díaz, C.; Rodríguez-Díaz, J. Pozzolanic activity of volcanic ashes produced by the eruption of the Tajogaite Volcano in La Palma, Canary Islands. *Constr. Build. Mat.* **2024**, *419*, 135498. [CrossRef]
5. Troncoso, L.; Bustillos, J.; Romero, J.E.; Guevara, A.; Carrillo, J.; Montalvo, E.; Izquierda, T. Hydrovolcanic ash emission between August 14 and 24, 2015 at Cotopaxi volcano (Ecuador): Characterization and eruption mechanisms. *J. Volcanol. Geotherm. Res.* **2017**, *341*, 228–241. [CrossRef]
6. Saputra, D.; Ratna, S.; Hairiah, K.; Widianto, D.; Noordwijk, M. Recovery after volcanic ash deposition: Vegetation effects on soil organic carbon, soil structure, and infiltration rates. *Plant Soil* **2022**, *474*, 163–179. [CrossRef]
7. Ligot, N.; Bogaert, P.; Biass, S.; Lobet, G.; Delmelle, P. Grain size modulates volcanic ash retention on crop foliage and potential yield loss. *Nat. Hazard. Earth Syst. Sci.* **2023**, *23*, 1355–1369. [CrossRef]
8. Pilatasig, D. Mapeo de las Unidades Productivas de seis Cultivos en las Comunidades de Cotopaxi, Mediante un Sistema de Información Geográfica. Bachelor's Thesis, Universidad Técnica de Cotopaxi, Latacunga, Ecuador, 2024. Available online: https://repositorio.utc.edu.ec/items/973a9e1f-46d7-48dd-8a54-5b44eab5c257 (accessed on 10 September 2024).
9. Díaz, M. Evaluación Agrnómica de Fréjol (*Phaseolus vulgaris* L.) Mixturiado Bajo un Sistema Agroecológico en la Granja Experimental, La Pradera. Bachelor's Thesis, Universidad Rpecnica del norte, Ibarra, Ecuador, 2021. Available online: https://repositorio.utn.edu.ec/handle/123456789/11526 (accessed on 10 September 2024).
10. Contreras, A. Desarrollo Fenológico del Frejol Panamito (*Phaseolus vulgaris*) a Base de Abono Orgánico en Ecuador. Bachelor's Thesis, Universidad Técnica de Babahoyo, Babahoyo, Ecuador, 2023. Available online: http://dspace.utb.edu.ec/bitstream/handle/49000/14820/E-UTBFACIAG0AGROP-000063.pdf?sequence=5&isAllowed=y (accessed on 10 September 2024).
11. Guzzon, F.; Arandia Rios, L.W.; Caviedes Cepeda, G.M.; Céspedes Polo, M.; Chavez Cabrera, A.; Muriel Figueroa, J.; Pixley, K.V. Conservation and use of Latin American maize diversity: Pillar of nutrition security and cultural heritage of humanity. *Agronomy* **2021**, *11*, 172. [CrossRef]
12. Gavilánez Luna, F.C.; Carabalí Vargas, A.A. Effect of three times of flooding in the flowering stage of three maize hybrids in Ecuador. *Agric. Res.* **2024**, *26*, 7–13. [CrossRef]
13. Ermolin, M.S.; Ivaneev, A.I.; Fedyunina, N.N.; Fedotov, P.S. Nanospeciation of metals and metalloids in volcanic ash using single particle inductively coupled plasma mass spectrometry. *Chemosphere* **2021**, *281*, 130950. [CrossRef]

14. Claros, P. Evaluacion de la Capacidad Antioxidante Total y Contenido de Polifenoles Totales del *Phaseolus vulgaris* "Frijol". Bachelor's Thesis, Universidad Nacional José Faustino Sánchez Carrión, Huacho, Peru, 2021. Available online: https://repositorio.unjfsc.edu.pe/handle/20.500.14067/5297 (accessed on 10 September 2024).
15. López-Froilán, R.; Hernández-Ledesma, B.; Cámara, M.; Pérez-Rodríguez, M. Evaluation of the Antioxidant Potential of Mixed Fruit-Based Beverages: A New Insight on the Folin-Ciocalteu Method. *Food Anal. Met.* **2018**, *11*, 2897–2906. [CrossRef]
16. Pekal, A.; Pyrzynska, K. Evaluation of aluminum complexation reaction for flavonoid content assay. *Food Anal. Met.* **2014**, *7*, 1776–1782. [CrossRef]
17. Agudo, L. Técnicas para la determinación de compuestos antioxidante en alimentos. *Rev. Educ. Ectremadura* **2010**, *8*, 27–34.
18. Ramírez, M. Determinación de la Capacidad Antioxidante Presente en las Semillas de Cordia Dentata por el Método ABTS y DPPH. Bachelor's Thesis, Universidad El Bosque, Bogota, Columbia, 2023. Available online: https://repositorio.unbosque.edu.co/bitstream/handle/20.500.12495/10599/DeterminacicondeacapacidadantioxidantepresenteenlassemillasdeCordiadentataporelmtodoABTSyDPPH?sequenceisAllowed (accessed on 10 September 2024).
19. Mendoza, M. Inducción de metabolitos de interés nutracéutico en germinados de frijol (*Phaseolus vulgaris* L.) y el efecto de su consumo en un modelo de dislipidemia. Bachelor's Thesis, Universidad Autónoma de Querétaro facultad de química, Santiago de Querétaro, Mexico, 2018.
20. Simonič, M.; Erjavec, A.; Volmajer ValhM, J. Application of ICP-OES for determination of mercury species in environmental samples. *Holist. Approach Environ.* **2024**, *14*, 101–108. [CrossRef]
21. Garcia, E.; Esteban, M. Validation of an ICP-OES Method for the Quantitative Determination of Trace Elements in Food Samples. *J. Anal. Atomic Spectr* **2016**, *31*, 550–558.
22. Tohma, H.; Köksal, E.; Kılıç, Ö.; Alan, Y.; Yılmaz, M.A.; Gülçin, İ.; Bursal, E.; Alwasel, S.H. RP-HPLC/MS/MS analysis of the phenolic compounds, antioxidant and antimicrobial activities of *Salvia* L. species. *Antioxidants* **2016**, *5*, 5040038. [CrossRef] [PubMed]
23. Irakli, M.; Skendi, A.; Bouloumpasi, E.; Chatzopoulou, P.; Biliaderis, C.G. LC-MS identification and quantification of phenolic compounds in solid residues from the essential oil industry. *Antioxidants* **2021**, *10*, 2016. [CrossRef]
24. Luskal, T.; Castillo, S.; Villar-Briones, A.; Orešič, M. MZmine 2: Modular framework for processing, visualizing, and analyzing mass spectrometry-based molecular profile data. *BMC Bioinform.* **2010**, *11*, 395.
25. Cellier, G.; Moreau, A.; Chabirand, A.; Hostachy, B.; Ailloud, F.; Prior, P. A Duplex PCR Assay for the Detection of *Ralstonia solanacearum* Phylotype II Strains in *Musa* spp. *PLoS ONE* **2015**, *10*, e0122182. [CrossRef]
26. Sánchez, E.; Vizcaino, G.; Mejía, F.; Cipriani, I. Análisis mineralógico y multielemental de la ceniza volcánica, producto de la erupción del Cotopaxi en 2015, por difracción de rayos X (XRD) y espectrometría de masas con plasma acoplado inductivamente (ICP-MS) y sus posibles aplicaciones e impactos. *infoANALÍTICA* **2018**, *6*, 9–23. [CrossRef]
27. Shoji, S.; Takahashi, T. Environmental and agricultural significance of volcanic ash soils. *Glob. Environ. Res.* **2002**, *6*, 113–135.
28. Cadena, E. Tendencias Estacionales del Sedimento de Ceniza Proveniente de Erupciones Vulcanianas del Volcán Tungurahua, Periodo 2015–2019. Bachelor's Thesis, Universidad San Francisco de Quito, Quito, Ecuador, 2019. Available online: https://repositorio.usfq.edu.ec/handle/23000/8988 (accessed on 10 May 2024).
29. Panico, A.M.; Garufi, F.; Nitto, S.; Di Mauro, R.; Longhitano, R.C.; Magrì, G.; De Guidi, G. Antioxidant activity and phenolic content of strawberry genotypes from *Fragaria x ananassa*. *Pharm. Biol.* **2009**, *47*, 203–208. [CrossRef]
30. Davis, M.M. Volcanic ash: A review of its effects on plant growth and agricultural productivity. *Agric. Ecosyst. Environ.* **2016**, *222*, 164–175.
31. Kokubun, T. Enhancement of phenolic compounds in plants exposed to stress. *J. Agric. Food Chem.* **2008**, *56*, 6003–6008.
32. Rice-Evans, C.A.; Miller, N.J.; Paganga, G. Structure-antioxidant activity relationships of flavonoids and phenolic acids. *Free Radic. Biol. Med.* **1996**, *20*, 933–956. [CrossRef]
33. Huang, H.; Ullah, F.; Zhou, D.-X.; Yi, M.; Zhao, Y. Mechanisms of ROS Regulation of Plant Development and Stress Responses. *Front. Plant Sci.* **2019**, *10*, 800. [CrossRef] [PubMed]
34. Rastogi, S.; Pandey, M.M.; Rawat, A.K.S. Ethnopharmacological uses, phytochemistry and pharmacology of genus *Adiantum*: A comprehensive review. *J. Ethnopharmacol.* **2018**, *215*, 101–119. [CrossRef]
35. Ciriminna, R.; Scurria, A.; Tizza, G.; Pagliaro, M. Volcanic ash as multi-nutrient mineral fertilizer: Science and early applications. *JSFA Rep.* **2022**, *2*, 528–534. [CrossRef]
36. Hu, L.; Wu, Z.; Robert, C.A.M.; Ouyang, X.; Züst, T.; Mestrot, A.; Xu, J.; Erb, M. Soil chemistry determines whether defensive plant secondary metabolites promote or suppress herbivore growth. *Proc. Natl. Acad. Sci. USA* **2021**, *118*, e2109602118. [CrossRef]
37. Liu, W.; Feng, Y.; Yu, S.; Fan, Z.; Li, X.; Li, J.; Yin, H. The Flavonoid Biosynthesis Network in Plants. *Int. J. Mol. Sci.* **2021**, *22*, 12824. [CrossRef]
38. Marschner, P. *Mineral Nutrition of Higher Plants*, 3rd ed.; Academic Press: Cambridge, MA, USA, 2012.
39. Fang, Y.; Zhao, Y.; Yang, J. Effects of volcanic ash on the mineral content of crops: A review. *J. Soil. Sci. Plant Nutr.* **2018**, *18*, 142–157.

40. Ravenscroft, P.; Brammer, H.; Richards, K. *Ground Water and Global Change: Trends, Impacts and Challenges*; Wiley-Blackwell: Toronto, ON, Canada, 2009.
41. Alloway, B.J. *Contaminated Land and Its Management: The Role of Soil Science*; Cambridge University Press: Cambridge, MA, USA, 2008.
42. Zhou, M.; Wang, H.; Zhang, Q. The Role of Heatmaps in Visualizing High-Dimensional Data. *Nat. Rev. Methods Primers* **2020**, *1*, 1–4.
43. González, M.T.; Aguirre, A.; Contreras, E. Micronutrient accumulation in leguminous plants grown in volcanic soils. *Plant Soil* **2017**, *412*, 303–312.
44. Ramirez, M.; Peñafiel, C.; León, A. Nutritional implications of volcanic soils in crop production. *Agron. J.* **2019**, *111*, 1747–1761.
45. Kumari, P.C.; Kim, S.; Kim, S. Antiplatelet effects of daphnetin mediated by TxA2 generation inhibition. *Int. J. Mol. Sci.* **2023**, *24*, 5779. [CrossRef]
46. Di Stasi, L.C. Natural Coumarin Derivatives Activating Nrf2 Signaling Pathway as Lead Compounds for the Design and Synthesis of Intestinal Anti-Inflammatory Drugs. *Pharmaceuticals* **2023**, *16*, 511. [CrossRef] [PubMed]
47. Counter, S.A.; Buchanan, L.H.; Ortega, F.; Amarasiriwardena, C.; Hu, H. Environmental Lead Contamination and Pediatric Lead Intoxication an Andean Ecuadorian Village. *Int. J. Occup. Environ. Health* **2000**, *6*, 169–176. [CrossRef]
48. Benavides, Á.; Romero, B.; Pérez-Almeida, I.; Pernía, B. Evaluation of the concentration of heavy metals in vegetables from Ecuador. *Revis. Bionatura* **2022**, *7*, 58. [CrossRef]

toxics

Article

Determination of Lead in Fruit Grown in the Vicinity of Tailings Dams of a Mine in Zacatecas, Mexico

Verónica Ávila Vázquez [1], Miguel Mauricio Aguilera Flores [1,*], Agali Naivy Veyna Robles [1], Lilia Elizabeth Solís Lerma [1], Omar Sánchez Mata [1] and Sergio Miguel Durón Torres [2]

[1] Interdisciplinary Professional Unit of Engineering, Campus Zacatecas, Instituto Politécnico Nacional, Blvd. del Bote 202 Cerro del Gato Ejido La Escondida, Col. Ciudad Administrativa, Zacatecas 98160, Mexico; vavila@ipn.mx (V.Á.V.); omsanchezm@ipn.mx (O.S.M.)

[2] Academic Unit of Chemical Sciences, Autonomous University of Zacatecas, Carr. Zacatecas-Guadalajara Km. 6, Ejido la Escondida, Zacatecas 98160, Mexico

* Correspondence: maguileraf@ipn.mx; Tel.: +52-555-729-6000 (ext. 83562)

Abstract: This study analyzed the lead concentrations in fruit grown near tailings dams of a mine in Zacatecas (Mexico) using electrochemical techniques. A 3 × 4 factorial design, with three levels of apple tree distance (low, medium, and high) and four levels of apple tree part (stem, leaf, pulp, and peel), was performed to predict the pathway for contamination (foliar or radicular). Samples of each apple tree part, soil, and irrigation water were collected. The lead concentrations were determined by anodic stripping voltammetry. The results showed lead concentrations of 172 ppm and 0.012 ppm for the soil and irrigation water, which were discarded as sources of contamination since they were below the allowable limits by the Mexican standards (400 ppm and 2 ppm, respectively). However, lead concentrations in the stem and leaf ranged from 6.6 ppm to 30.7 ppm, and pulp and peel exceeded 300 times the allowable limit by the *Codex Alimentarius* (0.1 ppm). The apple tree part was a significant factor in the experimental design. Hence, it was predicted that the pathway for contamination is by foliar absorption. The fruit is highly contaminated by its proximity to the mine. Therefore, mitigation actions must be performed to avoid health risks for the consumers of this fruit.

Keywords: anodic stripping voltammetry; Codex Alimentarius; foliar absorption; fruit contamination; health risks; lead concentrations; Mexican standards; radicular absorption

1. Introduction

Mining is one of the main economic sectors in Mexico that generates social benefits such as employment, foreign exchange, and cultural development. Mexico has remained one of the leading mining countries globally, ranking among the ten highest producers of 16 metals and minerals in 2022. This sector contributed 2.46% to the national Gross Domestic Product [1]. However, this activity has considerable negative impacts on the environment since it generates elevated amounts of waste (tailings dams), causing loss of biodiversity and vegetation cover, mass destruction of water bodies, land-use changes, air contamination, social conflicts, a high cost of living, and food insecurity [2,3].

Tailings dams are the principal source of contamination by heavy metals since they contaminate the soil, water, and air, and the heavy metals contained in them could be bioaccumulated and biomagnified in the food chain [2,4,5]. Arsenic, cadmium, chromium, mercury, and lead rank among the priority metals due to their high toxicity degree and public health significance [4,6]. The United States Agency for Toxic Substances and Disease

Registry classified these elements as human carcinogens and mutagens even at low levels of exposure [7]. Lead is extensively used in mining and other industrial processes. It is not easy to give up its use due to its properties. However, this element is non-biodegradable, and its continuous use accumulates its concentration in the environment with increasing hazards [8]. Therefore, lead has been considered a threat to human health [9,10].

Lead is released into the air by mines and factories, transferred to soil and surface water by rainwater, and introduced into vegetable and fruit cultivation by contaminated water, air, and soil [10,11]. Hence, humans, animals, and plants bioaccumulate this element when they coexist in a contaminated environment [2]. Lead is easily absorbed in the human body as a calcium analog, and children are more vulnerable than adults [10]. The World Health Organization recommends a blood lead concentration of 5 μg/dL as a trigger for a thorough review of how a person is being exposed to lead and for action to reduce or end this exposure based on the *Guideline for Clinical Management of Exposure to Lead* [12]. Lead exposure causes inflammation and cardiovascular, digestive, respiratory, neurologic, and urinary diseases [8].

On the other hand, the absorption of lead by plants is generally the first step in incorporating it into the food chain. Plant species obtain lead from irrigation water, the atmosphere, and soil–root transfer [13–15]. Although lead has no biological purpose in plants, few species have been reported as tolerant and hyperaccumulators since they accumulate concentrations of this metal at >1000 μg/g without harmful effects [16]. However, many plants are intolerant to it, showing fast inhibition of growth, blackening of the root system, chlorosis, photosynthesis inhibition, physiological activity disorders, and cell death at high concentrations [17].

It has been observed that several species of plants absorb lead and accumulate it in the roots, and only a minimum fraction is transported to aerial plant parts [18,19]. Therefore, root and tuber crops such as carrots, potatoes, cassava, and curcumin could remove lead from the soil and bioaccumulate it [18,20]. In this sense, agricultural production systems constitute a significant non-point source of heavy metal-type pollutants, and their use facilitates their accumulation in the soil and their transfer into the soil–plant–consumer chain [21]. Several cases of edible products that absorb lead have been found, such as cassava (19.92 mg/kg), curcumin (3.25 mg/kg) [20], beetroots (0.173 mg/kg), tomatoes (0.294 mg/kg), carrots (0.206 mg/kg), celery (0.259 mg/kg) [22], and pumpkin leaves (6.40 μg/g) [23].

Hence, it is necessary to analyze the food products cultivated at sites susceptible to lead contamination. Similarly, it is essential to identify the pathway by which this element enters the plant and fruit and to propose mitigation and control measures based on health and environmental regulations. The Food and Agriculture Organization of the United Nations (FAO) and the World Health Organization (WHO) issued the *Codex Alimentarius* ("Food Code"), which is a set of standards, guidelines, and codes of practice to protect the health of consumers. This standard establishes the maximum level of heavy metals in food without causing any adverse effects on human health; 0.1 ppm is the maximum lead level in apples [24]. Although Mexico has not emitted a regulation on lead concentration in fruits or crops, the Mexican standards establish a lead concentration limit in irrigation water of 2 ppm [25] and agricultural/residential/commercial and industrial soil types of 400 and 800 ppm, respectively [26].

Various analytical techniques are employed to determine the lead concentrations in different arrays, such as atomic absorption spectroscopy, flow injection analysis, and inductively coupled plasma mass spectrometry. However, these techniques become unattractive since they use expensive instruments that must be specially maintained and require tedious sample preparations, lengthy operational procedures, and unsuitability for online monitoring. Instead, electrochemical techniques, where the analyte is derived from the measurement of the current intensity as a function of the applied potential, show limits of detection in the order of ppb and are low-cost [27].

Therefore, this study aimed to analyze the lead concentrations in fruit grown near tailings dams of a mine in Zacatecas (Mexico) using electrochemical techniques. Furthermore, a 3 × 4 factorial design, with three levels of apple tree distance (low, medium, and high), four levels of apple tree part (stem, leaf, pulp, and peel), and two replicates, was performed to predict the pathway for contamination (foliar or radicular absorption). The results obtained could aid in taking actions based on the "Food Code" for the prevention and reduction of health risks of lead in consumers of this product.

2. Materials and Methods

An orchard where apple trees grow near tailings dams of a mine in Zacatecas (Mexico) was selected as a case study to determine lead concentrations. This mine operated from the 1950s until February 2006. It was subsequently closed for more than 11 years due to a strike. Then, it was rehabilitated in August 2018, and copper, silver, zinc, and lead were extracted. This study was conducted in three stages. The first stage implicated sample collection and the treatment of irrigation water, soil, stems, and leaves of apple trees, and apple pulp and peel. Sampling was carried out in September 2019. The apple harvest takes place mainly between July and October in Mexico, and September is usually the month with the highest production [28]. The second stage involved determining the lead concentrations in all samples. This stage was performed by electrochemical technique, specifically anodic stripping voltammetry (ASV), to determine the lead concentration in the different matrices. The third stage considered a statistical analysis of the results to predict if the pathway for contamination is by foliar or radicular absorption.

2.1. Stage 1. Sample Collection and Treatment

A representative water sample was taken from a surface water body used to irrigate apple trees. The surface water body is intermittent and is 0.8 km from the nearest apple orchard (Figure 1). The sample was obtained using a lead-free container immersed for two hours in nitric acid (at 3% v/v) and thoroughly rinsed with deionized water. The container was rinsed three times with irrigation water. Then, the sample was collected and stored at 4 °C until its use. The water sample was filtered to remove suspended particles and treated according to *Method 3051A Microwave Assisted Acid Digestion* with the following modification [29]. A total of 5 mL of nitric acid (at 65% v/v) and 5 mL of deionized water were added for sample treatment. This adjustment was performed to avoid electrode pitting due to the medium acidity [27].

A 3 × 4 factorial design with three levels of apple tree distance (low, medium, and high), four levels of apple tree part (stem, leaf, pulp, and peel), and two replicates were selected to study the effect of the distance between the apple trees and tailing dams of the mine (factor 1) and the tree part where it accumulates the lead (factor 2), as shown in Table 1. The lead concentration was the response analyzed in the experimental design.

Figure 1 shows the map of the study area where the sampling was carried out on the apple trees surrounding the tailings dams of the mine. Three apple crop plots were identified, taking the centroids of the area of each crop plot as a reference point (low, medium, and high) based on their distance from the tailings dams. Two apple trees 80 m away from the centroid in each crop were selected to take the samples. Therefore, samples of the stem, leaf, pulp, and peel of the apple trees chosen were collected according to the experimental design. Likewise, soil samples where apple trees grow were collected to rule out whether the contamination is foliar or reticular absorption. The distances for taking the samples were ensured to be as precise as possible.

Figure 1. Map of the study area.

Table 1. The 3 × 4 factorial design used in this study.

Assay Number	Factor 1: Apple Tree Distance [1]	Factor 2: Apple Tree Part	Apple Tree Number [2]
1	Low	Stem	1
2	Medium	Stem	3
3	High	Stem	5
4	Low	Leaf	1
5	Medium	Leaf	3
6	High	Leaf	5
7	Low	Pulp	1
8	Medium	Pulp	3
9	High	Pulp	5
10	Low	Peel	1
11	Medium	Peel	3
12	High	Peel	5
13	Low	Stem	2
14	Medium	Stem	4
15	High	Stem	6
16	Low	Leaf	2
17	Medium	Leaf	4
18	High	Leaf	6
19	Low	Pulp	2
20	Medium	Pulp	4
21	High	Pulp	6
22	Low	Peel	2
23	Medium	Peel	4
24	High	Peel	6

[1] Distance between tailings dams of the mine and the study area centroids: low (80 m), medium (343 m), and high (587 m). [2] The sampled apple tree location is shown in Figure 1 with black numbers (1–6).

Stems, leaves, and fruit were collected from each studied apple tree. The peel and pulp were separated from the fruit. The samples were cleaned with tap water and phosphate-free soap. Then, they were washed first with distilled water and deionized water. Later, the samples were dried at 70 °C in a convective flow stove FELISA FE-291AD (Feligneo, Zapopan, Mexico) for 48 h. Afterward, they were macerated with a mortar and placed in polyethylene bags at room temperature until digestion. The digestion was performed according to *Method 3051A Microwave Assisted Acid Digestion* with the mentioned modification [27,29].

The soil samples collected were associated with the apple tree selected in the collection of the samples. A soil sample of approximately 1 kg was taken from the closest part of the trunk of each sampled tree. A soil sampling auger was used, which allowed the soil to be probed to a depth of at least 15 cm. The soil samples were stored in polyethylene bags based on the Mexican Standard NMX-AA-132-SCFI-2016 [30]. Then, they were placed in aluminum trays and dried at 60 °C in a convective flow stove FELISA FE-291AD (Feligneo, Zapopan, Mexico) to remove the moisture. Then, they were sieved at 500 microns and placed in polyethylene bags at room temperature until their digestion. The digestion was performed according to *Method 3051A Microwave Assisted Acid Digestion* with the mentioned modification [27,29].

An apple tree and soil samples from a residential house were taken as blanks in Chalchihuites, Zacatecas, Mexico, remote from the study area (at 50 km) and without perceptible lead contamination sources. The site's geographic coordinates were—103.898931 West longitude and 23.396635 North latitude. The samples were equally treated and analyzed as described. This site was selected for the blank samples because it does not have an apparent source of lead contamination, and it would serve to validate the electrochemical techniques and compare the results between the blank samples and study matrices.

2.2. Stage 2. Measuring the Lead Concentrations in Matrices by Electrochemical Techniques

Lead concentrations in the stem, leaf, peel, pulp, and irrigation water were measured by the electrochemical technique of differential pulse anodic stripping voltammetry (DPASV) and the soil samples by linear potential sweep anodic stripping voltammetry (LSASV), using a Potentiostat/Galvanostat Model 283 (AMETEK, Oak Ridge, TN, USA) and a Rotating Ring-Disk Electrode System Model 636 (AMETEK, Oak Ridge, TN, USA). Glassy (vitreous) carbon, saturated calomel, and platinum were used as the working, reference, and counter electrodes (Figure 2).

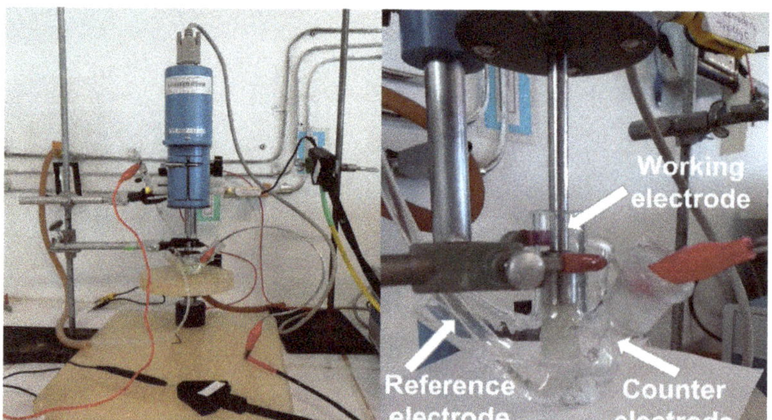

Figure 2. Three-electrode system for anodic stripping voltammetry analysis.

The limit of detection (LoD) and limit of quantitation (LoQ) were estimated using the standard addition method for each technique applied. This method consisted of adding aliquots of 1 or 0.5 microliters (μL) (depending on whether the initial current intensity was high or low) of a stock solution of lead (1000 ppm) in 2 wt% nitric acid (Sigma-Aldrich, San Luis, CA, USA). A standard curve for the low concentration was prepared from the lead stock standard (J.T. Baker, Phillipsburg, NJ, USA) to 0.1, 0.25, 0.5, 0.8, and 1 ppm. The estimated limits are shown in Table 2.

Table 2. Limit of detection (LoD) and limit of quantitation (LoQ) for the electrochemical techniques.

Electrochemical Technique	LoD (ppm)	LoQ (ppm)
DPASV [1]	0.002	0.002
LSASV [2]	1.530	3.850

[1] DPASV: differential pulse anodic stripping voltammetry. [2] LSASV: linear potential sweep anodic stripping voltammetry.

All the experiments were performed using a three-electrode conventional electrochemical cell fabricated with glass and a capacity of 10 mL (Figure 2). The solutions were prepared with a ratio of 1:2 (sample: cell volume), using 5 mL of amalgamating solution [100 ppm of mercury (II) nitrate monohydrate (J.T. Baker, Phillipsburg, NJ, USA) and 0.2 M potassium nitrate solution (Sigma-Aldrich, San Luis, CA, USA), gauged with nitric acid 10% v/v (J.T. Baker, Phillipsburg, NJ, USA)], and 5 mL of the digested sample. Each sample was placed in a process of purging with nitrogen gas for 10 min.

The lead concentration measurements by DPASV were performed using the standard addition method. This method consisted of adding aliquots of 1 or 0.5 μL (depending on whether the initial current intensity was high or low) of a stock solution of lead (1000 ppm) in 2 wt% nitric acid (Sigma-Aldrich, San Luis, CA, USA) until obtaining 3 to 5 additional peaks concerning the initial concentration peak. The operating conditions consisted of forming a mercury film and depositing the problem ions on the working electrode surface, applying a potential of −1.2 V for 435 s and agitation at 900 rpm. Then, the agitation was stopped for 45 s to homogenize the mercury amalgam. Subsequently, a sweep from −0.6 V to −0.4 V was applied at a velocity of 20 mV/s (a pulse amplitude of 5 mV every 250 ms), with pulse height and width of 50 mV and 50 ms, respectively. Each sample was analyzed in duplicate.

The lead concentration measurements by LSASV were performed similarly. For this case, 5 mL of amalgamating solution, 1.25 mL of digested sample, and 3.75 mL of deionized water were added. A sweep from −1.2 to 0.3 V was applied at a velocity of 20 mV/s (a pulse amplitude of 5 mV every 250 ms). Then, the agitation was stopped for 45 s. Each sample was analyzed in duplicate.

The results obtained in the voltammograms were analyzed with a data analysis program using the Origin Pro 8.0 software (OriginLab, Northampton, MA, USA). The baseline of the curves was eliminated. The maximum current intensity of the samples and the additions were plotted concerning the concentrations of the added aliquots of the stock solution of lead, resulting in a graph with an exponential growth trend (Figure 3a). Then, a linear regression of the data of the maximum current intensity obtained in the voltammogram versus the lead concentrations of each added aliquot was performed (Figure 3b). The value that is intersected with the ordinate corresponds to the sample concentration without dilution factors. Finally, the obtained value was multiplied by the dilution factors (125 of the digestion and 2 of the cell) to obtain the lead concentration.

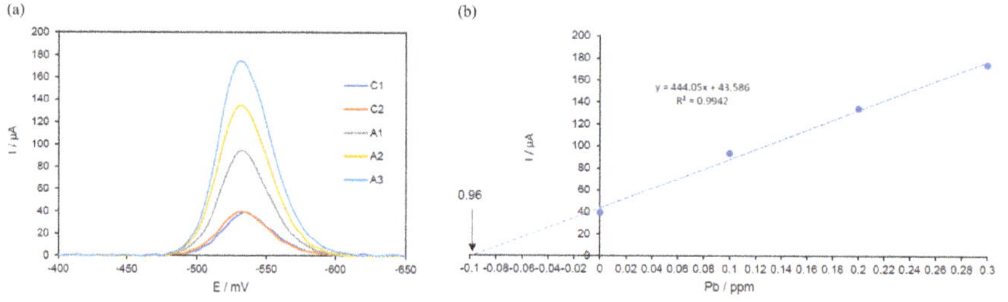

Figure 3. (**a**) Voltammograms of the leaf sample (assay number 6), where C1 and C2 are the initial sample readings and A1, A2, and A3 are the lead additions. (**b**) The maximum current intensity versus the lead concentration in the leaf sample (assay number 6).

2.3. *Stage 3. Statistical Analysis of the Results*

An analysis of variance (ANOVA) was performed from the experimental design using Design-Expert® version 12 software (Trial version) (Stat-Ease, Inc., Minneapolis, MN, USA). The significance of the model, the factors, and the lack of fit were established at a p-value of <0.05 [31]. The prediction of the pathway for contamination (foliar or radicular absorption) was determined from the results of the lead concentration, significant factors in the ANOVA, and environmental conditions (the elevation profile and predominant wind direction).

3. Results and Discussion

3.1. *Lead Concentrations in the Different Matrices*

Table 3 shows the lead concentrations in the stem, leaf, pulp, and peel of the apple trees and soil.

Table 3. Lead concentrations in apple trees and soil.

Assay Number	Factor 1: Apple Tree Distance [1]	Factor 2: Apple Tree Part	Lead Concentration (ppm)	
			Apple Tree	Soil
1	Low	Stem	15.0	135
2	Medium	Stem	18.3	172
3	High	Stem	6.6	142
4	Low	Leaf	16.7	135
5	Medium	Leaf	21.5	172
6	High	Leaf	24.2	142
7	Low	Pulp	15.5	135
8	Medium	Pulp	13.2	172
9	High	Pulp	9.0	142
10	Low	Peel	13.9	135
11	Medium	Peel	28.3	172
12	High	Peel	30.4	142
13	Low	Stem	16.0	161
14	Medium	Stem	10.7	164
15	High	Stem	11.7	140
16	Low	Leaf	21.6	161
17	Medium	Leaf	30.7	164
18	High	Leaf	19.4	140
19	Low	Pulp	14.5	161
20	Medium	Pulp	14.2	164
21	High	Pulp	10.4	140
22	Low	Peel	7.1	161
23	Medium	Peel	11.3	164
24	High	Peel	25.5	140

[1] Distance between tailings dams of the mine and the study area centroids: low (80 m), medium (343 m), and high (587 m).

On the one hand, lead concentrations between 6.6 ppm and 18.3 ppm were found in the stems of the apple trees, with the averages based on the distance of 15.5 ppm (low), 14.5 ppm (medium), and 9.2 ppm (high) (Table 3). Lead concentrations between 9 ppm and 15.5 ppm were found in the pulps of the apple, with the averages based on the distance of 15 ppm (low), 13.7 ppm (medium), and 9.7 ppm (high) (Table 3). The distance between the apple trees and the tailings dams of the mine could be a factor influencing the lead concentration in the apple trees since the higher lead concentrations were identified in apple trees of the low zone (80 m of tailings dams) and the lower ones in the high zone (587 m of tailings dams).

On the other hand, the highest lead concentrations were found in the leaves of the apple trees and the peels of the apples (Table 3). Lead concentrations between 16.7 ppm and 30.7 ppm, with averages based on the distance of 19.1 ppm (low), 26.1 ppm (medium), and 21.8 ppm (high), were found in the leaves of the apple trees (Table 3). Lead concentrations between 11.3 ppm and 30.4 ppm were found in the peels of the apples, with averages based on the distance of 15.5 ppm (low), 19.8 ppm (medium), and 28 ppm (high) (Table 3). It is noted that the distance between apple trees and the tailings dams of the mine was not a factor influencing the lead concentration since the higher lead concentration was estimated in the apple trees of the medium zone (at 343 m of tailings dams). Therefore, the apple tree distance (low, medium, and high) and apple tree part (stem, leaf, pulp, and peel) were the factors studied in the experimental design, evaluating their influence on the response (lead concentration) in the analysis of variance ANOVA.

Lead concentrations between 135 ppm and 172 ppm were found in the soil samples, with the averages based on the distance of 148 ppm (low), 168 ppm (medium), and 141 ppm (high) (Table 3). It can be noted that there is no clear trend in which the lead concentration has a direct relationship with the distance between apple trees and the tailings dams of the mine. The Mexican Standard NOM-147-SEMARNAT/SSA1-2004 establishes criteria for determining the remediation concentrations of the soils contaminated by lead and other metals, specifying lead reference values of 400 ppm and 800 ppm for agricultural/residential/commercial and industrial use [26]. None of the soil samples exceeded the reference value for soil of agricultural use. Therefore, the soil of the apple tree plots does not require remediation for lead contamination. However, the soil of the apple tree plots has been contaminated since lead is naturally present in all soils at concentrations of 15 to 40 ppm [32], exceeding these values up to 4.3 times the analyzed soil samples in this study. Hence, tailings dams of the mine could be a contamination source of lead in these apple tree plots since lead is one of the minerals extracted by the mine.

Additionally, future contamination should not be ruled out if the control techniques are not maintained at the tailings dams since Félix et al. [27] characterized the soil around the tailings dams of the mine in this study in 2018, finding lead concentrations of 1115 ppm. This value was 2.87 times higher than the reference value stipulated in the Mexican Standard [26]. Therefore, these results suggest that the lead is dispersed by wind erosion and chemical wear of the tailings dams until it reaches the apple tree plots.

Irrigation water was taken from a surface water body and was discarded as a contamination source of lead in the soil of the apple tree plots since a lead concentration of 0.012 ppm was found. This value was 416.7 and 166.7 times lower than the maximum concentrations for irrigation stipulated by the United States Environmental Protection Agency (5 ppm) [33] and Mexican Standard NOM-001-SEMARNAT-2021 (2 ppm) [25]. Apple trees are irrigated weekly with this water. However, this practice is considered suitable since it does not represent a contamination source due to the very low lead concentration reported in this study.

The samples, considered blanks, showed 0.012 ppm, 0.100 ppm, 0.013 ppm, and 0.090 ppm in the stem, leaf, pulp, and peel, respectively. The average values of the study area exceeded 13.1, 22.3, 12.8, and 21.1 times the blank values, respectively. Likewise, the blank soil sample showed a lead concentration of 24 ppm. The average value of the soil samples of the apple tree plots was 152.3 times higher than the blank value. Therefore, it is evident that the apple tree plots are strongly contaminated by lead. The results suggest that the leaves and peels of apple trees have a higher capacity to accumulate this contaminant.

The Food and Agriculture Organization of the United Nations and the World Health Organization emitted the General Standard for Contaminants and Toxins in Food and Feed through the *Codex Alimentarius CXS 193-1995*, stipulating a maximum lead level in apples of 0.1 ppm [24]. The apples analyzed in this study exceeded 128 and 211 times this limit, considering average values, and the highest values in the apple peel and pulp samples exceeded 245 and 300 times, respectively. Hence, fruit represents health risks related to lead in consumers of this product.

Table 4 shows studies on lead accumulation in crops near contaminated soils.

Table 4. Studies on lead accumulation in crops near contaminated soils.

Crops	Lead Concentration (ppm)	Reference
Oat	45.3	[34]
Broad bean	55.0	[34]
Maize	66.6–6166.0	[35]
Vegetables	0.1–0.3	[36]
Rice	2.0–16.0	[37]
Ipomoea	76.9	[38]
Apple	6.6–30.7	This study

García-Gallegos et al. [34] found that oat and broad bean crops can absorb lead from the soil through their roots and translocate it to the aerial part (stems and leaves), showing tolerance to the contamination by lead. They demonstrated that the plant height, root volume, and total dry biomass are not affected by lead contamination. Therefore, the plants did not show any phytotoxic effects by direct contact with lead-contaminated soil. However, Ruiz-Huerta et al. [35] also studied the accumulation of arsenic and heavy metals (among them lead) in maize near mine tailings. They identified visible affectations in the plant, such as chlorosis, thinner leaves, and growth inhibition. In addition, lead was one of the heavy metals with the highest concentration in maize (Table 4). Therefore, comparing both studies with the results obtained in this work, it is evident that plants cultivated in zones close to tailings dams of the mines accumulate heavy metals, causing visible and non-visible toxic effects.

Zhuang et al. [36] investigated the contamination levels of lead in soils, vegetables, and rice grown near the Dabaoshan mine in China. Generally, the vegetables did not exceed the maximum permissible level recommended for fresh-leaf vegetables in China (Table 4). However, the lead concentrations in different parts of the rice plant were in the order of straw > hull > grain, with average values of 16 ppm, 5 ppm, and 2 ppm, respectively. The values exceeded 7.2 times the permissible values (0.2 ppm) for cereals in China. The authors [36] did not predict the pathway for contamination. However, they highlighted the importance of investigating the status of heavy metal concentrations in food crops grown near tailings dams.

Liu et al. [37] studied the heavy metal contamination of soils and crops affected by a Chenzhou lead/zinc mine in Hunan, China. Cereal (rice, maize, and sorghum), pulses (soybean, Adzuki bean, mung bean, and peanut), vegetables (ipomoea, capsicum, taro, and string bean), and the rooted soils were the matrices studied. In general, edible leaves or stems of crops showed higher contamination than seeds or fruits. Ipomoea was the most severely contaminated crop with lead (Table 4), finding concentrations in its leaves 8.5 times higher than the maximum permitted level (9 ppm). The results coincide with those obtained in this work, where apple tree leaves showed the highest concentrations (average values of 22.3 ppm, 21.1 ppm, 13.1 ppm, and 12.8 ppm in leaves, peels, stems, and pulps, respectively).

Bioconcentration and translocation factors have been proposed to demonstrate the behavior of heavy metals in plants. The bioconcentration factor evaluates the content of heavy metals in the plant. It is calculated by dividing the metal concentration in the plant part by the metal concentration in the soil. The translocation factor measures the quantity of the heavy metals transferred from one organ to another. It is calculated by dividing the metal concentration in the plant shoot by the metal concentration in the plant root and shoot systems [39]. Bioconcentration factors were calculated for this study, showing values of 0.45 and 0.43, 0.47 and 0.41, and 0.49 and 0.47 for the low, medium, and high zones, respectively. These values are less than 1, indicating more lead in the environment than in the apple trees [39]. Therefore, the lead in the environment is again associated with mine waste (tailings dams) as a contamination source.

The translocation factors could not be calculated since the lead concentration in apple tree roots was outside the scope of this study. However, it could be analyzed as future work to rule out that apple trees are hyperaccumulating plants and act as phytoremediators of the contaminated site. It is possible that the translocation factors would also result in values less than one based on the values obtained in the bioconcentration factors. Therefore, apple trees would be ruled out as hyperaccumulators or phytoremediators.

3.2. Statistical Validation

Table 5 shows the analysis of variance ANOVA of the 3 × 4 factorial design.

Table 5. Analysis of variance ANOVA of the 3 × 4 factorial design.

Source	Sum of Squares	Degree of Freedom	Mean Square	F Value	p-Value
Model	837.31	11	76.12	3.15	0.0302
Factor 1 [1]	49.38	2	24.69	1.02	0.3891
Factor 2 [2]	406.01	3	135.34	5.60	0.0123
Factor 1 × Factor 2	381.93	6	63.65	2.64	0.0722
Pure Error	289.83	12	24.15		
Cor Total	1127.15	23			

[1] Factor 1: Distance between apple trees and tailings dams of the mine. [2] Factor 2: Apple tree part.

It can be noted that the model of the experimental design is significant according to the analysis of variance ANOVA since the p-value is less than 0.05 (Table 5). The distance between apple trees and the tailings dams of the mine (factor 1) turned out to be a non-significant factor since its p-value is higher than 0.05 (Table 5). This result coincides with the results shown in Table 3, where it was demonstrated that the distance between apple trees and the tailings dams of the mine did not influence the lead concentration since the higher lead concentration was estimated in apple trees of the medium zone (at 343 m of tailings dams).

On the other hand, the apple tree part (factor 2) is a significant factor since its p-value is less than 0.05 (Table 5). The different parts of the apple trees showed average values of lead concentration of 22.3 ppm, 21.1 ppm, 13.1 ppm, and 12.8 ppm in the leaves, peels, stems, and pulps, respectively, from highest to lowest, highlighting significant differences in the average value of each part of the apple trees. Likewise, the interaction of both factors (factor 1 × factor 2) is also significant, with a p-value less than 0.05. Therefore, combining these two factors influences the lead concentration in apple trees. Although the main effect of distance was not significant, it becomes relevant when interacting with the tree part. Hence, the distance between apple trees and the tailings dams of the mine could not be completely ruled out. Therefore, the pathway for contamination (foliar or radicular absorption) was analyzed and predicted.

3.3. Pathway for Contamination (Atmospheric Deposition)

Figure 4 shows the elevation profile from the tailings dams of the mine to the apple orchard. It can be observed that the average elevation profiles of the tailings dams and apple orchards are 2530–2550 and 2490–2530 m above sea level, respectively. Although the elevation difference between the tailings dams and the apple orchard would promote the conditions for leaching heavy metals in the soil, the orography as a climatic factor of the terrain prevents leachates from moving towards the apple orchard, generating a natural barrier that surrounds the surface of the nearby soil between the crop area and the mine tailings.

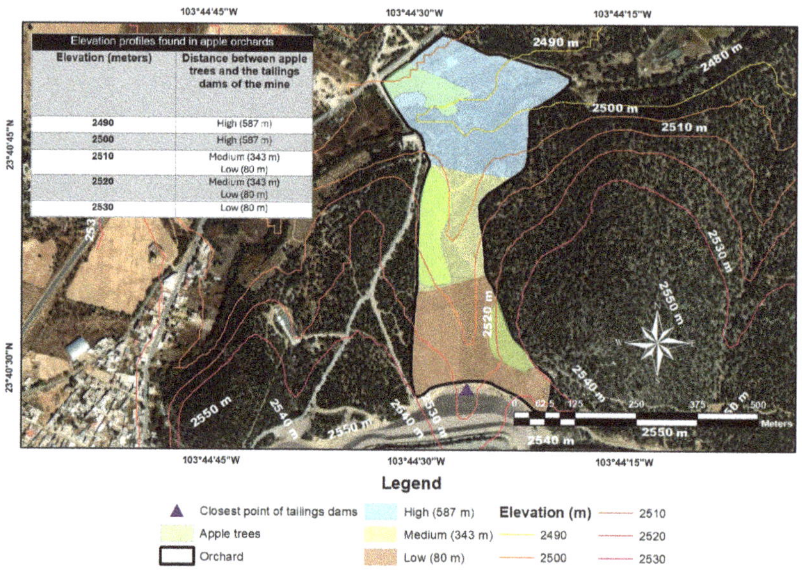

Figure 4. Elevation profile of the study area.

Wang et al. [38] analyzed the contamination of farmland and orchards with heavy metals close to the tailings dams of an abandoned lead–zinc mine in Zixing City, China. The residents primarily focus on cultivating fruit trees for commercial purposes, with cultivating vegetables and staple crops as a secondary activity. However, the design and construction of the tailings dams lacked proper standards. Wang et al. [38] predicted that contamination is mainly caused by the migration and diffusion of the metals downstream since tailings dams were in the upstream region. This condition did not apply to the study area of this work.

Figure 5 shows the predominant wind direction of the study area. It can be noted that the flow vector of the predominant wind direction that goes from the tailings dams to the apple orchard is North-Northeast (NNE) and North (N), with an average velocity between 5.7 and $\geq$11.1 m/s. Both climatic elements (direction and velocity) promote the conditions for a possible wind dispersion of particles to the crop plot. Based on the results of the ANOVA, shown in Table 5, the distance between apple trees and the tailings dams was not a significant factor. Hence, fruit contamination could be attributed to the dispersion of particulate material with lead content towards the apple orchard plots (atmospheric deposition) caused by the predominant wind direction and velocity (Figure 5). Therefore, the pathway for contamination was predicted as foliar absorption.

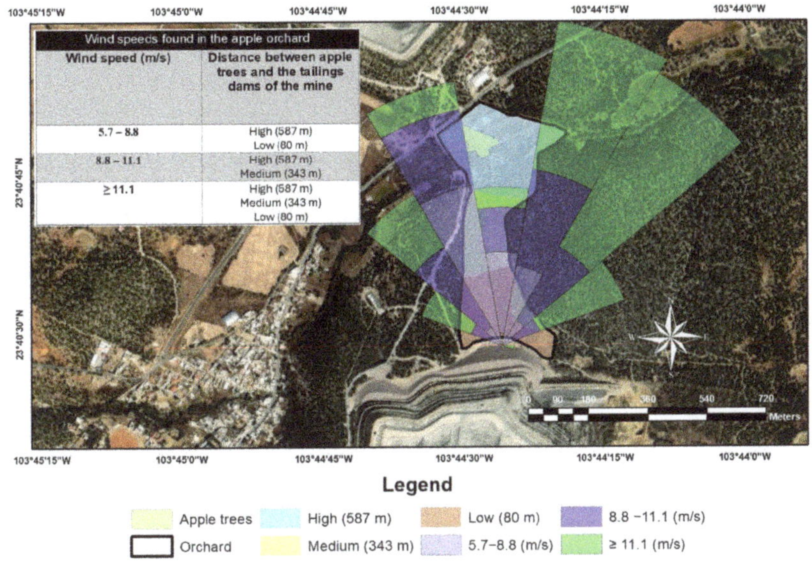

Figure 5. Predominant wind direction of the study area.

Zhu et al. [40] compared the mechanisms of heavy metal absorption from soil in the leaves and roots of different crops. They identified that the highest absorption is in the roots. However, they also found that there is foliar absorption of heavy metals present in the air. The highest heavy metal concentrations in the leaves were lead, cadmium, arsenic, and chromium. According to this study, it can be associated that the pathway for lead contamination of the apple trees (leaves and fruit) is by atmospheric deposition (wind erosion and chemical wear of the tailings dams of the mine). Therefore, it is verified that lead accumulation in apple trees is caused by foliar absorption.

An evaluation of the influence of the seasonal period on lead concentrations found in apple trees was outside the scope of this study. However, it is essential to consider it as future work. This study was carried out during the months of rainfall on the site (September 2019 presented a precipitation of 69 mm [41]), so this factor could have favored the atmospheric deposition of lead on the aerial part of apple trees and promoted the absorption and accumulation of this heavy metal. Yu et al. [42] demonstrated that atmospheric deposition by sedimentation or dragging of raindrops positively impacts heavy metal enrichment in crops. Therefore, the precipitation was a factor that affected the lead concentrations found in the apple trees of this study.

3.4. Mitigation Measures in the Control of Contamination by Lead

Use of natural barriers as endemic trees to break or interrupt winds, as well as chemical or biological stabilization with biosolids, the use of synthetic or vegetative covers (phytostabilization), protection of the surface with recovered soil or materials that allow the fixation of plant species native to the region, and carrying out reforestation and restoration programs to stabilize the slopes of the main barrier to keep the surfaces protected from erosion by wind and rain action are the mitigation measures proposed in the control of tailings dams of the mine. These measures could prevent solid particles from being emitted into the atmosphere caused by the loss of moisture from the tailings dams' surface or the slope of the containment curtain and the formation of runoffs that affect surface and underground water bodies [43].

The Food and Agriculture Organization of the United Nations (FAO) and the World Health Organization (WHO) established the Code of Practice CAC/RCP 56-2004 for the prevention and reduction of the presence of lead in foods and to reduce its content in crops [44]. The proposed recommendations are outlined below.

- Periodically monitor the lead content in soils near crop plots and in the crop plots to discharge that which exceeds the limits established in the environmental regulations.
- Do not use compounds containing lead, such as pesticides based on lead arsenate, or substances that could be contaminated with lead, such as copper fungicides or improperly prepared phosphate fertilizers.
- Do not use machines or equipment that use gasoline with lead as fuel, such as dryers.
- The product (apples) must be protected from lead contamination, for example, exposure to lead from air, soil, or water contamination.
- Organic and inorganic amendments such as compost, biosolids, or manure can be incorporated into the soil, and protective sheets can be used to reduce soil contact deposition on trees and to prevent lead from becoming available to trees.
- Water for irrigation could be protected from sources of lead contamination, and its lead content must be monitored to prevent or mitigate contamination of crops by lead.
- Periodically analyze the lead content in the fruit (apples) to ensure that its concentration does not exceed the limits established by the environmental regulations.

It is essential to ensure the safety and stability of tailings dams, controlling the environmental impacts under the applicable regulations since their omission can cause severe damage to the population and the environment and high economic losses. Such was the case of the collapse of the tailings in Guangdong, China, in 2010. This event caused economic losses of 460 million yuan, the death of 22 people, damage to 6370 houses, and an affected crop area of 72.6 km^2. The causes were associated with the climatological conditions (heavy rain) and neglectful work of the mining departments (design, supervision, and construction) [45]. Therefore, government authorities must pay attention to this activity to avoid damage to the environment that causes high economic losses and damage to the population by lack of control and stability of the tailings dams.

4. Conclusions

The results showed the possibility of potential lead exposure in consumers of apples grown in the vicinity of the tailings dams of a mine in Zacatecas (Mexico) since the apple peel and pulp samples exceeded 245 and 300 times, respectively—the allowed limit by the *Codex Alimentarius* (0.1 ppm). The lead concentrations in soil (from 135 ppm to 172 ppm) and irrigation water (0.012 ppm) were below the allowed limits by the Mexican standards NOM-147-SEMARNAT/SSA1-2004 (<400 ppm) and NOM-001-SEMARNAT-2021 (2 ppm), respectively, being discarded as contamination sources. However, lead concentrations in the stem and leaf from 6.6 ppm to 30.7 ppm were found, predicting that the pathway for contamination is by foliar absorption due to the wind dispersion of tailings dams that arrive at the apple orchard. Therefore, the results of this study provide information on the contamination by lead in fruit grown near tailings dams of a mine and the possibility of human exposure to lead. In addition, the results support taking actions based on the Code of Practice CAC/RCP 56-2004 for preventing and reducing lead contamination in foods emitted by the FAO/WHO.

Author Contributions: Conceptualization, V.Á.V., M.M.A.F. and L.E.S.L.; methodology, V.Á.V., M.M.A.F., O.S.M. and S.M.D.T.; software, M.M.A.F., O.S.M. and S.M.D.T.; validation, V.Á.V., M.M.A.F., O.S.M. and S.M.D.T.; formal analysis, A.N.V.R. and L.E.S.L.; investigation, V.Á.V., M.M.A.F., A.N.V.R., L.E.S.L. and O.S.M.; resources, V.Á.V., M.M.A.F. and S.M.D.T.; data curation, V.Á.V., M.M.A.F., A.N.V.R., L.E.S.L. and O.S.M.; writing—original draft preparation, V.Á.V. and M.M.A.F.; writing—review and editing, V.Á.V., M.M.A.F., A.N.V.R., L.E.S.L., O.S.M. and S.M.D.T.; visualization, M.M.A.F.; supervision, M.M.A.F.; project administration, M.M.A.F.; funding acquisition, M.M.A.F. All authors have read and agreed to the published version of the manuscript.

Funding: This research was funded by the Instituto Politécnico Nacional, grant number SIP-20190214, and by Consejo Zacatecano de Ciencia, Tecnología e Innovación.

Institutional Review Board Statement: Not applicable.

Informed Consent Statement: Not applicable.

Data Availability Statement: The datasets generated and/or analyzed during the current study are available from the corresponding author upon reasonable request.

Acknowledgments: The authors thank the Renewable Energies Laboratory staff at the Autonomous University of Zacatecas for their technical and analytical support.

Conflicts of Interest: The authors declare no conflicts of interest.

References

1. CAMIMEX. Sustainability Report 2023. *Mining and Sustainability: A Commitment to Mexico.* Available online: https://camimex.org.mx/sostenibilidad2023/index.html (accessed on 25 December 2024).
2. Kumar, A.; Kumar, A.; Cabral-Pinto, M.M.S.; Chaturvedi, A.K.; Shabnam, A.A.; Subrahmanyam, G.; Mondal, R.; Gupta, D.K.; Malyan, S.K.; Kumar, S.S.; et al. Lead Toxicity: Health Hazards, Influence on Food Chain, and Sustainable Remediation Approaches. *Int. J. Environ. Res. Public Health* **2020**, *17*, 2179. [CrossRef]
3. Worlanyo, A.S.; Jiangfeng, L. Evaluating the environmental and economic impact of mining for post-mined land restoration and land-use: A review. *J. Environ. Manag.* **2021**, *279*, 111623. [CrossRef]
4. Ekere, N.R.; Ugbor, M.C.J.; Ihedioha, J.N.; Ukwueze, N.N.; Abugu, H.O. Ecological and potential health risk assessment of heavy metals in soils and food crops grown in abandoned urban open waste dumpsite. *J. Environ. Health Sci. Eng.* **2020**, *18*, 711–721. [CrossRef] [PubMed]
5. Techane, G.; Sahilu, G.; Alakangas, L.; Mulat, W.; Kloos, H. Assessment of heavy metal pollution associated with tailing dam in gold mining area, southern Ethiopia. *Geosyst. Eng.* **2023**, *26*, 1–11. [CrossRef]
6. Tchounwou, P.B.; Yedjou, C.G.; Patlolla, A.K.; Sutton, D.J. Heavy metal toxicity and the environment. *Exp. Suppl.* **2012**, *101*, 133–164. [PubMed]
7. Agency for Toxic Substances and Disease Registry. Guidance for the Preparation of Toxicological Profiles. Available online: https://www.atsdr.cdc.gov/toxprofiles/guidance/profile_development_guidance.pdf (accessed on 6 January 2025).
8. Wani, A.L.; Ara, A.; Usmani, J.A. Lead toxicity: A review. *Interdiscip. Toxicol.* **2015**, *8*, 55–64. [CrossRef] [PubMed]
9. Boskabady, M.; Marefati, N.; Farkhondeh, T.; Shakeri, F.; Farshbaf, A.; Boskabady, M.H. The effect of environmental lead exposure on human health and the contribution of inflammatory mechanisms, a review. *Environ. Int.* **2018**, *120*, 404–420. [CrossRef]
10. Collin, M.S.; Venkatraman, S.K.; Vijayakumar, N.; Kanimozhi, V.; Arbaaz, S.M.; Sibiya Stacey, R.G.; Anusha, J.; Choudhary, R.; Lvov, V.; Tovar, G.I.; et al. Bioaccumulation of lead (Pb) and its effects on human: A review. *J. Hazard. Mater. Adv.* **2022**, *7*, 100094. [CrossRef]
11. Liao, X.; Li, Y.; Miranda-Avilés, R.; Puy-Alquiza, M.J.; Bian, J.; Hernández-Anguiano, J.H.; Serafín-Muñoz, A.H.; Datta, S.; Zha, X.; Liu, J.; et al. Assessments of Pollution Status and Human Health Risk of Potentially Toxic Elements in Primary Crops and Agricultural Soils in Guanajuato, Mexico. *Water Air Soil. Pollut.* **2023**, *234*, 670. [CrossRef]
12. World Health Organization. WHO Guideline for Clinical Management of Exposure to Lead. Available online: https://iris.who.int/bitstream/handle/10665/347360/9789240037045-eng.pdf?sequence=1 (accessed on 6 January 2025).
13. Bi, X.; Feng, X.; Yang, Y.; Li, X.; Shin, G.P.Y.; Li, F.; Qiu, G.; Li, G.; Liu, T.; Fu, Z. Allocation and source attribution of lead and cadmium in maize (*Zea mays* L.) impacted by smelting emissions. *Environ. Pollut.* **2009**, *157*, 834–839. [CrossRef]
14. Uzu, G.; Sobanska, S.; Sarret, G.; Muñoz, M.; Dumat, C. Foliar Lead Uptake by Lettuce Exposed to Atmospheric Fallouts. *Environ. Sci. Technol.* **2010**, *44*, 1036–1042. [CrossRef] [PubMed]

15. Tong, G.; Wu, S.; Yuan, Y.; Li, F.; Chen, L. Modeling of Trace Metal Migration and Accumulation Processes in a Soil-Wheat System in Lihe Watershed, China. *Int. J. Environ. Res. Public Health* **2018**, *15*, 2432. [CrossRef]
16. Reeves, R.D.; Baker, A.J.M.; Jaffré, T.; Erskine, P.D.; Echevarria, G.; van der Ent, A. A global database for plants that hyperaccumulate metal and metalloid trace elements. *New Phytol.* **2018**, *218*, 407–411. [CrossRef]
17. Nas, F.S.; Ali, M. The effect of lead on plants in terms of growing and biochemical parameters: A review. *MOJ Eco. Environ. Sci.* **2018**, *3*, 265–268.
18. Kumar, B.; Smita, K.; Flores, L.C. Plant mediated detoxification of mercury and lead. *Arab. J. Chem.* **2017**, *10*, S2335–S2342. [CrossRef]
19. Collin, S.; Baskar, A.; Geevarghese, D.M.; Ali, M.N.V.S.; Bahubali, P.; Choudhary, R.; Lvov, V.; Ibrahin Tovar, G.I.; Senatov, F.; Koppala, S.; et al. Bioaccumulation of lead (Pb) and its effects in plants: A review. *J. Hazard. Mater. Lett.* **2022**, *3*, 100064. [CrossRef]
20. Nobuntou, W.; Parkpian, P.; Kim Oanh, N.T.; Noomhorm, A.; Delaune, R.D.; Jugsujinda, A. Lead distribution and its potential risk to the environment: Lesson learned from environmental monitoring of abandon mine. *J. Environ. Sci. Health Part A* **2010**, *45*, 1702–1714. [CrossRef]
21. Kabata-Pendias, A. *Trace Elements in Soils and Plants*, 4th ed.; CRC Press: Boca Raton, FL, USA, 2010; pp. 1–548.
22. Rusin, M.; Domagalska, J.; Rogala, D.; Razzaghi, M.; Szymala, I. Concentration of cadmium and lead in vegetables and fruits. *Sci. Rep.* **2021**, *11*, 11913. [CrossRef] [PubMed]
23. Nwachukwu, J.I.; Clarke, L.J.; Symeonakis, E.; Brearley, F.Q. Assessment of human exposure to food crops contaminated with lead and cadmium in Owerri, South-eastern Nigeria. *J. Trace Elem. Miner.* **2022**, *2*, 100037. [CrossRef]
24. Food and Agriculture Organization of the United Nations and World Health Organization. General Standard for Contaminants and Toxins in Food and Feed CXS 193-1995. Available online: https://www.fao.org/fao-who-codexalimentarius/sh-proxy/es/?lnk=1&url=https%253A%252F%252Fworkspace.fao.org%252Fsites%252Fcodex%252FStandards%252FCXS+193-1995%252FCXS_193e.pdf (accessed on 18 January 2025).
25. Ministry of the Environment and Natural Resources. Mexican Official Standard NOM-001-SEMARNAT-2021, Which Establishes the Permissible Limits of Pollutants in Wastewater Discharges in Receiving Bodies Owned by the Nation. Available online: https://www.dof.gob.mx/nota_detalle.php?codigo=5645374&fecha=11/03/2022#gsc.tab=0 (accessed on 18 January 2025).
26. Ministry of the Environment and Natural Resources and Ministry of Health. Mexican Official Standard NOM-147-SEMARNAT/SSA1-2004, Which Establishes Criteria to Determine the Remediation Concentrations of Soils Contaminated by Arsenic, Barium, Beryllium, Cadmium, Hexavalent Chromium, Mercury, Nickel, Silver, Lead, Selenium, Thallium, and/or Vaadium. Available online: https://www.profepa.gob.mx/innovaportal/file/1392/1/nom-147-semarnat_ssa1-2004.pdf (accessed on 18 January 2025).
27. Félix, L.T.; Durón, S.M.; Ávila, V.; Correa, H.C.; Aguilera, M.M. Determination of Pb in *Brickellia Veronicifolia* for Anodic Stripping Voltammetry. *ECS Trans.* **2018**, *84*, 297. [CrossRef]
28. Covarrubias-Ramírez, J.M.; Parga-Torres, V.M.; Martínez-Rodríguez, J.G. Postharvest Management and Marketing of Apples in Mexico. In *Fuit Industry*, 1st ed.; Kahramanoğlu, I., Wan, C., Eds.; IntechOpen: London, UK, 2022; Volume 1, pp. 208–554.
29. United States Environmental Protection Agency. Method 3051A: Microwave Assisted Acid Digestion of Sediments, Sludges, Soils, and Oils. Available online: https://www.epa.gov/sites/default/files/2015-12/documents/3051a.pdf (accessed on 1 December 2024).
30. Ministry of Economy. Mexican Standard NMX-AA-132-SCFI-2016, Soil Sampling for Metals and Metalloids Identification and Quantification, and Sample Handling. Available online: http://www.economia-nmx.gob.mx/normas/nmx/2010/nmx-aa-132-scfi-2016.pdf (accessed on 1 December 2024).
31. Montgomery, D.C. *Design and Analysis of Experiments*, 10th ed.; John Wiley & Sons, Inc.: Hoboken, NJ, USA, 2019; pp. 1–114.
32. Center for Agriculture, Food, and the Environment of the University of Massachusetts Amherst. Soil Lead Fact Sheet, Soil Lead Contamination. Available online: https://ag.umass.edu/soil-plant-nutrient-testing-laboratory/fact-sheets/soil-lead-fact-sheet (accessed on 25 December 2024).
33. United States Environmental Protection Agency. Guidelines for Water Reuse 2012. Available online: https://www.epa.gov/sites/default/files/2019-08/documents/2012-guidelines-water-reuse.pdf (accessed on 25 December 2024).
34. García-Gallegos, E.; Hernández-Acosta, E.; García-Nieto, E.; Acevedo-Sandoval, O.A. Lead content and translocation in oats (*Avena sativa*, L.) and broad bean (*Vicia faba*, L.) from contaminated soil. *Rev. Chapingo Ser. Cienc. For. Ambiente* **2011**, *17*, 19–29.
35. Ruiz-Huerta, E.A.; Armienta-Hernández, M.A. Accumulation of arsenic and heavy metals in maize near mine tailings. *Rev. Int. Contam. Ambient.* **2012**, *28*, 103–117.
36. Zhuang, P.; Zou, B.; Li, N.Y. Heavy metal contamination in soils and food crops around Dabaoshan mine in Guangdong, China: Implication for human health. *Environ. Geochem. Health* **2009**, *31*, 707–715. [CrossRef] [PubMed]
37. Liu, H.; Probst, A.; Liao, B. Metal contamination of soils and crops affected by the Chenzhou lead/zinc mine spill (Hunan, China). *Sci. Total Environ.* **2005**, *339*, 153–166. [CrossRef]

38. Wang, Q.; Cai, J.; Gao, F.; Li, Z.; Zhang, M. Pollution Level, Ecological Risk Assessment and Vertical Distribution Pattern Analysis of Heavy Metals in the Tailings Dam of an Abandon Lead–Zinc Mine. *Sustainability* **2023**, *15*, 11987. [CrossRef]
39. Ba, V.N.; Thien, B.N.; Phuong, H.T.; Hong-Loan, T.T.; Anh, T.T. Bioconcentration and translocation of elements from soil to vegetables and associated health risk. *J. Food Compost. Anal.* **2024**, *132*, 106296.
40. Zhu, Z.; Yang, X.D.; Xu, Z.Q.; Fei, J.C.; Peng, J.W.; Rong, X.M.; Huang, Y.; Yang, X.E. Foliar uptake, translocation and accumulation of heavy metals from atmospheric deposition in crops. *J. Plant Nutr. Soil. Sci.* **2021**, *27*, 332–345.
41. National Water Commission. National Meteorological System of Mexico. *National Climatological Database*. Available online: https://smn.conagua.gob.mx/tools/RESOURCES/Normales_Climatologicas/Mensuales/zac/mes32054.txt (accessed on 24 February 2025).
42. Yu, P.; Shao, X.; Wang, M.; Zhu, Z.; Tong, Z.; Peng, J.; Deng, Y.; Huang, Y. Effects of atmospheric deposition on heavy metal contamination in paddy field systems under different functional areas in ChangZhuTan, Hunan Province, China. *Sci. Total Environ.* **2024**, *933*, 172953. [CrossRef]
43. Ministry of the Environment and Natural Resources. Mexican Official Standard NOM-141-SEMARNAT-2003, Which Establishes the Procedure for Characterizing Tailings, as Well as the Specifications and Criteria for the Characterization and Preparation of the Site, Project, Construction, Operation, and Post-Operation of Tailings Dams. Available online: https://www.profepa.gob.mx/innovaportal/file/1317/1/nom-141-semarnat-2003.pdf (accessed on 28 January 2025).
44. Food and Agriculture Organization of the United Nations (FAO) and the World Health Organization (WHO). CAC/RCP 56-2004 Code of Practice for the Prevention and Reduction of Lead in Food. Available online: https://www.fao.org/input/download/standards/10099/CXP_056s.pdf (accessed on 30 January 2025).
45. Lyu, Z.; Chai, J.; Xu, Z.; Qin, Y.; Cao, J. A Comprehensive Review on Reasons for Tailings Dam Failures Based on Case History. *Adv. Civil. Eng.* **2019**, *2019*, 4159306. [CrossRef]

toxics

MDPI

Article

Comparative Evaluation of Inductively Coupled Plasma Mass Spectrometry (ICP-MS) and X-Ray Fluorescence (XRF) Analysis Techniques for Screening Potentially Toxic Elements in Soil

Ilaria Guagliardi [1,*], Nicola Ricca [1] and Domenico Cicchella [2]

1 Institute for Agriculture and Forest Systems in the Mediterranean (CNR-ISAFOM), National Research Council of Italy, 87036 Rende, Italy; nicola.ricca@cnr.it
2 Department of Science and Technology, University of Sannio, 82100 Benevento, Italy; domenico.cicchella@unisannio.it
* Correspondence: ilaria.guagliardi@cnr.it; Tel.: +39-0984841427

Abstract: Soil contamination by potentially toxic elements (PTEs) poses a major environmental concern. The distribution and concentration of these elements can vary significantly in polluted areas, making detailed assessments crucial. A comprehensive analysis is essential to accurately characterise contamination patterns, as a foundation for effective site evaluation and remediation efforts. This study evaluates the effectiveness and reliability of X-ray fluorescence (XRF) and inductively coupled plasma mass spectrometry (ICP-MS) for determining PTEs in soil samples. Statistical analyses reveal significant differences between the two techniques for Sr, Ni, Cr, V, As, and Zn, likely due to variations in detection sensitivity, calibration methods, or matrix effects. Pb exhibits a weaker difference, suggesting a potential, yet statistically insignificant, difference between methods. Correlation analyses indicate a strong linear relationship for Ni and Cr, while Zn and Sr display high variability, limiting direct comparability. Bland–Altman plots highlight systematic biases, particularly for V, where XRF consistently underestimates concentrations compared to ICP-MS. These findings underscore the importance of selecting the appropriate analytical technique based on detection limits, sample characteristics, and measurement reliability. While both methods provide valuable insights for environmental monitoring, carefully considering their limitations is crucial for accurate contamination assessment.

Keywords: inductively coupled plasma mass spectrometry (ICP-MS); X-ray fluorescence (XRF); soil; analytical difference

1. Introduction

Soil contamination by potentially toxic elements (PTEs) is a significant environmental issue [1]. Polluted areas often show considerable variability in the distribution and concentration of PTEs [2,3]. Therefore, a thorough and detailed area assessment is essential for a comprehensive geospatial understanding of soil contamination [4]. This understanding is a prerequisite for effective site characterisation and reclamation [5,6].

Measuring trace metals is crucial [7], especially in complex environmental matrices that require analytical techniques to provide information about spatially distributed elements and allow examination of their geochemical dynamics [8].

The comparison of screening techniques for PTE assessment is a critical area of research, particularly in environmental monitoring and health risk assessments. Different methods have distinct advantages and limitations that influence their applicability in various contexts [9]. These differences can affect operational principles, sensitivity, and specific application scenarios. Inductively coupled plasma mass spectrometry (ICP-MS) and X-ray fluorescence (XRF) are two of the most commonly used techniques in environmental monitoring, food safety, and health assessments. However, they each have distinct advantages and limitations that determine their suitability for different analytical scenarios.

Inductively coupled plasma mass spectrometry (ICP-MS) is a highly sensitive and precise analytical technique used to quantitatively determine trace and ultra-trace elements in various sample types, including soils, water, and biological materials, providing essential data for risk assessments [10]. It is widely recognised for its exceptional sensitivity and low detection limits, often achieving levels as low as parts per trillion (ppt) for many elements, multi-element analysis, and wide dynamic range. This makes it the preferred method for trace elemental analysis, particularly for PTEs [11–13]. It is widely applied in environmental, geological, and industrial analyses due to its ability to detect multiple elements simultaneously with low detection limits. This technique employs high-temperature plasma to ionise the elements, followed by mass spectrometric analysis to quantify the ions produced. The ability of ICP-MS to analyse multiple elements simultaneously [14] and its high throughput capacity are additional advantages that enhance its utility in complex sample matrices [11,15]. For instance, studies have demonstrated that ICP-MS can accurately detect a wide range of elements, including PTEs such as lead, cadmium, and arsenic, with high accuracy and precision [16]. Some limitations can be achieved for matrix effects, sample preparation, which can be time-consuming, instrument cost, and maintenance.

X-ray fluorescence (XRF) is a non-destructive analytical technique that uses X-ray excitation to induce fluorescence in a sample. This process allows identifying and quantifying solid samples' elemental composition, including soil, rocks, and sediments [17,18], based on their characteristic X-ray emissions. The method relies on the interaction between high-energy X-rays and the sample material to excite atomic electrons and generate characteristic secondary (fluorescent) X-rays. XRF allows for qualitative and quantitative sample composition analysis [19].

While XRF typically has higher detection limits than ICP-MS, it offers rapid analysis and minimal sample preparation advantages. This makes it particularly well suited for field applications and real-time monitoring [20,21]. Recent advancements in XRF technology, such as the development of portable XRF analysers, have further enhanced its applicability in various settings, including environmental assessments and food safety inspections [20,22]. Despite these advantages, XRF has some limitations. The penetration depth for fluorescence emission in soil, particularly for lead, is restricted to the surface layer due to mass absorption coefficients, soil density, and instrument characteristics [23]. As a result, a homogeneous sample is required, which may need to be achieved through acid dissolution into liquids, grinding, or preparing pressed pellets [24]. Additionally, XRF instruments are relatively expensive and less accessible than other techniques, and their detection limits are higher than those of inductively coupled plasma (ICP) methods.

One of the critical differences between these two techniques lies in their sensitivity to sample matrix effects. ICP-MS requires sample digestion, often involving acid treatments, which can introduce uncertainties if the digestion is incomplete or if certain elements are not fully released from the matrix [25]. In contrast, XRF can analyse solid samples directly without extensive preparation, which can be advantageous when dealing with heterogeneous materials or when preserving the integrity of the sample is paramount [21,22]. However, the non-destructive nature of XRF can lead to challenges in quantification, particularly for elements present at low concentrations, as the technique may suffer from matrix effects that can skew results [21].

Moreover, the operational costs associated with each technique differ significantly. ICP-MS instruments are typically more expensive and require skilled personnel for operation and maintenance, which can limit their accessibility for routine analyses [12,22]. On the other hand, XRF instruments, especially portable versions, are generally more affordable and user-friendly, allowing for broader adoption in various sectors, including environmental monitoring and public health [20,21]. This cost-effectiveness makes XRF an attractive option for preliminary screening, where rapid results are needed, and the highest precision is not critical.

In terms of elemental coverage, both techniques have their strengths. ICP-MS can analyse a wide range of elements, including those difficult to detect with XRF, such as certain lanthanides and actinides [11,12]. Conversely, XRF can measure a broader spectrum of elements in a single analysis, which can be particularly useful in applications where multiple elements must be assessed simultaneously [21,22]. However, the elements analysed can vary based on calibration standards and instrument settings, which necessitates careful method validation for accurate quantification [26].

The integration of both techniques can also provide complementary data, enhancing the overall understanding of PTEs in various matrices. For instance, using XRF for initial screening followed by confirmatory analysis with ICP-MS can help validate results and provide a more comprehensive assessment of elemental concentrations [27]. This multi-method approach is particularly beneficial in complex environmental studies, where the presence of multiple contaminants may complicate the analysis.

The choice between ICP-MS and XRF ultimately depends on specific analytical requirements, including the type of sample, the elements of interest, the required detection limits, and the available resources. For instance, in environmental studies where trace levels of toxic metals are of concern, ICP-MS may be the preferred technique due to its superior sensitivity [11,15]. Conversely, for rapid screening in field conditions or when analysing bulk materials, XRF may be more appropriate due to its speed and ease of use [20,21]. In terms of specific applications, both techniques have been employed to assess the presence of toxic elements in various contexts. For example, studies have utilised ICP-MS to evaluate the concentrations of heavy metals in bone samples, revealing significant disparities in results compared to XRF [28]. Similarly, XRF has been effectively used to screen for toxic elements in consumer products, demonstrating its utility in regulatory contexts [25]. The choice between ICP-MS and XRF often depends on the specific requirements of the analysis, including the desired sensitivity, the nature of the sample, and the available resources.

In conclusion, while ICP-MS and XRF are valuable tools for PTE analysis, their distinct operational characteristics, sensitivity, and application contexts necessitate careful consideration when selecting the appropriate technique for a given analytical task. Integrating both methods in a complementary manner could also be beneficial, allowing for the strengths of each technique to be leveraged in comprehensive analytical workflows.

This study aims to compare the effectiveness and reliability of two analytical techniques—X-ray fluorescence (XRF) and inductively coupled plasma mass spectrometry (ICP-MS)—in determining the concentrations of PTEs in soil samples. Soil from urban and peri-urban areas of the municipalities of Cosenza and Rende, in Calabria, southern Italy, was analysed to assess these methods' consistency, accuracy, and applicability, providing insights into their suitability for environmental monitoring and contamination assessment.

2. Materials and Methods

2.1. Study Area

The study area, covering 92 km^2, is situated in the Cosenza–Rende territory in the Calabria region (southern Italy) and is characterised by high vehicular traffic. It lies within the Crati Basin graben, which the Sila Massif borders to the east and the Coastal Chain to the west [29]. The Crati Basin is an NW-oriented tectonic depression that formed during the Pliocene and has been influenced by faults associated with the horst–graben system [30–32].

Geologically, the area features a thick succession of Pliocene sediments, Pleistocene to Holocene alluvial sands, and Miocene carbonate rock outcrops [33–35], which overlie a Paleozoic intrusive–metamorphic complex formed by paragneiss, biotite schists, and grey phyllitic schists with quartz, chlorite, and muscovite, which, in some cases, are in the process of weathering. The predominant soil types include Fluvisols, Leptosols, Arenosols, Cambisols, Calcisols, Umbrisols, and Phaeozems [36], as well as Vertisols, Luvisols, Entisols, Inceptisols, Mollisols, and Alfisols [37], as mapped in the Calabria region's 1:250,000-scale soil map [38].

The area experiences a temperate continental climate, marked by significant daily and annual temperature variations [39]. The local soils primarily derive from sediments originating from the surrounding mountain massifs [40].

2.2. Soil Sampling

In this study, a collection of 50 soil samples from both residual and non-residual topsoil across various locations, including gardens, parks, flowerbeds, and agricultural fields within the study area, was performed. Additionally, duplicate pairs from every 10th site, which were then split in the laboratory to create replicates, were collected. Before sampling, surface litter was removed from each site.

Topsoil samples (0–10 cm in depth) were collected from five locations—one at each corner and one in the centre of a 20 $\times$ 20 m square—using a hand auger. These samples were combined to create a bulk sample. The collected soil was thoroughly mixed, and any foreign materials, such as roots, stones, pebbles, and gravel, were removed. The final volume of the sample ranged from 1 to 1.5 kg, which was further reduced by half through quartering.

2.3. Determination of PTE Contents Using XRF and ICP-MS Techniques

Soil samples were dried at 40 °C in the laboratory to eliminate moisture, ensuring a water-free reference for elemental analysis. The soil was then homogenized and sieved to obtain fine soil particles ($\leq$2 mm). All subsequent analyses were conducted on this fine soil, serving as a standard interstudy comparison reference.

Each prepared soil sample was analysed using X-ray fluorescence spectrometry (XRF) and inductively coupled plasma mass spectrometry (ICP-MS) to determine the concentrations of arsenic (As), chromium (Cr), nickel (Ni), lead (Pb), strontium (Sr), vanadium (V), and zinc (Zn).

In this study, the XRF instrument used was the wave dispersive X-ray fluorescent spectrometer Philips PW 1480 (Amsterdam, The Netherlands), with a rhodium anti-cathode tube. It was calibrated using certified reference materials to ensure accurate quantification, precisely the reference rock powders JA-1 andesite, JB-1a basalt, JF-1 feldspar, and JGb-1 gabbro issued by the Geological Survey of Japan [41]. Fine soil samples were compressed into pellets and placed in the spectrometer, and an X-ray beam was directed at the surface. The emitted fluorescence signals were recorded, and the software processed the spectral data to identify and quantify the elements.

In this study, for the ICP-MS method, PTEs were extracted using microwave-assisted digestion in closed Teflon (TFM) vessels with automated pressure and temperature control. The digestion process employed 14 mL of a ternary acid mixture consisting of 6 mL of perchloric acid ($HClO_4$), 5 mL of hydrofluoric acid (HF), and 3 mL of nitric acid (HNO_3) added to a sample quantity of 0.100 g. After the application of the thermal program, the vessels were cooled and opened, and the digested solutions were transferred into

50 mL PFTE containers and placed on a plate at 200 °C to allow their reduction to gel. An evaporation process was then planned. The dissolution procedure was repeated a second time by adding 2 mL of $HClO_4$. After the evaporation of this second phase, 5 mL of HNO_3 was added, and the containers were then left to cool. The residue is dissolved in 2 mL HNO_3 and 3–5 mL ultra-pure water, heated at 60–80 °C overnight, transferred to a polycarbonate container, and diluted to 100 mL (1000× dilution of sample) before analysis. Analysis of the soil samples was carried out using an ICP-MS PerkinElmer SCIEX—ELAN 6100 DRC-e (MDS Inc., Concord, ON, Canada). The measurement conditions for ICP-MS are presented in Table 1.

Table 1. Measurement conditions for ICP-MS (Perkin-Elmer Sciex ELAN 6100 DRC-e).

ICP-MS Parameter	Value
RF power—W	1200
Argon plasma gas flow—L min^{-1}	15
Nebulizer gas flow—L min^{-1}	0.82–0.86
Auxiliary gas flow—L min^{-1}	1.15
Lens voltage—V	6.00
Nebulizer	Cross flow
Plasma torch	quartz
Sample uptake—mL min^{-1}	1
Scanning mode	Peak hop
Dwell time—ms	50
Sweeps/Reading	20
Number of replicates	3
Read delay time—s	15
Cell gas	CH_4
DRC gas flow—mL min^{-1}	0.7
RPq (rejection parameter q)	0.65
Isotope ratio precision (RSD for Ag-107/Ag-109)	<0.2%
LOD—μg kg^{-1}	
Sr	0.24
Ni	0.27
Cr	5.02
V	1.9
As	0.7
Pb	0.31
Zn	2.5

Elemental concentrations were then determined according to established international standard methodologies (Table 2). Accuracy of the data was determined as <3% through analysis of the U.S. Geological Survey standards AGV-1, BCR-1, BR, DR-N, GA, GSP-1, and NIM-G, described in the Geochemical Database for Reference Materials and Isotopic Standards (GeoReM, http://georem.mpch-mainz.gwdg.de/, accessed on 18 December 2024) at the rate of 1 per each batch of 20 samples. Duplicate analyses were included in each batch at a frequency of one in twenty samples. Measurement errors were assessed using the relative standard deviation (<5%), calculated from three random sample replicates.

Single-element standard solutions of Cr, Ni, Pb, V, and Zn with an initial concentration of 10 μg/mL were mixed and used for calibration after appropriate dilution to obtain the following concentrations: 0.5, 1.0, 5.0, 10.0, 25.0, and 50.0 ng/mL.

All solutions were prepared with double deionized water (Millipore Purification System Synergy, Molsheim, France).

Table 2. Mean concentrations (mg kg^{-1}) obtained for the U.S. Geological Survey standards.

U.S.G.S. Standards	Sr	Ni	Cr	V	As	Pb	Zn
AGV-1	0.51	0.01	0.008	0.12	0.09	0.03	0.08
BCR-1	0.31	0.01	0.13	0.4	0.61	0.01	0.12
BR	1.32	0.2	0.03	0.32	0.02	0.05	0.16
DR-N	0.4	0.01	0.04	0.22	0.003	0.05	0.14
GA	0.31	0.007	0.02	0.03	0.001	0.03	0.08
GSP-1	0.36	0.008	0.01	0.05	0.0001	0.05	0.1
NIM-G	-	0.02	0.01	0.003	0.02	0.04	0.03
Accuracy (%)	1.1	2.4	2	2.8	1.1	2.8	1.4
RPD (%)	2.3	1.3	1.8	1.1	1.1	2.7	1.5

2.4. Statistical Analyses

Statistical analyses were performed on each dataset. The paired *t*-test was used to compare the means of two related samples to assess whether they had a significant difference. The Wilcoxon signed-rank test was used to determine statistical similarities or differences for non-parametric datasets containing two related samples. Regression analysis was conducted using Microsoft Excel to determine if a relationship existed between datasets.

3. Results and Discussion

Table 3 summarises the statistics of the soil samples determined by both the XRF and ICP-MS techniques.

Table 3. Descriptive statistics on PTEs (mg kg^{-1}) in soil samples, determined using XRF and ICP-MS techniques.

	Sr		Ni		Cr		V	
	XRF	**ICP-MS**	**XRF**	**ICP-MS**	**XRF**	**ICP-MS**	**XRF**	**ICP-MS**
N. samples	50	50	50	50	50	50	50	50
Min	120.00	105.19	19.00	15.76	47.00	45.71	64.00	96.48
Max	384.00	490.39	82.00	68.11	309.00	231.56	197.00	257.03
Median	233.50	209.70	33.00	25.93	85.50	72.26	100.50	135.73
Mean	231.42	208.94	35.12	27.99	91.36	76.83	107.18	143.93
St. dev.	67.22	68.79	11.07	8.97	37.84	27.63	28.40	36.29
CV (%)	29	32.9	31.5	32	41.4	36	26.5	25.2
Skewness	0.348	1.447	1.582	1.951	3.859	3.693	1.211	1.009
Kurtosis	−0.203	4.172	4.748	6.250	20.03	17.92	1.456	0.687
Difference mean	−22.48 (ICP < XRF)		−7.12 (ICP < XRF)		−14.53 (ICP < XRF)		+36.75 (ICP > XRF)	
Largest absolute difference	109.39		14.95		77.44		159.03	
	As		**Pb**		**Zn**			
	XRF	**ICP-MS**	**XRF**	**ICP-MS**	**XRF**	**ICP-MS**		
N. samples	50	50	50	50	50	50		
Min	4.00	4.43	12.00	14.85	43.00	78.36		
Max	14.00	18.98	247.00	250.93	871.00	689.48		
Median	7.00	8.73	32.50	36.29	116.50	160.72		
Mean	7.58	9.16	65.38	70.01	181.40	211.39		
St. dev.	2.74	3.25	71.02	70.97	172.81	138.08		
CV (%)	36.2	35.5	108.6	101.4	95.3	65.3		
Skewness	0.735	1.238	1.467	1.459	2.328	1.986		
Kurtosis	−0.532	1.615	0.673	0.530	5.258	3.697		
Difference mean	+1.58 (ICP > XRF)		+4.63 (ICP > XRF)		+29.99 (ICP > XRF)			
Largest absolute difference	14.98		62.15		355.22			

The statistical analysis of elemental distributions reveals that most elements exhibit a right-skewed pattern. This means that while most measurements are clustered around lower values, a few samples show notably higher concentrations. These high-concentration outliers can influence overall variability and must be carefully considered when comparing analytical techniques.

When comparing the two methods, ICP-MS generally shows broader distributions, suggesting more significant measurement variability than XRF [42–45]. This could be attributed to its higher sensitivity, lower detection limits, or potential matrix effects affecting certain elements. In contrast, XRF measurements tend to be more tightly clustered, exhibiting lower variability but potentially underrepresenting extreme values [43,46]. Notably, XRF demonstrates higher kurtosis for some elements, indicating the presence of more extreme values, while ICP-MS typically has lower variability—except for Zn, which displays significant dispersion, suggesting inconsistencies in its measurement across samples [42].

3.1. Specific-Element Variability and Trends

Examining variability in more detail, the coefficient of variation (CV) highlights Pb and Zn as the elements with the highest variability, indicating substantial fluctuations in concentration measurements across both methods. Additionally, Cr stands out with the highest skewness and kurtosis, emphasizing its strongly asymmetric distribution and the presence of extreme values. An observation of individual elements reveals method-specific trends:

- Strontium (Sr): ICP-MS and XRF produce similar distributions, though XRF reports a slightly higher median value.
- Nickel (Ni): The ICP-MS distribution appears more skewed, with a few samples displaying significantly higher concentrations than the rest.
- Chromium (Cr): The XRF distribution has a wider tail on the higher concentration side, suggesting a tendency for overestimation in some samples.
- Vanadium (V): ICP-MS and XRF exhibit noticeable differences, with XRF yielding a tighter distribution and less variability.
- Arsenic (As): XRF measurements appear more uniformly distributed, whereas ICP-MS results include a few higher outliers, suggesting greater sensitivity in detecting lower-concentration variations.
- Lead (Pb): Both methods demonstrate high variability, but ICP-MS reports more extreme values, potentially due to its greater sensitivity at lower concentrations.

The observed higher variability in ICP-MS data may result from its greater sensitivity and lower detection limits, which can amplify measurement fluctuations, particularly for trace elements. Additionally, systematic biases in elements such as Cr and Pb suggest potential influences from factors like calibration differences, instrument-specific detection efficiencies, or matrix effects that alter how elements interact with the measurement process. In contrast, the generally tighter distributions seen in the XRF data imply more reliable measurements, though this reliability could come at the cost of missing some extreme values that ICP-MS captures. This difference in measurement behaviour underscores the importance of understanding each technique's strengths and limitations when interpreting the results. These findings align with those of previous studies. Significant differences have been reported between ICP-MS and XRF for elements such as strontium, nickel, chromium, arsenic, and zinc, with lead showing a weaker yet notable variation [47–49].

3.2. Statistical Validation: Paired t-Test and Wilcoxon Signed-Rank Test Analysis

A paired t-test was conducted to formally assess whether ICP-MS and XRF produce significantly different measurements. This test evaluates whether the mean differences between paired ICP-MS and XRF values are statistically significant. The calculations were performed between the mean values from each dataset, the standard deviation of each group, and the number of data values. The results of the paired t-test are summarized in Table 4. The statistical findings further reinforce the observed trends, confirming that some elements exhibit strong measurement discrepancies between techniques.

Table 4. Paired t-test analysis. The results were calculated as mean ± SD for each technique across all samples.

Element	t-Statistic	p-Value
Sr	5.63	8.80×10^{-7}
Ni	13.40	5.22×10^{-18}
Cr	5.84	4.14×10^{-7}
V	−7.66	6.39×10^{-10}
As	−3.35	1.56×10^{-3}
Pb	−1.99	5.17×10^{-2}
Zn	−3.10	3.23×10^{-3}

The p-value is a key statistical metric used to determine whether the differences observed between the XRF and ICP-MS measurements are statistically significant. A p-value below 0.05 indicates strong evidence that the differences between the two methods are unlikely to be due to random variation alone, suggesting a statistically significant discrepancy. Conversely, a p-value above 0.05 implies that any observed differences could be due to chance, meaning that the two methods yield comparable results for that element.

In this analysis, the elements Sr, Ni, Cr, V, As, and Zn exhibit p-values below 0.05, indicating significant differences between the measurements obtained using XRF and ICP-MS. These differences could be attributed to variations in detection sensitivity, calibration methods, or matrix effects specific to each technique [50].

Lead presents a p-value of 0.0517, slightly above the conventional significance threshold of 0.05. This suggests that the difference between XRF and ICP-MS measurements for Pb is relatively weak and does not meet the strict criteria for statistical significance. However, given its proximity to the threshold, it may still indicate a potential difference that could be relevant in practical applications, especially if additional factors such as sample heterogeneity or instrumental precision are considered.

Overall, these statistical results highlight method-dependent variations in elemental measurements, emphasizing the importance of understanding the strengths and limitations of each analytical technique when interpreting soil composition data.

The Wilcoxon signed-rank test is a non-parametric statistical method used to determine whether there are significant differences between two related datasets—in this case, ICP-MS and XRF measurements. Unlike parametric tests, such as the paired t-test, which assumes normally distributed differences, the Wilcoxon test does not require normality and is, therefore, more reliable when the data exhibit skewness, outliers, or other non-normal characteristics [51–53]. This makes it particularly useful for comparing element concentrations when distributions are irregular or highly skewed.

By applying the Wilcoxon signed-rank test and the paired t-test, valuable insights can be gained regarding the nature of differences between ICP-MS and XRF measurements. This feature is critical in the context of ICP-MS and XRF measurement comparisons, where underlying data may be influenced by various factors, including measurement noise and variability inherent in analytical procedures [54].

If both p-values from the Wilcoxon test and the t-test are low ($p < 0.05$), this indicates that ICP-MS and XRF measurements differ systematically, meaning that the observed differences are unlikely to be due to random variation alone. In this case, the two methods may not be interchangeable for certain elements, and the source of systematic discrepancies—such as calibration issues, matrix effects, or differences in sensitivity—should be further investigated.

If the Wilcoxon test is significant ($p < 0.05$), but the paired t-test is not, this suggests that the difference between the two methods is likely due to non-normal behaviour in the data rather than a simple shift in mean values. Such a result implies that the dataset contains skewed distributions, outliers, or heavy tails, which influence the Wilcoxon test but not the t-test. This outcome underscores the importance of considering distribution shape when comparing analytical techniques.

If neither test is significant ($p > 0.05$), this suggests that ICP-MS and XRF are in strong agreement, with no statistically meaningful differences in their distributions. In such cases, the methods can be considered interchangeable for the elements under study, at least within the given measurement conditions.

3.3. Bar Chart

The bar chart comparing the mean differences (ICP-MS—XRF) and the largest absolute difference for each element provides a visual representation of how the two measurement techniques (ICP-MS and XRF) compare in terms of both average differences and extreme values (Figure 1). The chart includes blue bars representing the mean differences and red bars representing the largest absolute differences.

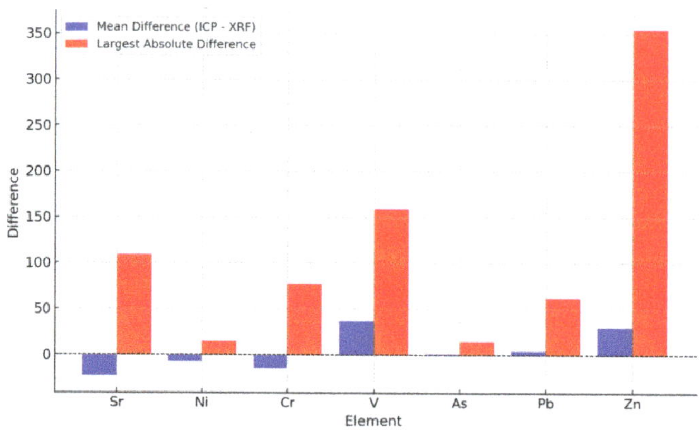

Figure 1. Bar chart values determined using XRF and ICP-MS techniques.

The blue bars reflect the mean differences between each element's ICP-MS and XRF measurements. These values are calculated by subtracting the XRF measurement from the ICP-MS measurement for each paired observation. Positive mean differences indicate that ICP-MS tends to report higher concentrations than XRF for the respective elements. This could suggest that ICP-MS is more sensitive or has a lower detection limit for that element, leading to higher readings. Negative mean differences suggest that XRF tends to overestimate the concentrations, reporting higher values compared to ICP-MS.

The red bars represent the largest absolute difference observed between the ICP-MS and XRF measurements for each element. This value highlights the maximum difference observed in any individual pair of measurements, providing insight into outliers or extreme variations. Large values for the red bars indicate that while the average difference between the two techniques might be small, there are individual measurements where the difference is much more substantial. These extreme differences may be due to sample heterogeneity, instrumental calibration issues, or matrix effects.

The element-specific observations suggest that the mean difference for V is +36.75, indicating that ICP-MS consistently reports higher concentrations than XRF. This significant positive value suggests that ICP-MS might have better sensitivity for V or is better equipped to detect lower concentrations, leading to its higher values compared to XRF.

Zinc shows the largest absolute difference at 355.22, meaning that there are instances where the differences between ICP-MS and XRF are very large for certain samples. This suggests that some individual samples have extreme variation, potentially due to outliers, instrumental inconsistencies, or matrix effects that impact the measurements from both techniques differently.

Strontium, nickel, and chromium show negative mean differences, indicating that XRF tends to overestimate concentrations compared to ICP-MS. For these elements, the negative values suggest systematic overreporting by XRF, which could be due to factors such as calibration bias, matrix effects, or differences in the sensitivity of the two techniques.

Arsenic and lead show relatively small mean differences, meaning that the overall concentrations measured using ICP-MS and XRF are fairly similar on average. However, there is still some variation, reflected in the differences between individual samples. This suggests that while the methods agree in general, there are still occasional differences that could be due to instrumental or environmental factors influencing the results.

The overall agreement vs. variability suggests that both the general trend (mean difference) and the potential for outliers (absolute difference) make it clear that while ICP-MS and XRF may produce similar results for some elements, there are cases where significant differences arise. Understanding the sources of these differences—whether due to instrument-specific factors, sample heterogeneity, or calibration differences—is essential for making informed decisions about which method to use for each element [55,56].

3.4. Regression Trends

The scatter plots in Figure 2 visually represent the relationship between the measurements obtained using ICP-MS and XRF for each element, performed by a Pearson correlation analysis, whose results are reported in Table 5. Each point on the plot corresponds to a paired observation for a specific element, with the x-axis representing the XRF measurement and the y-axis representing the ICP-MS measurement. The green line in each plot represents the ideal 1:1 correlation, where the measurements from both techniques would be identical, meaning that for any given value of XRF, ICP-MS would report the same value. The closer the data points are to this line, the stronger the agreement between the two techniques.

Some elements, such as Ni and Cr, show a clear linear relationship between ICP-MS and XRF measurements, meaning that the two methods tend to produce similar results across the measured concentration range. In these cases, the data points closely follow the 1:1 line, suggesting that there is a good correlation between the two techniques.

For elements like Zn and Sr, the scatter plots reveal significant dispersion (scatter of points), indicating high variability between the measurements obtained using ICP-MS and XRF. This suggests that while the two methods might provide some consistent readings, there are substantial fluctuations and discrepancies in the measurements for these elements, reducing the reliability of direct comparisons.

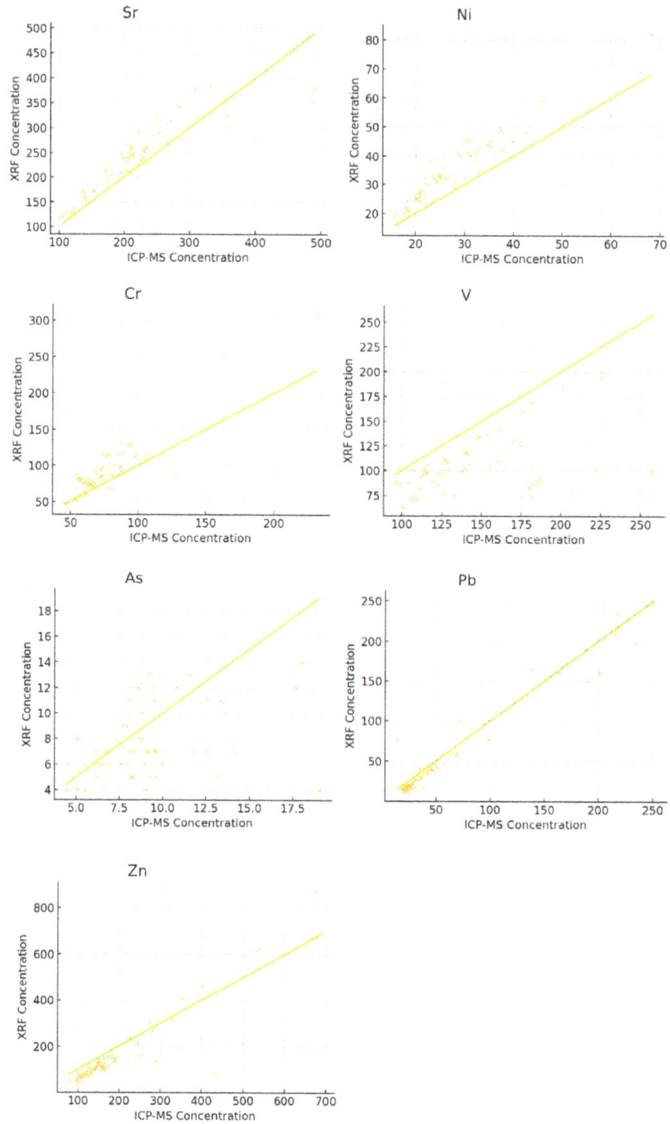

Figure 2. Scatter plots.

Table 5. Correlation analysis.

Element	Pearson Correlation (r)	*p*-Value
Sr	0.91 (strong)	2.01×10^{-20}
Ni	0.95 (very strong)	3.65×10^{-26}
Cr	0.90 (strong)	4.01×10^{-19}
V	0.47 (moderate)	5.46×10^{-4}
As	0.39 (weak)	4.69×10^{-3}
Pb	0.97 (very strong)	2.50×10^{-32}
Zn	0.93 (very strong)	4.34×10^{-22}

In the case of V, there is an observable trend where XRF underestimates concentrations compared to ICP-MS. The data points for V tend to fall below the 1:1 line, suggesting that XRF consistently reports lower concentrations than ICP-MS for this element.

3.5. Correlation Analysis

Table 4 shows the Pearson correlation analysis results. The correlation coefficient (r), which ranges from −1 to +1, quantifies the degree of association between the two methods: r = 1 indicates perfect positive correlation, r = −1 indicates perfect negative correlation, and r = 0 indicates no correlation.

For Ni, Pb, Zn, Sr, and Cr, the correlation coefficients exceed 0.90, suggesting a very strong correlation between ICP-MS and XRF and highlighting that both ICP-MS and XRF perform similarly in terms of precision and accuracy. This strong relationship indicates that one method does not systematically overestimate or underestimate the concentrations compared to the other. Given the strong correlation for these elements, it may be possible to use either ICP-MS or XRF interchangeably when measuring them, without a significant risk of differences. The strong relationship also suggests that the methods likely measure the same underlying phenomenon, with any differences likely being within acceptable limits for analytical purposes.

For V and As, the correlation coefficients are below 0.50, indicating a weaker correlation between ICP-MS and XRF, implying that the measurements from ICP-MS and XRF do not consistently align. This inconsistency may suggest that either or both techniques are influenced by factors that are not equally addressed by the two methods. These differences could arise from differences in sensitivity, detection limits, or matrix effects specific to these elements. The weak correlation between ICP-MS and XRF for V and As may indicate that one method tends to underestimate or overestimate the concentration of these elements compared to the other. In this case, the choice of method may need to be based on the specific requirements of the analysis. For example, if high precision is critical, it may be important to rely on the method that shows the most consistent results for these elements, or further optimization and validation might be required to align the two methods.

3.6. Bland–Altman Plot Analysis

The Bland–Altman plot is a powerful tool for assessing the agreement between the two measurement techniques. Rather than simply evaluating correlation, it provides a visual representation of systematic biases and identifies outliers where the two methods diverge significantly [57].

In the Bland–Altman plot, the x-axis represents the mean concentration for each sample, calculated as

$$\text{Mean} = \frac{\text{ICP-MS} + \text{XRF}}{2}$$

The y-axis represents the difference between ICP-MS and XRF for each sample:

$$\text{Difference} = \text{ICP-MS} - \text{XRF}$$

In the Bland–Altman plot shown in Figure 3, the green line represents the mean difference, indicating whether one method systematically reports higher values than the other. The grey dashed lines represent the limits of agreement (±1.96 standard deviations from the mean), capturing 95% of the data. Points outside this range are considered outliers, representing significant discrepancies between the two techniques.

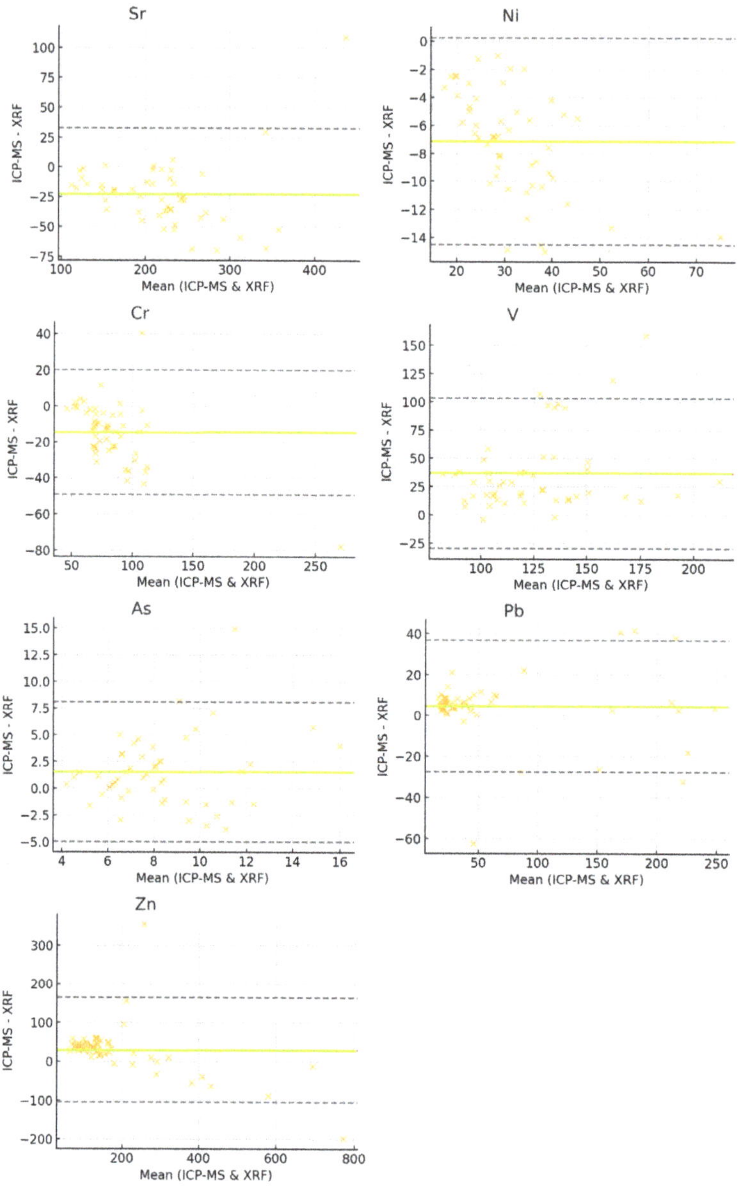

Figure 3. Bland–Altman plots.

Certain samples exhibit differences exceeding ±1.96 standard deviations, suggesting major disagreements between ICP-MS and XRF. These outliers may arise from matrix effects, instrument limitations, calibration errors, or sample inhomogeneity.

For Sr,

- the sample CSRN137 exhibited an extreme difference of +109.39, with ICP-MS reporting 490.39 and XRF reporting 381.

For Ni,

- the samples CSRN020, CSRN024, and CSRN126 showed a ~15 unit lower value in ICP-MS compared to XRF.

For Cr,

- the sample CSRN074 showed an ICP-MS value of 231.56 and an XRF value of 309 (−77.44 difference)
- the sample CSRN097 showed an ICP-MS value of 127.61 vs. an XRF value of 87 (+40.61 difference).

For V, the largest outliers were represented by

- the sample CSRN135, with a +159.03 difference (ICP-MS 257.03 vs. XRF 98),
- the sample CSRN042, with a +107.58 difference,
- the sample CSRN133 with a +120.03 difference.

As displayed, the largest difference was for

- the sample CSRN042 (+14.98), with ICP-MS 18.97 and XRF 4.

Pb,

- with the sample CSRN113 showed a difference of -62.15, with ICP-MS being 14.85 and XRF 77.

Zn exhibited an extreme outlier with

- the sample CSRN067: a difference of +355.22, with being ICP-MS 434.22 and XRF 79.

Several factors could explain the observed differences and outliers between ICP-MS and XRF measurements.

The ICP-MS technique is highly sensitive but subject to matrix suppression, where certain sample components interfere with ionization efficiency. In contrast, the XRF technique is a surface-sensitive technique, meaning that non-homogeneous samples can lead to errors.

Moreover, element-specific interference can occur; for example, for vanadium, ICP-MS consistently reports much higher values than XRF since XRF may not efficiently detect V at low concentrations due to its excitation energy constraints. For nickel, for some samples, XRF consistently reports higher values than ICP-MS. This can be due to background correction errors in XRF affecting Ni detection. XRF may underestimate vanadium concentrations due to its low sensitivity in certain energy ranges.

In addition, XRF analyses only the surface layer, while ICP-MS analyses a fully digested sample. If a sample is not ground evenly, different elemental distributions may be measured using each technique.

XRF has higher detection limits than ICP-MS, making it less sensitive to low concentrations. Background noise in XRF can cause misclassification of elements.

The Bland–Altman approach focuses on individual sample-level agreement, identifying whether measurements from XRF and ICP-MS consistently align. On the other hand, the *t*-test method evaluates differences across the entire dataset rather than at the sample level. Outliers influence Bland–Altman more strongly, as it highlights large deviations in specific cases, while the *t*-test accounts for overall distribution trends, capturing method-dependent variations more broadly. Thus, while Bland–Altman suggests that the methods generally agree, it also reveals cases of major discrepancies. Meanwhile, the *t*-test demonstrates that despite individual agreements, systematic biases exist, reinforcing the need to carefully interpret the results based on the strengths and limitations of each technique.

4. Conclusions

This study compared the effectiveness and reliability of X-ray fluorescence (XRF) and inductively coupled plasma mass spectrometry (ICP-MS) for determining potentially toxic elements in soil samples from urban and peri-urban areas of Cosenza and Rende, Calabria, southern Italy. The analysis revealed both the strengths and limitations of each technique, emphasizing their suitability for different applications in environmental monitoring and contamination assessment.

ICP-MS demonstrated higher sensitivity and a broader dynamic range, making it more effective for detecting low concentrations of PTEs. However, it also exhibited greater variability, likely due to matrix effects and instrumental calibration differences. XRF, while providing rapid and non-destructive analysis, tended to underestimate certain elements (e.g., V) and exhibited lower accuracy in highly heterogeneous samples. The statistical analysis showed significant differences between the two methods for Sr, Ni, Cr, V, As, and Zn, while Pb presented a weaker but notable discrepancy.

These findings emphasise the importance of understanding the limitations and strengths of each technique when interpreting soil composition data. While ICP-MS and XRF may provide comparable results for some elements, significant differences arise in others, particularly due to matrix effects, calibration inconsistencies, and instrumental sensitivities. Therefore, careful consideration should be given to the choice of analytical method depending on the element of interest, ensuring accurate and reliable assessments of elemental concentrations in environmental and geochemical studies.

Author Contributions: I.G.: conceptualisation, methodology, validation, formal analysis, investigation, resources, data curation, writing—original draft preparation, writing—review and editing, visualisation, project administration; N.R.: validation, formal analysis, investigation; D.C.: conceptualisation, methodology, supervision. All authors have read and agreed to the published version of the manuscript.

Funding: This work was funded by the Next Generation EU—Italian NRRP, Mission 4, Component 2, Investment 1.5, call for the creation and strengthening of 'Innovation Ecosystems', building 'Territorial R&D Leaders' (Directorial Decree n. 2021/3277)—project Tech4You—Technologies for climate change adaptation and quality of life improvement, n. ECS0000009.

Institutional Review Board Statement: Not applicable.

Informed Consent Statement: Not applicable.

Data Availability Statement: The original contributions presented in this study are included in the article; further inquiries can be directed to the corresponding authors.

Conflicts of Interest: The authors declare no conflicts of interest. This work reflects only the authors' views and opinions; neither the Ministry for University and Research nor the European Commission can be considered responsible for them.

References

1. Ambrosino, M.; Albanese, S.; De Vivo, B.; Guagliardi, I.; Guarino, A.; Lima, A.; Cicchella, D. Identification of Rare Earth Elements (REEs) distribution patterns in the soils of Campania region (Italy) using compositional and multivariate data analysis. *J. Geochem. Explor.* **2022**, *243*, 107112. [CrossRef]
2. Yu, B.; Lu, X.; Fan, X.; Fan, P.; Zuo, L.; Yang, Y.; Wang, L. Analyzing environmental risk, source and spatial distribution of potentially toxic elements in dust of residential area in Xi'an urban area, China. *Ecotoxicol. Environ. Saf.* **2021**, *208*, 111679. [CrossRef] [PubMed]
3. Liu, W.; Yuan, C.; Hassan Farooq, T.; Chen, P.; Yang, M.; Ouyang, Z.; Fu, Y.; Yuan, Y.; Wang, G.; Yan, W.; et al. Pollution index and distribution characteristics of soil heavy metals among four distinct land use patterns of Taojia River Basin in China. *Gondwana Res.* **2024**, *135*, 198–207. [CrossRef]

4. Su, C.; Yang, Y.; Jia, M.; Yan, Y. Integrated framework to assess soil potentially toxic element contamination through 3D pollution analysis in a typical mining city. *Chemosphere* **2024**, *359*, 142378. [CrossRef]
5. Buttafuoco, G.; Tarvainen, T.; Jarva, J.; Guagliardi, I. Spatial variability and trigger values of arsenic in the surface urban soils of the cities of Tampere and Lahti, Finland. *Environ. Earth Sci.* **2016**, *75*, 896. [CrossRef]
6. Huang, J.; Wu, Y.; Sun, J.; Li, X.; Geng, X.; Zhao, M.; Fan, Z. Health risk assessment of heavy metal(loid)s in park soils of the largest megacity in China by using Monte Carlo simulation coupled with Positive matrix factorization model. *J. Hazard Mater.* **2021**, *415*, 125629. [CrossRef]
7. Xiang, P.; Han, Y.-H.; Li, H.-B.; Gao, P. Editorial: Advanced analytical techniques for heavy metals speciation in soil, crop, and human samples. *Front. Chem.* **2023**, *11*, 1151371. [CrossRef]
8. Guagliardi, I. Editorial for the Special Issue "Potentially Toxic Elements Pollution in Urban and Suburban Environments". *Toxics* **2022**, *10*, 775. [CrossRef]
9. Haghighizadeh, A.; Rajabi, O.; Nezarat, A.; Hajyani, Z.; Haghmohammadi, M.; Hedayatikhah, S.; Delnabi Asl, S.; Aghababai Beni, A. Comprehensive analysis of heavy metal soil contamination in mining Environments: Impacts, monitoring Techniques, and remediation strategies. *Arab. J. Chem.* **2024**, *17*, 105777. [CrossRef]
10. Gonçalves, D.A.M.; Pereira, W.V.d.S.; Johannesson, K.H.; Pérez, D.V.; Guilherme, L.R.G.; Fernandes, A.R. Geochemical background for potentially toxic elements in forested soils of the state of Pará, Brazilian Amazon. *Minerals* **2022**, *12*, 674. [CrossRef]
11. Ahmed, A.Y.; Abdullah, M.P.; Wood, A.K.; Hamza, M.S.; Othman, M. Determination of some trace elements in marine sediment using ICP-MS and XRF (a comparative study). *Orient. J. Chem.* **2013**, *29*, 645–653. [CrossRef]
12. Elkadi, M.; Pillay, A.; Manuel, J.; Khan, M.R.; Stephen, S.; Molki, A. Sustainability study on heavy metal uptake in neem biodiesel using selective catalytic preparation and hyphenated mass spectrometry. *Sustainability* **2014**, *6*, 2413–2423. [CrossRef]
13. Vrdoljak, G.; Palmer, P.T.; Jacobs, R.M.; Moezzi, B.; Viegas, A. Comparison of XRF, TXRF, and ICP-MS methods for determination of mercury in face creams. *J. Med. Radiat. Sci.* **2021**, *9*, 1–8. [CrossRef]
14. Aarab, I.; Chakir, E.; Maazouzi, Y.; Bounouira, H.; Didi, A.; Amsil, H.; Badague, A. Major and trace elements content in cannabis sativa–l cultivated in north of Morocco and heavy metals health risk assessment. *E3S Web Conf.* **2023**, *469*, 00027. [CrossRef]
15. Lohmeier, S.G.; Lottermoser, B.G.; Schirmer, T.; Gallhofer, D. Copper slag as a potential source of critical elements—A case study from Tsumeb, Namibia. *J. South. Afr. Inst. Min. Metall.* **2021**, *121*, 129–142. [CrossRef]
16. Bowler, R.M.; Beseler, C.L.; Gocheva, V.; Colledge, M.; Kornblith, E.; Julian, J.R.; Lobdell, D.T. Environmental exposure to manganese in air: Associations with tremor and motor function. *Sci. Total Environ.* **2016**, *541*, 646–654. [CrossRef]
17. Delgado, M.; Jason, P.; Humberto, G.; Alba, C.; Gustavo, C.; Alfredo, C.; Alberto, D.; Jorge, G. Comparison of ICP-OES and XRF performance for Pb and As analysis in environmental soil samples from Chihuahua City, Mexico. *Phys. Rev. Res. Int.* **2011**, *1*, 29–44.
18. Congiu, A.; Perucchini, S.; Cesti, P. Trace metal contaminants in sediments and soils: Comparison between ICP and XRF quantitative determination. *E3S Web Conf.* **2013**, *1*, 09004. [CrossRef]
19. Kalnicky, D.J.; Singhvi, R. Field portable XRF analysis of environmental samples. *J. Hazard. Mater.* **2001**, *83*, 93–122. [CrossRef]
20. Roberts, A.A.; Guimarães, D.; Tehrani, M.W.; Lin, S.; Parsons, P.J. A field-based evaluation of portable XRF to screen for toxic metals in seafood products. *X-Ray Spectrom.* **2023**, *53*, 506–519. [CrossRef]
21. Guimarães, D.; Praamsma, M.L.; Parsons, P.J. Evaluation of a new optic-enabled portable X-ray fluorescence spectrometry instrument for measuring toxic metals/metalloids in consumer goods and cultural products. *Spectrochim. Acta B* **2016**, *122*, 192–202. [CrossRef] [PubMed]
22. McComb, J.; Rogers, C.; Han, F.X.; Tchounwou, P.B. Rapid screening of heavy metals and trace elements in environmental samples using portable X-ray fluorescence spectrometer, a comparative study. *Water Air Soil Pollut.* **2014**, *225*, 2169. [CrossRef] [PubMed]
23. Gauglitz, G.; Vo-Dinh, T. *Handbook of Spectroscopy*; WILEY-VCH Verlag GmbH & Co. KGaA: Weinheim, Germany, 2003. [CrossRef]
24. Fukai, M.; Kikawada, Y.; Oi, T. Determination of lanthanoids in seawater by inductively coupled plasma mass spectrometry after pre-concentration with a chelating resin disk. *Nat. Sci.* **2016**, *8*, 431–441. [CrossRef]
25. Palmer, P.T.; Jacobs, R.M.; Yee, S.; Li, T.; Reed, C.; Nebenzahl, D.; Douglas, M. Best practices for the use of portable X-ray fluorescence analyzers to screen for toxic elements in fda-regulated products. *J. Regul. Sci.* **2019**, *7*, 1–11. [CrossRef]
26. Almusawi, A.A.H.; Jebril, N.; AL-Auhaimid, A.A.H.; Hammoodi, Z.H.K. Carpal tunnel syndrome and calcium deposit in the surgically transacted transverse carpal ligament. *Sci. World J.* **2022**, *2022*, 2864485. [CrossRef]
27. McIver, D.J.; VanLeeuwen, J.; Knafla, A.L.; Campbell, J.A.; Alexander, K.; Gherase, M.R.; Fleming, D. Evaluation of a novel portable X-ray fluorescence screening tool for detection of arsenic exposure. *Physiol. Meas.* **2015**, *36*, 2443–2459. [CrossRef]
28. Little, N.C.; Florey, V.; Molina, I.; Owsley, D.W.; Speakman, R.J. Measuring heavy metal content in bone using portable X-ray fluorescence. *J. Open Archaeol.* **2014**, *2*, 19–21. [CrossRef]
29. Ricca, N.; Guagliardi, I. Evidences of Soil Consumption Dynamics over Space and Time by Data Analysis in a Southern Italy Urban Sprawling Area. *Land* **2023**, *12*, 1056. [CrossRef]

30. Iovine, G.; Guagliardi, I.; Bruno, C.; Greco, R.; Tallarico, A.; Falcone, G.; Lucà, F.; Buttafuoco, G. Soil-gas radon anomalies in three study areas of Central-Northern Calabria (Southern Italy). *Nat. Hazards* **2018**, *91*, 193–219. [CrossRef]
31. Tansi, C.; Muto, F.; Critelli, S.; Iovine, G. Neogene-Quaternary strike-slip tectonics in the central Calabrian Arc (southern Italy). *J. Geodyn.* **2007**, *43*, 393–414. [CrossRef]
32. Van Dijk, J.P.; Bello, M.; Brancaleoni, G.P.; Cantarella, G.; Costa, V.; Frixa, A.; Golfetto, F.; Merlini, S.; Riva, M.; Torricelli, S. A regional structural model for the northern sector of the Calabrian Arc (Southern Italy). *Tectonophysics* **2000**, *324*, 267–320. [CrossRef]
33. Gaglioti, S.; Infusino, E.; Caloiero, T.; Callegari, G.; Guagliardi, I. Geochemical Characterization of Spring Waters in the Crati River Basin, Calabria (Southern Italy). *Geofluids* **2019**, *2019*, 3850148. [CrossRef]
34. Guagliardi, I.; Astel, A.M.; Cicchella, D. Exploring Soil Pollution Patterns Using Self-Organizing Maps. *Toxics* **2022**, *10*, 416. [CrossRef] [PubMed]
35. Infusino, E.; Guagliardi, I.; Gaglioti, S.; Caloiero, T. Vulnerability to Nitrate Occurrence in the Spring Waters of the Sila Massif (Calabria, Southern Italy). *Toxics* **2022**, *10*, 137. [CrossRef]
36. IUSS Working Group WRB. *World Reference Base for Soil Resources 2006*; First Update 2007; World Soil Resources Reports No. 103; FAO: Rome, Italy, 2007.
37. Soil Survey Staff. *Keys to Soil Taxonomy*, 11th ed.; USDA-NRCS: Washington DC, USA, 2010.
38. R.S.S.A. (Agenzia Regionale per lo Sviluppo e per i Servizi in Agricoltura). I suoli della Calabria. Carta dei suoli in scala 1: 250000 della Regione Calabria. In *Monografia Divulgativa: Programma Interregionale Agricoltura-Qualità—Misura 5, ARSSA, Servizio Agropedologia*; Rubbettino Editore: Catanzaro, Italy, 2003.
39. Buttafuoco, G.; Caloiero, T.; Guagliardi, I.; Ricca, N. Drought assessment using the reconnaissance drought index (RDI) in a southern Italy region. In Proceedings of the 6th IMEKO TC19 Symposium on Environmental Instrumentation and Measurements, Reggio Calabria, Italy, 24–26 June 2016; pp. 52–55.
40. Buttafuoco, G.; Caloiero, T.; Ricca, N.; Guagliardi, I. Assessment of drought and its uncertainty in a southern Italy area (Calabria region). *Measurement* **2018**, *113*, 205–210. [CrossRef]
41. Imai, N.; Terashima, S.; Itoh, S.; Ando, A. 1994 compilation values for GSJ reference samples and GSJ reference samples, 'Igneous rock series'. *Geochem. J.* **1195**, *29*, 91–95. [CrossRef]
42. Balcaen, L.; Bolea-Fernández, E.; Resano, M.; Vanhaecke, F. Inductively coupled plasma—Tandem mass spectrometry (ICP-MS/MS): A powerful and universal tool for the interference-free determination of (ultra)trace elements—A tutorial review. *Anal. Chim. Acta* **2015**, *894*, 7–19. [CrossRef]
43. Greenberg, I.; Sawallisch, A.; Stelling, J.; Vohland, M.; Ludwig, B. Optimization of sample preparation and data evaluation techniques for X-ray fluorescence prediction of soil texture, pH, and cation exchange capacity of loess soils. *Soil Sci. Soc. Am. J.* **2023**, *88*, 27–42. [CrossRef]
44. Al Maliki, A.; Al-lami, A.K.; Hussain, H.M.; Al-Ansari, N. Comparison between inductively coupled plasma and X-ray fluorescence performance for Pb analysis in environmental soil samples. *Environ. Earth Sci.* **2017**, *76*, 433. [CrossRef]
45. Schmidt, K.; Autenrieth, D.; Nagisetty, R. A comparison of field portable X-ray fluorescence (FP XRF) and inductively coupled plasma mass spectrometry (ICP-MS) for analysis of metals in the soil and ambient air. *Res. Sq.* **2024**, *rs.3*, rs-3849271. [CrossRef]
46. Chen, Q.; Kissel, C.; Govin, A.; Liu, Z.; Xie, X. Correction of interstitial water changes in calibration methods applied to XRF core-scanning major elements in long sediment cores: Case study from the South China Sea. *Geochem. Geophys. Geosyst.* **2016**, *17*, 1925–1934. [CrossRef]
47. Javadi, S.; Mouazen, A. Data fusion of XRF and Vis-nir using outer product analysis, granger–Ramanathan, and least squares for prediction of key soil attributes. *Remote Sens.* **2023**, *13*, 2023. [CrossRef]
48. Gebregiorgis, D.; Giosan, L.; Hathorne, E.; Anand, P.; Nilsson-Kerr, K.; Plaß, A.; Frank, M. What can we learn from X-ray fluorescence core scanning data? A paleomonsoon case study. *Geochem. Geophys. Geosyst.* **2020**, *21*, e2019GC008414. [CrossRef]
49. Lopresti, M.; Mangolini, B.; Milanesio, M.; Caliandro, R.; Palin, L. Multivariate versus traditional quantitative phase analysis of X-ray powder diffraction and fluorescence data of mixtures showing preferred orientation and microabsorption. *J. Appl. Crystallogr.* **2022**, *55*, 837–850. [CrossRef]
50. Colomban, P.; Gironda, M.; Franci, G.; D'abrigeon, P. Distinguishing genuine imperial Qing dynasty porcelain from ancient replicas by on-site non-invasive XRF and Raman spectroscopy. *Materials* **2022**, *15*, 5747. [CrossRef]
51. Haidous, N.; Sawilowsky, S. Robustness and power of the Kornbrot rank difference, signed ranks, and dependent samples *t*-test. *Am. J. Appl. Math. Stat.* **2013**, *1*, 99–102. [CrossRef]
52. Wiedermann, W.; Alexandrowicz, R. A modified normal scores test for paired data. *Methodology* **2011**, *7*, 25–38. [CrossRef]
53. Shieh, G.; Jan, S.; Randles, R. Power and sample size determinations for the Wilcoxon signed-rank test. *J. Stat. Comput. Simul.* **2007**, *77*, 717–724. [CrossRef]
54. Ang, J.; Zhang, S. Evaluating long-horizon event study methodology. *SSRN Electron. J.* **2011**. [CrossRef]

55. Bhatia, M.; Specht, A.; Vallabhuni, R.; Sulaiman, D.; Konda, M.; Balcom, P.; Qureshi, A. Portable X-ray fluorescence as a rapid determination tool to detect parts per million levels of Ni, Zn, As, Se, and Pb in human toenails: A south India case study. *Environ. Sci. Technol.* **2021**, *55*, 13113–13121. [CrossRef]
56. Fleming, D.; Madani, N.; Kaiser, M.; Kim, J.; Keltie, E.; Drage, N.; Brough, L. Portable X-ray fluorescence of zinc and selenium with nail clippings–mother and infant nutrition investigation (mini). *PLoS ONE* **2024**, *19*, e0310845. [CrossRef] [PubMed]
57. Kim, J.; Lee, J.H. A novel graphical evaluation of agreement. *BMC Med. Res. Methodol.* **2022**, *22*, 51. [CrossRef] [PubMed]

toxics

MDPI

Article

Characterization of Volcanic Ash Influence on the Nutritional Quality and Biological Traits in Potato Crops of the Cotopaxi Region

Raluca A. Mihai [1,*], Ramiro Fernando Vivanco Gonzaga [2], Nathaly Raquel Romero Balladares [2] and Rodica D. Catana [3]

1 CIAM, Department of Life Science and Agriculture, Universidad de Las Fuerzas Armadas-ESPE, Av. General Ruminahui s/n y, Sangolqui 171103, Ecuador

2 Department of Life Science and Agriculture, Universidad de Las Fuerzas Armadas-ESPE, Av. General Ruminahui s/n y, Sangolqui 171103, Ecuador; rfvivanco2@espe.edu.ec (R.F.V.G.); nrromero@espe.edu.ec (N.R.R.B.)

3 Institute of Biology Bucharest of Romanian Academy, 296 Splaiul Independentei, 060031 Bucharest, Romania; rodica.catana@ibiol.ro

* Correspondence: rmihai@espe.edu.ec

Abstract: This study investigates the impact of volcanic ash from Cotopaxi Volcano on the nutritional quality and biological traits of potato tubers (*Solanum tuberosum* L.) cultivated in the Cotopaxi region. Methods: Samples collected from exposed and unexposed areas were used to characterize the volcanic ash influence on the metabolic aspects of the potato crop. The colorimetric method; DPPH, ABTS, and FRAP assays; and ICP-OES were used to better understand potatoes' reaction to the stress. Results: Antioxidant activity was significantly higher (4.80 ± 2.38 µmol Trolox g^{-1} DW-DPPH assay; 11.05 ± 2.57 µmol Trolox g^{-1} DW-ABTS assay; and 11.96 ± 4.57 µmol Fe^{2+} g^{-1} DW-FRAP assay) in ash-exposed samples, suggesting enhanced stress responses. The bioactive compounds studied followed a comparable trend, with high content in the exposed tubers. Also, significant changes in elemental composition were registered: Potassium levels decreased in unexposed samples, while magnesium and iron levels increased. Metallic elements (zinc; lithium; boron; manganese; barium; lead; nickel; chromium; indium) were in concentrations <0.01 mg/kg. Conclusion: These findings demonstrate that volcanic ash alters the metabolic and antioxidant profiles of potato tubers, enhancing nutraceutical properties while posing food safety risks due to heavy metals. This dual impact highlights the challenges and opportunities for agriculture in volcanic regions like Cotopaxi.

Keywords: antioxidant activity; flavonoids; heavy metals; inductively coupled plasma (ICP); phenolic compounds; *Solanum tuberosum* L.; volcanic ash

1. Introduction

Potatoes are among the most widely grown crops worldwide, due to their high yield potential and ability to grow in diverse climatic conditions [1]; they are recognized as a vital resource in nutrition [2] and agricultural versatility.

For the Andean region, potato tubers are one of the most important crops, where diverse environmental factors contribute to their development and quality. The Andean region is a treasure trove of potato genetic diversity, which is crucial for ensuring the resilience of potato crops to pests, diseases, and climate change [3].

Comprehending the interplay between volcanic ash and plant development is important to assess the risks to agriculture and public health in volcanic regions. The Cotopaxi Volcano, one of the most active volcanoes in the world, emits ash characterized by a unique composition of minerals and heavy metals that significantly influences the surrounding agricultural lands, with both beneficial and detrimental aspects [4]. According to Smith and Lee, 2020 [5], volcanic ash contains silica, aluminum, iron, calcium, and magnesium, with reported increases of 20–30% in phenolic content in crops exposed to volcanic soils [6]. We hypothesize that volcanic ash from Cotopaxi increases the antioxidant profiles of potatoes, improving their nutraceutical quality, although it could increase food safety risks due to heavy metal accumulation. These elements can alter soil properties, potentially affecting the morphological and chemical characteristics of crops grown in ash-affected regions. Heavy metals pose particular concerns, as they may be absorbed by potato tubers, thus entering the food chain and potentially compromising food safety [7].

Few studies are available on the influence of volcanic ash on the nutritional quality of potatoes. Ligot et al., 2025 [8], point out that current models for predicting the impact of ash on crops use only ash thickness (or mass load) as a variable, without taking into account other factors (plant traits; growth stage; metabolic profiles; etc.); their results have been used to develop new vulnerability functions for estimating yield loss in potatoes exposed to an ash fall event. Other studies have shown that crops cultivated in soil improved with volcanic ash, showing an increased yield, crude protein, starch content in potato tubers, and also better resistance to potato diseases [9,10]. In this context, our study examines the impact of ash originating from the Cotopaxi Volcano on potatoes cultivated under two different conditions: exposed and unexposed to volcanic ash. By analyzing the antioxidant content and examining the overall impact of volcanic ash on potato tubers, this research provides crucial insights into how volcanic activity affects local agriculture. It highlights not only the challenges faced by farmers in volcanic regions but also the potential opportunities to enhance the functional and health-promoting qualities of staple crops like potatoes. However, it also addresses the risk of compromised food security at a national level due to the presence of heavy metals in volcanic ash. These findings are particularly significant for the Cotopaxi region, where volcanic activity continues to influence both the environment and the livelihoods of its inhabitants.

This study emphasizes the need for further research into optimizing agricultural practices in volcanic regions, balancing the advantages of improved crop quality with potential risks to food security. By understanding the complex interactions between soil properties and plant metabolism, agricultural systems in volcanic areas can be better managed to ensure both productivity and crop security.

2. Materials and Methods

Potato tubers (*Solanum tuberosum* L.) were collected from two distinct areas in the Cotopaxi region, Ecuador: areas exposed to volcanic ash (CC1 and CC2) and unexposed areas (CS1 and CS2). The areas are separated by 20 km and are approximately 60 km from the Cotopaxi volcano. The exposed samples (CC1 and CC2) came from agricultural fields in the parish of Guasaganda, affected by emissions from Cotopaxi, while the unexposed samples (CS1 and CS2) were collected from fields in the parish of Mulaló, without direct influence of recent volcanic ash. The collection was carried out during the harvest season, selecting mature tubers from plants cultivated under traditional local agricultural practices.

Sampling areas were randomly selected within representative agricultural fields with recent ash exposure (post 2015). Exposed soils had an average pH of 6.2, 3.5% organic matter, and high mineral load (e.g., Fe: 120 mg/kg), versus pH 6.8 and 2.8% in unexposed soils. ICP-OES analysis had detection limits of 0.01 mg/kg and recoveries of 95–105% for multielement standards.

Bioactive compound extraction was conducted according to the procedure outlined by Claros, 2021 [11], with modifications adapted for *S. tuberosum*. Fresh and mature potato tubers were processed by grinding them in a mortar to obtain a homogeneous paste. Subsequently, 1 g of the paste was precisely weighed using an analytical balance and subjected to maceration with 96% ethanol (10 mL) in 15 mL Falcon tubes. The mixture was manually stirred using a glass rod and then refrigerated at 5 °C for 72 h to optimize the extraction. The absorbance was quantified using a UV-Vis spectrophotometer (Thermo Fisher Scientific, Waltham, MA, USA). All biochemical experiments (extractions and determinations) were carried out in triplicate.

The phenolic content in samples collected from both sites with and without volcanic ash *S. tuberosum* tubers was quantified using the Folin–Ciocalteu colorimetric method, following the protocol described by López-Froilán et al., 2018 [12]. To determine the total phenolic compounds, a volume of 1.0 mL diluted potato extract (1:10) or standard solution was mixed with Folin–Ciocalteu reagent (same volume) and a small volume of distilled water. In total, 4 mL of Na_2CO_3 (100 mg/L) was added after 4 min, and the volume was adjusted to 25 mL using distilled water. A spectrophotometer was used to detect absorbance at 750 nm after the solution was incubated for 90 min at room temperature in the dark. Ethanol was used as a blank, with results given in mg GAE/L, using a calibration curve based on gallic acid (0–250 mg GAE/L).

For flavonoid quantification, Pekal et al.'s [13] method was used, which consists of mixing crude potato extract (1 mL), solvent (1.5 mL), CH_3COONa 1 M (100 µL), $AlCl_3$ 10% *v*/*v* (100 µL), and distilled water (2.3 mL). The mixture was left to rest for 40 min at room temperature before measuring absorbance at 435 nm.

The antioxidant capacity evaluation through FRAP assays was based on the methodology of Agudo, 2010 [14]. As amount of 593 nm was used to measure the absorbance. FRAP reagent was prepared by mixing acetate buffer 300 mM, pH 3.6 (100 mL), $FeCl_3$ 20 mM (10 mL), and distilled water (12 mL). Each reaction consisted of FRAP reagent (300 µL), potato extract (100 µL), and distilled water (300 µL), incubated for 4 min, in the dark, at room temperature. Ethanol was used as a blank, with a standard curve $y = 0.5981x - 0.0082$ with $R^2 = 0.9989$.

The Ramírez, 2023 [15] protocol was used for DPPH assay, which consists of mixing DPPH reagent (2.9 mL) with sample (0.1 mL) and maintaining the mixture for 30 min in the dark. As amount of 517 nm was used to measure the absorbance. Ethanol was used as a blank, with a standard curve $y = 18.073x + 1.2252$ with $R^2 = 0.9868$.

For the ABTS assay, the methodology described by Mendoza, 2018 [16], was followed, generating the ABTS•+ radical, and adjusted at an absorbance of 0.76 ± 0.1 at 754 nm. A mixture of ABTS•+ solution (980 µL) and extract (20 µL) was incubated for 7 min at room temperature, and absorbance was measured at 754 nm. A positive control represented by Trolox was used, with a standard curve $y = 34.102x + 9.2946$ with $R^2 = 0.9612$.

For the Inductively Coupled Plasma (ICP) analysis, dry samples of *S. tuberosum* tubers collected from soil control (SC) and post-Cotopaxi eruption conditions (PCCs) were prepared by weighing 0.5 g of finely ground plant material and transferring it into high-resistance Teflon tubes. A reagent mixture composed of nitric acid (HNO_3), hydrogen peroxide (H_2O_2), and ultrapure water was added to each tube. The samples underwent a microwave-assisted digestion process following a standardized program for plant materials, as described in [16,17]. After digestion, the resulting solutions were filtered and diluted with ultrapure water in volumetric flasks, preparing them for trace element analysis. A 10 mL volume was used for the following analysis. Trace element quantification was conducted using an ICP-OES Thermo Fisher Scientific 7400 Duo (Waltham, MA, USA) (Inductively Coupled Plasma Optical Emission Spectrometer). A multielement ICP Mix 33 standard (0.01–7.5 mg/L) and 5 ppm yttrium (as an internal standard) were used for calibration and to account for potential instrumental variations and enhance analytical precision [18,19].

With a significance level of $p < 0.05$, two-factor ANOVA was used to evaluate significant differences between experimental groups in the statistical analysis, which was carried out using RStudio software version 4.3.2. Every experiment was carried out in triplicate, and the mean ± standard deviation (SD) was used to express the results. Pearson's correlation coefficient was used to assess the relationship between antioxidant capacity and secondary metabolite concentration. Bar graphs with error bars used to visualize data were generated using the ggplot2 package, ensuring an accurate and clear representation of the observed variations. Additionally, a heatmap was constructed to illustrate the distribution of trace elements obtained from the ICP-OES analysis, facilitating comparative evaluation across samples affected and unaffected by volcanic ash.

3. Results

The TPC values in samples collected from the ash-exposed place (CC1) varied between 2.206 and 2.270 mg GAE/g DW. Similarly, CC2 samples displayed TPC values varying between 2.183 and 2.211 mg GAE/g DW. In contrast, the samples collected from the non-ash-exposed place (CS1) exhibited lower TPC values with an average of 1.070 ± 0.054 mg GAE/g DW. The CS2 samples showed a similar trend, with TPC values with a mean of 1.053 ± 0.039 mg GAE/g DW (Figure 1).

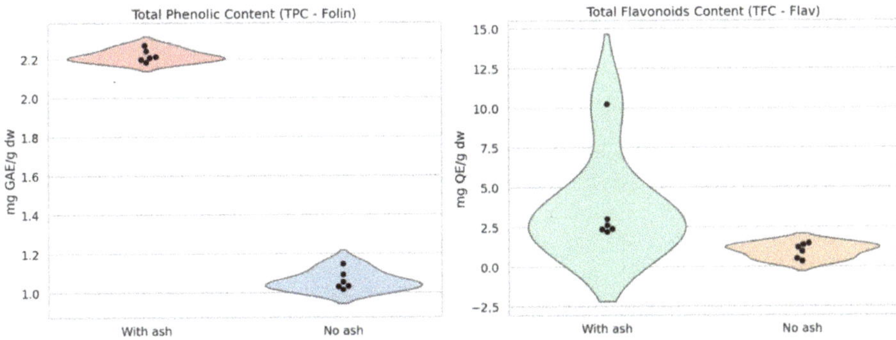

Figure 1. Bioactive compounds analyzed in the potato samples cultivated from exposed and unexposed areas in the Cotopaxi region. The black dots represent individual measurements, providing insight into the variability within each group.

TFC values followed a distinct pattern among the samples. CC1 samples exhibited TFC values ranging from 2.229 to 2.635 mg QE/g DW, with an average of 2.428 ± 0.168 mg QE/g DW. In CC2, a notable variation was observed, with TFC values ranging from 2.387 to 10.248 mg QE/g DW, averaging 5.884 ± 4.011 mg QE/g DW, suggesting significant metabolic fluctuations potentially influenced by volcanic ash exposure. CS1 samples presented TFC values between 0.980 and 1.464 mg QE/g DW (1.272 ± 0.200 mg QE/g DW average), while CS2 exhibited the lowest TFC values (0.372–1.216 mg QE/g DW), with an average of 0.698 ± 0.436 mg QE/g DW. The variability in TFC of CC2 (2.387 to 10.248 mg QE/g DW) could reflect soil heterogeneity or differential metabolic responses, with no statistical outliers identified (Grubbs test, $p > 0.05$).

The negative DPPH value in CS1 (−0.12 μmol Trolox g^{-1} DW) indicates possible assay interference, adjusted to zero for further analysis (Figure 1).

The antioxidant capacity of potato samples cultivated under volcanic ash influence in the Cotopaxi region, obtained using three different assays, showed that samples collected from ash-exposed areas (CC1 and CC2) exhibited significantly higher antioxidant activity across all assays compared to samples collected from non-ash places (CS1 and CS2). In the DPPH assay, CC2 recorded the highest activity (4.80 ± 2.38 μmol Trolox g^{-1} DW), while CS1 displayed negative values. Similarly, ABTS results confirm higher antioxidant potential in CC2 (11.05 ± 2.57 μmol Trolox g^{-1} DW), whereas CS1 and CS2 exhibited significantly lower activity. The FRAP assay, which measures reducing power, also showed a clear difference between samples. CC1 had the highest reducing capacity (11.96 ± 4.57 μmol Fe^{2+} g^{-1} DW), followed closely by CC2, while CS1 and CS2 presented the lowest values (Figure 2).

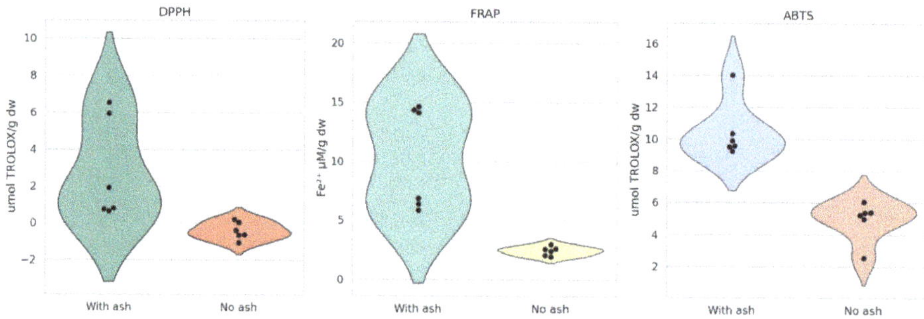

Figure 2. Antioxidant activity in potato samples cultivated with and without volcanic ash in the Cotopaxi region. Legend: DPPH—α-diphenyl-α-picrylhydrazyl free radical scavenging; FRAP—ferric-reducing antioxidant power; ABTS—free radical scavenging activity. The black dots represent individual measurements, providing insight into the variability within each group.

The correlation analysis between the antioxidant properties of potato samples cultivated in the Cotopaxi region (Figure 3) revealed distinct patterns among the different antioxidant assays (DPPH, FRAP, ABTS) and bioactive compound contents (TPC and TFC).

Our results show a strong and significant correlation between FRAP and DPPH (r = 0.986 ***), suggesting a consistent interplay between these two antioxidant capacities. Additionally, a high correlation among ABTS and TFC (r = 0.911 ***) was noted, reinforcing the implication of flavonoids in the scavenging activity captured by ABTS assay. Interestingly, DPPH also exhibited a robust positive correlation with TFC (r = 0.775 **), while FRAP showed a weaker but still significant correlation with TPC (r = 0.637 *), indicating phenolic compounds' role in ferric reducing antioxidant potential.

However, some negative correlations stood out, such as the relationship between TFC and FRAP (r = −0.088) and between TPC and TFC (r = −0.045), suggesting a complex interaction in which flavonoid content may inversely affect certain antioxidant properties under specific conditions. These results are consistent across affected (CC1 and CC2) and unaffected (CS1 and CS2) samples, highlighting the influence of volcanic ash on metabolic and antioxidant dynamics in the potato crop.

Subgroup analyses revealed additional data. For example, the correlation between TPC and ABTS was particularly pronounced in unaffected samples (CS1: r = 0.963; CS2: r = 0.791) compared to affected samples (CC1: r = −0.384; CC2: r = −0.334). This suggests that volcanic ash may alter the relationship between phenolic content and antioxidant capacity. Furthermore, flavonoid content showed a stronger relationship with antioxidant capacity in ash-treated samples (e.g., TFC and ABTS, CC1: r = 0.659; CC2: r = 0.649), possibly reflecting an adaptive metabolic response to volcanic ash exposure.

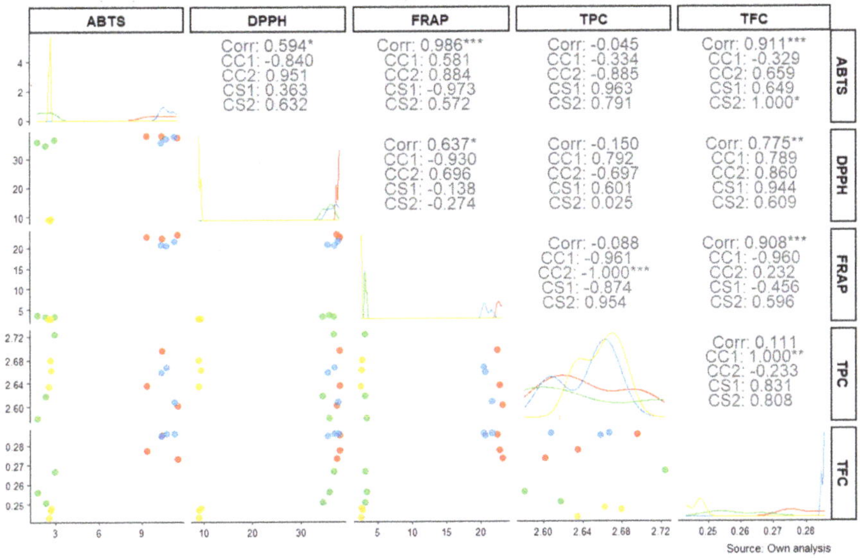

Figure 3. Correlation matrix between bioactive compounds and antioxidant capacity in potato cultivated in Cotopaxi region. The plot includes density distributions along the diagonal and scatterplots with regression lines for each pairwise comparison. Legend: CC1, CC2—samples collected from volcanic ash-exposed place; CS1, CS2—samples collected from non-ash-exposed place; TPC—total phenolic content; TFC—total flavonoid content; DPPH—α-diphenyl-α-picrylhydrazyl free radical scavenging; FRAP—ferric-reducing antioxidant power; ABTS—free radical scavenging activity; r—correlation coefficient; * $p < 0.05$; ** $p < 0.01$; *** $p < 0.001$. Different colors indicate data for each sample; lines represent the fitted correlation trend for the respective assays.

Inductive Plasma Coupled

The ICP analysis of potato samples from the Cotopaxi region revealed notable differences in elemental concentrations between tubers collected from unexposed soil (SC) and exposed soil (PCC). Potassium (K) was the most abundant element, with 17,215.34 mg/kg in control samples (CS1 and CS2) and 14,525.76 mg/kg in post-Cotopaxi eruption conditions (CC1 and CC2), followed by magnesium (Mg) at 661.28 mg/kg (SC) and 855.87 mg/kg (PCC). Calcium (Ca) and sodium (Na) showed moderate concentrations, while iron (Fe) increased from 104.47 mg/kg in soil control to 180.82 mg/kg in post-Cotopaxi eruption conditions. Other elements like boron (B) and copper (Cu) decreased post eruption, while aluminum (Al) showed a significant rise from 26.59 mg/kg in SC to 88.56 mg/kg in PCC. Metallic elements such as zinc, lithium, boron, manganese, barium, lead, nickel, chromium, and indium showed concentrations below 0.01 mg/kg.

Overall, volcanic ash deposition influenced the elemental composition of potatoes, increasing Fe, Al, and Mn while reducing Cu, Ni, and B levels. These changes may impact the nutritional and metabolic properties of potato crops in the region (Figure 4).

The ICP analysis of potato samples from the Cotopaxi region, visualized through a heatmap, confirmed that potassium (K) was the most abundant element, with concentrations of 17,215.34 mg/kg in soil-conditioned (SC) samples and 14,525.76 mg/kg in post-cultivation condition (PCC) samples, showing a noticeable decrease after cultivation. This trend is clearly reflected in the heatmap, where K appears as the most intense element. Magnesium (Mg) and calcium (Ca) followed in concentration, while elements like bismuth (Bi), cobalt (Co), and thallium (Tl) were nearly undetectable (<0.01 mg/kg). Transition metals such as iron (Fe) and zinc (Zn) showed moderate levels, with Fe increasing in PCC samples. Sodium (Na) and strontium (Sr) exhibited notable decreases, suggesting element mobilization during cultivation. The heatmap effectively illustrates the dominance of K and the minimal presence of other elements, reinforcing the influence of volcanic ash on elemental distribution and its potential implications for plant metabolism and nutritional quality (Figure 5).

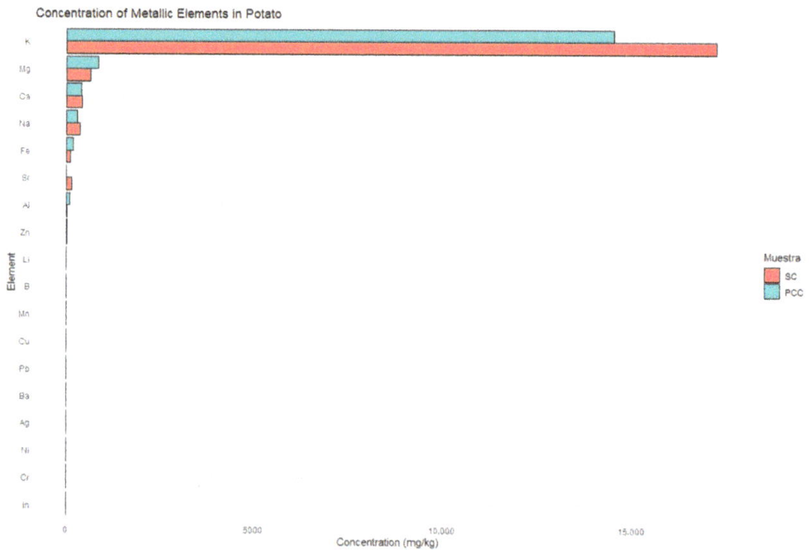

Figure 4. Concentration of metallic elements in potato samples. Legend: SC—tubers from the control soil; PCC—post-Cotopaxi eruption conditions. The *x*-axis represents element concentrations, while the *y*-axis lists the detected elements. Concentrations below 0.01 mg/kg of metallic elements were excluded from the visualization.

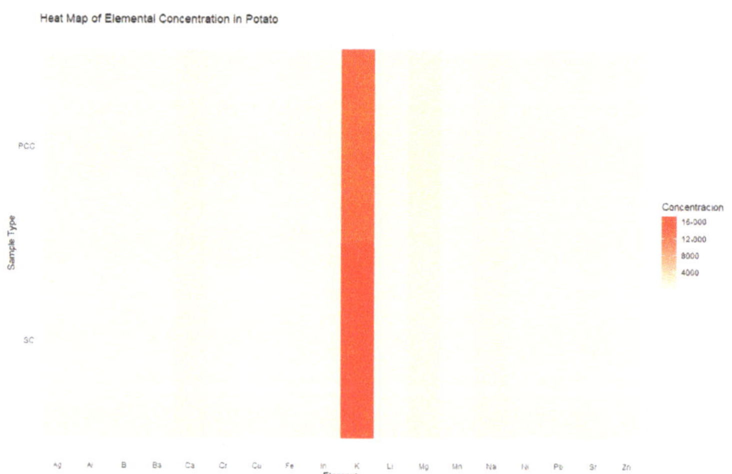

Figure 5. Heatmap of elemental concentration in potato samples (ICP). This heatmap visually represents the concentration levels of various metallic elements (mg/kg) in potato samples from the Cotopaxi region. The color gradient highlights the most and least concentrated elements, with potassium (K) standing out as the most abundant, followed by magnesium (Mg) and calcium (Ca). Elements with concentrations below 0.01 mg/kg were nearly undetectable and are not prominently displayed.

4. Discussion

The present study demonstrates that potatoes cultivated in volcanic ash-influenced soils exhibit enhanced bioactive profiles compared to those grown in non-volcanic substrates.

The significantly higher total phenolic content in exposed samples (CC1 and CC2) suggests that the unique mineral composition and physicochemical properties of volcanic ash may act as abiotic stressors, triggering the biosynthesis of phenolic compounds. These compounds are well known for their antioxidant activity and implication in plant defense mechanisms, which can ultimately contribute to improved nutritional quality and stress resilience [20,21].

Furthermore, the marked variation in total flavonoid content observed, particularly in the CC2 samples, indicates potential metabolic adaptation to the heterogeneous nature of volcanic ash deposition. This variability might reflect the differential activation of stress response pathways that modulate flavonoid biosynthesis to counteract environmental fluctuations, such as changes in soil pH or the presence of heavy metals. Such adaptive metabolic responses could not only enhance the antioxidant capacity of the tubers but also elucidate the intricate relationship between plant secondary metabolism and soil chemistry [22].

Another crucial aspect explored in this study is the antioxidant content of potatoes, which reflects the plant's defense mechanisms against external stressors. Antioxidants are critical for neutralizing reactive oxygen species, which can form under stressful conditions such as exposure to volcanic ash [9]. Interestingly, while volcanic ash represents a potential threat to food safety due to heavy metal contamination, its impact on plant stress responses may also have positive implications. The stress induced by ash exposure can potentially enhance the production of secondary metabolites, including antioxidants, especially phenolic compounds such as flavonoids, which contribute to the nutraceutical value of the tubers. These dual effects—both risks to food safety and potential benefits to nutritional quality—highlight the complexity of volcanic ash interactions with agricultural systems. The increase in antioxidant activity could be due to the upregulation of phenylalanine ammonia lyase (PAL) under ash-induced oxidative stress, promoting phenol synthesis [23]. However, this study is limited by the lack of transcriptomic data, detailed soil controls, and varietal specificity, which restricts the generalizability of the results.

The antioxidant assays (DPPH, ABTS, and FRAP) clearly demonstrate that volcanic ash-enriched soils significantly enhance the potatoes' antioxidant activity. Specifically, the CC2 samples exhibited the highest radical scavenging activity in both the DPPH (4.80 ± 2.38 µmol Trolox g^{-1} DW) and ABTS (11.05 ± 2.57 µmol Trolox g^{-1} DW) assays, while CC1 showed superior reducing power in the FRAP assay (11.96 ± 4.57 µmol Fe^{2+} g^{-1} DW). These results imply that the unique volcanic ash mineral composition may stimulate the biosynthesis of bioactive compounds, thereby improving the overall antioxidant profile of the crop. Such results align with previous studies that have reported enhanced phytochemical accumulation in plants grown in volcanic substrates [24].

Furthermore, the marked differences in antioxidant activities between the samples underscore the complex interplay between soil properties and plant metabolic responses. The negative DPPH values observed in the unexposed samples (CS1), in contrast with the robust activity in the exposed samples (CC1 and CC2), indicate that potatoes grown without volcanic ash might lack sufficient defense mechanisms against oxidative stress. This disparity points to an adaptive response where the stress imposed by volcanic ash environments triggers the upregulation of antioxidant pathways, ultimately contributing to improved nutritional quality. Similar adaptive metabolic responses have been documented in the recent literature, emphasizing the potential of volcanic ash to serve as a natural enhancer of crop quality [23,25].

The correlation analysis reveals complex interactions among antioxidant capacity and bioactive compounds in potato cultivated in the Cotopaxi region. A remarkably positive correlation between FRAP and DPPH (r = 0.986, $p < 0.001$) indicates consistent interplay between reducing power and radical scavenging activity. In addition, the significant association between ABTS and TFC (r = 0.911, $p < 0.001$) reinforces the crucial role of flavonoids in scavenging free radicals, while the moderate correlation between DPPH and TFC (r = 0.775, $p < 0.01$) further supports the contribution of these compounds to antioxidant defense mechanisms. These findings align with recent studies emphasizing the pivotal role of secondary metabolites in enhancing plant antioxidant systems [22,23].

Interestingly, the analysis also revealed negative correlations between flavonoids and FRAP (r = −0.088) and between TPC and TFC (r = −0.045), suggesting a complex interaction that may be modulated by environmental conditions such as volcanic ash exposure. Subgroup analyses showed that, in samples cultivated without volcanic ash, the correlation between TPC and ABTS was particularly strong (CS1: r = 0.963; CS2: r = 0.791), whereas ash-treated samples exhibited inverse relationships (CC1: r = −0.384; CC2: r = −0.334). This divergence implies that volcanic ash could alter the synthesis or interaction of these bioactive compounds, potentially as an adaptive metabolic response to abiotic stress. Furthermore, the enhanced correlation between TFC and ABTS in ash-treated samples (CC1: r = 0.659; CC2: r = 0.649) underscores the possibility that volcanic ash may stimulate specific metabolic pathways involved in flavonoid biosynthesis [23,26].

Although the property of volcanic ash as a multi-nutrient mineral fertilizer has been demonstrated since 2010, in areas of Russia and Indonesia (two areas of the world that host very active volcanoes), its mechanisms remain monthly unknown [26,27].

The ICP analysis reveals that volcanic ash deposition in the Cotopaxi region significantly alters the elemental composition of potatoes. Notably, potassium (K) levels decrease in tubers collected from unexposed soil (SC) to 14,525.76 mg/kg in tubers collected from exposed soil (PCC), suggesting nutrient mobilization or altered uptake dynamics during cultivation. In contrast, the concentrations of magnesium (Mg) and iron (Fe) increase in PCC samples (from 661.28 mg/kg to 855.87 mg/kg and from 104.47 mg/kg to 180.82 mg/kg, respectively), indicating that volcanic ash may enhance the bioavailability of certain micronutrients by modifying soil pH and cation exchange capacity [25]. The significant rise in aluminum (Al) from 26.59 mg/kg to 88.56 mg/kg further underscores the transformative impact of volcanic ash on soil mineralogy, potentially influencing both plant stress responses and metabolic functions [25]. The levels of Fe (180.82 mg/kg) and Al (88.56 mg/kg) in exposed samples are below the FAO/WHO toxicity thresholds (Fe: 425 mg/kg; Al: 1000 mg/kg in tubers), suggesting safety for human consumption under current conditions.

These shifts in elemental profiles have critical implications for the nutritional and metabolic quality of potato crops. The enhanced levels of Fe and Mg in PCC samples may improve the crop's nutritional value and support vital enzymatic processes associated with energy production and antioxidant defense [22]. However, the observed decreases of copper (Cu) and boron (B) (essential elements) raise concerns about potential deficiencies that could affect plant growth and metabolic balance. The heatmap visualization further corroborates these findings by illustrating the dominance of potassium alongside fluctuating trace element distributions, thereby highlighting the intricate relationship between volcanic ash inputs and nutrient uptake in potato cultivation. In our determinations, heavy metals appear to be in concentrations below 0.01 mg/kg, suggesting that heavy metals from the environment did not negatively influence the quality of potatoes in the area.

5. Conclusions

This study highlights the significant influence of volcanic ash on the nutritional and biological properties of potato tubers cultivated in the Cotopaxi region. Samples grown in volcanic ash-enriched soils demonstrated enhanced bioactive profiles. These findings suggest that the unique mineral composition of volcanic ash acts as an abiotic stressor, triggering adaptive metabolic responses that enhance the biosynthesis of secondary metabolites. The increased antioxidant activity observed in ash-treated samples, particularly in DPPH, ABTS, and FRAP assays, underscores the potential of volcanic ash to improve the nutraceutical qualities of crops. Elemental analysis further revealed that volcanic ash deposition alters the elemental composition of soils and tubers. While potassium (K) levels decreased, the increased concentrations of magnesium (Mg), iron (Fe), and aluminum (Al) indicate enhanced nutrient bioavailability and changes in soil chemistry. These shifts may contribute to improved nutritional quality but also raise concerns about potential heavy metal accumulation and food safety risks. The high variability in TFC observed in the CC2 samples reflects the heterogeneous nature of volcanic ash deposition and its potential to activate specific metabolic pathways in response to environmental stress.

Despite the potential benefits of volcanic ash exposure, such as increased antioxidant capacity, the findings raise critical questions regarding its long-term impact on crop safety and soil sustainability.

Author Contributions: Conceptualization, R.A.M.; methodology, R.A.M.; formal analysis, R.F.V.G. and N.R.R.B.; investigation, R.F.V.G. and N.R.R.B.; resources, R.A.M.; writing—original draft preparation, R.A.M.; writing—review and editing, R.D.C.; supervision, R.A.M.; project administration, R.A.M.; funding acquisition, R.A.M. All authors have read and agreed to the published version of the manuscript.

Funding: This research was funded by Universidad de Las Fuerzas Armadas-ESPE, grant number CV-GNP-0066-2020, and the Institute of Biology Bucharest, Romanian Academy RO1567-IBB06/2025. The APC was funded by Universidad de Las Fuerzas Armadas-ESPE.

Institutional Review Board Statement: Not applicable.

Informed Consent Statement: Not applicable.

Data Availability Statement: The original contributions presented in the study are included in the article; further inquiries can be directed to the corresponding author due to privacy.

Acknowledgments: The authors would like to express their sincere gratitude to the communities of Cotopaxi Province, particularly the members of Jatari Unancha College in Guasaganda, for their invaluable assistance during the plant collection process and also to the Universidad de Las Fuerzas Armadas-ESPE for financial support.

Conflicts of Interest: The authors declare no conflicts of interest. The funders had no role in the study's design; in the collection, analyses, or interpretation of data; in the writing of the manuscript; or in the decision to publish the results.

References

1. Ahmed, F.M.; Mondal, M.A.; Babul Akter, B. Organic Fertilizers Effect on Potato (*Solanum tuberosum* L.) Tuber Production in Sandy Loam Soil. *Int. J. Plant Soil Sci.* **2019**, *29*, 1–11. [CrossRef]
2. McGill, C.R.; Kurilich, A.C.; Davignon, J. The role of potatoes and potato components in cardiometabolic health: A review. *Ann. Med.* **2013**, *45*, 467–473. [CrossRef] [PubMed]
3. Calliope, S.R.; Lobo, M.O.; Sammán, N.C. Biodiversity of Andean potatoes: Morphological, nutritional and functional characterization. *Food Chem.* **2018**, *238*, 42–50. [CrossRef]
4. Mihai, R.A.; Espinoza-Caiza, I.A.; Melo-Heras, E.J.; Cubi-Insuaste, N.S.; Pinto-Valdiviezo, E.A.; Catana, R.D. Does the Mineral Composition of Volcanic Ashes Have a Beneficial or Detrimental Impact on the Soils and Cultivated Crops of Ecuador? *Toxics* **2023**, *11*, 846. [CrossRef]

5. Smith, V.C.; Costa, A.; Aguirre-Díaz, G.; Pedrazzi, D.; Scifo, A.; Plunkett, G.; Poret, M.; Tournigand, P.Y.; Miles, D.; Dee, M.W.; et al. The timing and impact of the large Tierra Blanca Joven eruption of Ilopango, El Salvador. *Proc. Natl. Acad. Sci. USA* **2020**, *117*, 26061–26068. [CrossRef]
6. Wang, L.; Zhou, Y. The role of volcanic substrates in enhancing plant secondary metabolism. *Ecol. Soil Chem.* **2021**, *19*, 210–225.
7. Song, Q.; Zheng, Y.-J.; Xue, Y.; Sheng, W.-G.; Zhao, M.-R. An evolutionary deep neural network for predicting morbidity of gastrointestinal infections by food contamination. *Neurocomputing* **2017**, *226*, 16–22. [CrossRef]
8. Ligot, N.; Barthelemi, L.; Falys, H.; Godin, B.; Bogaert, P.; Delmelle, P. Plant traits, growth stage, and ash mass load control the vulnerability of potato, corn, and wheat crops to volcanic ashfall. *Volcanica* **2025**, *8*, 81–94. [CrossRef]
9. Zakharikhina, L.V.; Litvinenko, Y.S.; Gainatulina, V.V. Volcanic Ash Application in Agricultural Practice. *Univers. J. Agric. Res.* **2022**, *10*, 77–87. [CrossRef]
10. Budiman, M.; Dian, F.; Hairiah, K.; van Noordwijk, M. Applying volcanic ash to croplands—The untapped natural solution. *Soil Secur.* **2021**, *3*, 100006.
11. Claros, P. Evaluacion de la Capacidad Antioxidante Total y Contenido de Polifenoles Totales Del *Phaseolus vulgaris* "Frijol". Universidad Nacional José Faustino Sánchez Carrión, Huacho. 2021. Available online: https://repositorio.unjfsc.edu.pe/handle/20.500.14067/5297 (accessed on 3 March 2025).
12. López-Froilán, R.; Hernández-Ledesma, B.; Cámara, M.; Pérez-Rodríguez, M. Evaluation of the Antioxidant Potential of Mixed Fruit-Based Beverages: A New Insight on the Folin-Ciocalteu Method. *Food Anal. Met.* **2018**, *11*, 2897–2906. [CrossRef]
13. Pekal, A.; Pyrzynska, K. Evaluation of aluminum complexation reaction for flavonoid content assay. *Food Anal. Met.* **2014**, *7*, 1776–1782. [CrossRef]
14. Agudo, L.M. Técnicas para la determinación de compuestos antioxidante en alimentos. *Autodidacta* **2010**, 27–34.
15. Ramírez Silva, M.P. Determinación de la Capacidad Antioxidante Presente en las Semillas de Cordia Dentata por el Método ABTS y DPPH. Bachelor's Thesis, Universidad El Bosque, Bogota, Columbia, 2023. Available online: https://repositorio.unbosque.edu.co/bitstream/handle/20.500.12495/10599/DeterminacicondeacapacidadantioxidantepresenteenlassemillasdeCordiadentataporelmtodoABTSyDPPH?sequenceisAllowed (accessed on 10 October 2024).
16. Mendoza, M. Inducción de Metabolitos de Interés Nutracéutico en Germinados de Frijol (Phaseolus vulgaris L.) y el Efecto de su Consumo en un Modelo de Dislipidemia. Ph.D. Thesis, Faculty of Chemistry, Querétaro, Mexico, 2018.
17. Tzvetkov, N.T.; Dinev, T.; Milanov, G. Application of ICP-OES for the Determination of Metal Elements in Soil and Plant Samples. *Environ. Monit. Assess.* **2018**, *190*, 1–10.
18. Ravenscroft, P.; Brammer, H.; Richards, K. *Groundwater and Global Change: Trends, Impacts and Challenges*; Wiley-Blackwell: Hoboken, NJ, USA, 2009.
19. Garcia, E.; Esteban, M. Validation of an ICP-OES Method for the Quantitative Determination of Trace Elements in Food Samples. *J. Anal. At. Spectr.* **2016**, *31*, 550–558.
20. Alloway, B.J. *Contaminated Land and Its Management: The Role of Soil Science*; Cambridge University Press: Cambridge, UK, 2008.
21. Chan, Y.S.; Lee, T.Y. Impact of volcanic ash on soil properties and plant secondary metabolites. *Ecol. Chem. Let.* **2022**, *20*, 205–218.
22. Kumar, P.; Patel, D. Environmental influences on phenolic and flavonoid synthesis in crop plants. *Plant Physiol. Rep.* **2023**, *29*, 150–165.
23. Khawula, S.; Gokul, A.; Niekerk, L.A.; Basson, G.; Keyster, M.; Badiwe, M.; Klein, A.; Nkomo, M. Insights into the Effects of Hydroxycinnamic Acid and Its Secondary Metabolites as Antioxidants for Oxidative Stress and Plant Growth under Environmental Stresses. *Curr. Issues Mol. Biol.* **2023**, *46*, 81–95. [CrossRef]
24. Garcia, R.M.; Santos, J.P.; Lopez, M. Impact of volcanic ash on phytochemical accumulation in crops. *J. Agric. Sci.* **2022**, *58*, 345–356.
25. Fernandez, A.L.; Mendoza, C.F.; Rodriguez, E. Volcanic ash influence on plant nutrient uptake and elemental composition. *J. Environ. Sci. Technol.* **2023**, *57*, 123–137.
26. Ciriminna, R.; Scurria, A.; Tizza, G.; Pagliaro, M. Volcanic ash as multi-nutrient mineral fertilizer: Science and early applications. *ChemRxiv* **2022**. [CrossRef]
27. El-Desoky, A.I.; Hassan, A.Z.A.; Mahmoud, A.M. Volcanic Ash as a Material for Soil Conditioner and Fertility. *J. Soil Sci. Agric. Eng.* **2018**, *9*, 491–495. [CrossRef]

toxics

MDPI

Article

Hydrochemical Characteristics, Controlling Factors, and High Nitrate Hazards of Shallow Groundwater in an Urban Area of Southwestern China

Chang Yang [1,2], Si Chen [1,*], Jianhui Dong [3], Yunhui Zhang [3,4,5,6], Yangshuang Wang [3,4,5,6], Wulue Kang [1], Xingjun Zhang [1], Yuanyi Liang [1], Dunkai Fu [4,5], Yuting Yan [4,5] and Shiming Yang [4,5]

[1] Chongqing Institute of Geology and Mineral Resources, Chongqing 401120, China
[2] Chongqing Huadi Resources Environment Technology Co., Ltd., Chongqing 401120, China
[3] Sichuan Provincial Engineering Research Center of City Solid Waste Energy and Building Materials Conversion and Utilization Technology, Chengdu 610106, China
[4] Yibin Research Institute, Southwest Jiaotong University, Yibin 644000, China
[5] Faculty of Geosciences and Environmental Engineering, Southwest Jiaotong University, Chengdu 611756, China
[6] Sichuan Province Engineering Technology Research Center of Ecological Mitigation of Geohazards in Tibet Plateau Transportation Corridors, Chengdu 611756, China
* Correspondence: cs9605@tom.com; Tel.: +86-13637822030

Abstract: Groundwater nitrate (NO_3^-) contamination has emerged as a critical global environmental issue, posing serious human health risks. This study systematically investigated the hydrochemical processes, sources of NO_3^- pollution, the impact of land use on NO_3^- pollution, and drinking water safety in an urban area of southwestern China. Thirty-one groundwater samples were collected and analyzed for major hydrochemical parameters and dual isotopic composition of NO_3^- ($\delta^{15}N$-NO_3^- and $\delta^{18}O$-NO_3^-). The groundwater samples were characterized by neutral to slightly alkaline nature, and were dominated by the Ca-HCO_3 type. Hydrochemical analysis revealed that water–rock interactions, including carbonate dissolution, silicate weathering, and cation exchange, were the primary natural processes controlling hydrochemistry. Additionally, anthropogenic influences have significantly altered NO_3^- concentration. A total of 19.35% of the samples exceeded the Chinese guideline limit of 20 mg/L for NO_3^-. Isotopic evidence suggested that primary sources of NO_3^- in groundwater include NH_4^+-based fertilizer, soil organic nitrogen, sewage, and manure. Spatial distribution maps indicated that the spatial distribution of NO_3^- concentration correlated strongly with land use types. Elevated NO_3^- levels were observed in areas dominated by agriculture and artificial surfaces, while lower concentrations were associated with grass-covered ridge areas. The unabsorbed NH_4^+ from nitrogen fertilizer entered groundwater along with precipitation and irrigation water infiltration. The direct discharge of domestic sewage and improper disposal of livestock manure contributed substantially to NO_3^- pollution. The nitrogen fixation capacity of the grassland ecosystem led to a relatively low NO_3^- concentration in the ridge region. Despite elevated NO_3^- and F^- concentrations, the entropy weighted water quality index (EWQI) indicated that all groundwater samples were suitable for drinking. This study provides valuable insights into NO_3^- source identification and hydrochemical processes across varying land-use types.

Keywords: nitrate; groundwater; land use; stable isotopes; drinking water quality

https://doi.org/10.3390/toxics13060516

1. Introduction

Groundwater serves as a vital global source of drinking water, yet its quality is facing escalating threats, especially in agricultural regions [1–3]. A survey demonstrates that approximately one-third of the world's population relies on groundwater as their primary drinking water supply, with agrarian areas exhibiting significantly higher contamination risks [4]. Among various pollutants, nitrate contamination is the most widespread issue linked to intensive agricultural processes [5–7]. For instance, previous studies revealed severe nitrate pollution hotspots: over 30 regions in Africa, 20 areas in Asia, and nine regions in Europe had average groundwater nitrate concentrations exceeding the WHO safety guideline (50 mg/L) [8], while in China, approximately 20% of shallow groundwater samples surpassed the national standard (20 mg/L) [9]. The primary driver of this contamination is the excessive use of nitrogen-based fertilizers and manure in agriculture [10–12], which disrupts natural hydrogeochemical cycles. Moreover, long-term exposure to nitrate-rich groundwater greatly threatens human health [13–15], such as gastric cancer. Despite these documented risks, systematic investigations into pollution sources and hydrochemical controls remain limited in many high-risk agricultural zones [16]. Therefore, comprehensive assessments of groundwater chemistry and anthropogenic influences are urgently needed to guide targeted water resource protection policies.

Previous studies have utilized conventional hydrochemical analytical approaches, including major ion ratio analysis and basic graphical methods, including Piper and Gibbs diagrams, to investigate groundwater contamination in agricultural areas [17–19]. These methodologies have indicated that nitrate contamination predominantly stems from chemical fertilizers, manure, and wastewater [20–22]. For example, Yang et al. [23] demonstrated that excessive nitrogen fertilizer application leads to nitrate release into soils, consequently elevating nitrate concentrations in the groundwater of China's Ordos Basin. Jin et al. [24] identified manure as a significant contributor to elevated nitrate levels in groundwater systems. On the other hand, to assess the implications of nitrate-contaminated groundwater, researchers have conducted comprehensive water quality evaluations [25–28]. Various quantitative methods have been employed to determine drinking water suitability, including the single-factor index [29], Nemerow index [30,31], and water quality index (WQI) [32–34]. The single-factor index offers a rapid and straightforward assessment of parameter exceedances, whereas the Nemerow index provides a holistic evaluation by integrating multiple pollutant concentrations [35–37]. The WQI, meanwhile, simplifies complex water quality data into an interpretable numerical score, facilitating informed decision-making [38–40]. Collectively, these methods yield critical insights for groundwater quality protection and remediation strategies.

While these conventional approaches have proven effective in elucidating nitrate pollution sources and assessing groundwater suitability, they exhibit notable methodological limitations [41–43]. A primary constraint lies in the frequent ambiguity of source identification due to overlapping ionic ratios among different nitrogen sources [44–46]. Furthermore, most existing studies inadequately consider potential hydrochemical processes that may modify or obscure source signatures. Similarly, current water quality assessment methods face critical shortcomings, particularly in their inability to quantitatively evaluate drinking water suitability or provide entirely objective assessment outcomes [47,48]. However, recent methodological advancements demonstrate promising solutions to these limitations through the integration of dual nitrate isotopes ($\delta^{15}N$ and $\delta^{18}O$) and the entropy weighted water quality index (EWQI) [49–51]. The isotopic approach leverages distinct source-specific signatures, minimizing uncertainties associated with mixed pollution sources. Concurrently, the EWQI enhances objectivity by quantifying assessment results and applying information entropy theory to derive statistically robust weighting coefficients [52,53].

Anthropogenic activities have influenced the groundwater chemistry to some extent [54]. Different land use types reflect different natural environments and human influences, which significantly impact groundwater quality [55]. Several studies have demonstrated that nitrate pollution strongly correlates with land use types [56]. For example, Iqbal found that land use types, particularly cropland and rural areas, strongly correlate with nitrate concentration [57]. Choi found that 23%, 43%, and 67% of samples in the cropping area, the cropping–livestock farming complex area, and the residential area exceeded the national nitrate concentration standard [56]. Exploring the impact of different land use types on nitrate concentrations is crucial for the sustainable agricultural management and precise pollution control.

As a vital area for economic development, the southwestern region of China has seen increasingly prominent nitrate pollution in groundwater due to intensified agricultural activities and accelerated urbanization in recent years [58,59]. Consequently, research on this region has grown progressively more comprehensive and in-depth. However, systematic studies on groundwater in the southwestern region remain insufficient, leaving residents in some areas poorly informed about local groundwater conditions. This situation contradicts China's commitment to sustainable development and significantly hinders regional economic growth. The study area is located in southwestern China, where agricultural development and drinking water quality are critical for economic development. Nevertheless, research on hydrochemical evolution and groundwater quality remains scarce. To address this limitation, this study employs hydrochemical analysis, ion ratios, and isotopic characteristics to identify nitrate sources, while using the EWQI to assess drinking water quality. Accordingly, the objectives of this study are as follows: (1) determine the general characteristics and spatial distribution of hydrochemical parameters in the study area, (2) analyze the sources of major ions and nitrate in groundwater, (3) identify the impact of land use type on NO_3^- pollution, and (4) evaluate groundwater quality for drinking and its spatial distribution in the study area. The findings of this research will provide valuable insights for groundwater management worldwide.

2. Study Area

The study area is located in the southwest part of Chongqing city, southwestern China, spanning from 106°13′ to 106°26′ E and 29°25′ to 29°38′ N (Figure 1). The geomorphology is characterized by parallel ridges and valleys trending NE-NNE. The area experiences a humid subtropical monsoon climate, with an average annual rainfall of approximately 1188 mm, and an average annual temperature of 18.63 °C.

Geologically, the valley regions are primarily underlain by the Suining Formation (J_3sn) and the Shaximiao Formation (J_2s) of the Jurassic strata. The left-side ridges are mainly composed of the Feixianguan Formation (T_1f), Jialingjiang Formation (T_1j), and Leikoupo Formation (T_2l) of the Triassic strata. In contrast, the right-side ridges are composed of the Longxing Formation (P_2l), Changxing Formation (P_2c), and Maokou Formation (P_1m) of the Permian strata. The valley regions are dominated by sandstone, mudstone, and shale, whereas the ridge regions are predominantly composed of dolomite and limestone.

Groundwater in the study area occurs mainly in two forms: (1) bedrock fracture water within the weathering zones of red beds, and (2) fracture water within carbonate rocks. Bedrock fracture water is mainly found in valley regions, typically at depths less than 20 m. Carbonate fracture water is distributed in a banded pattern along the axes of anticlines. Groundwater is primarily recharged by precipitation and is discharged through springs and underground rivers.

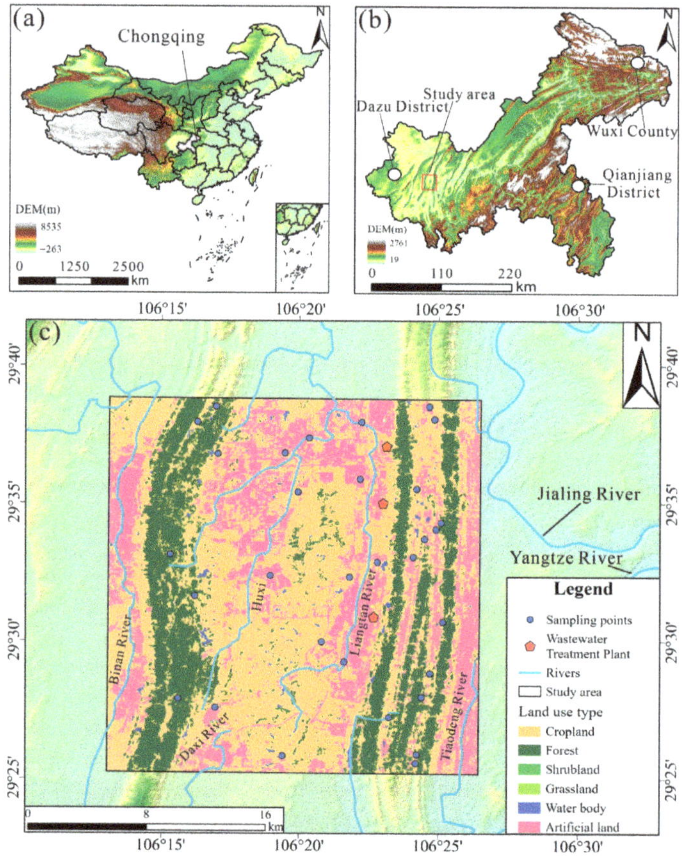

Figure 1. The location of the groundwater samples: (**a**) location of Chongqing City; (**b**) location of the study area in Chongqing City; (**c**) land use map of the study area and the location map of the groundwater samples.

Land use in the study area includes six categories: cropland, forest, shrubland, grassland, water body, and artificial land (Figure 1c). Forested areas are mainly distributed along ridge regions, while cultivated land is concentrated in the valley areas between ridges. Artificial surfaces refer to impervious land covers such as asphalt, concrete, and building structures, which are primarily focused near ridges and in the northern section of the central valley. The study area boasts abundant industrial resources, featuring two major industrial parks. It enjoys unique port advantages with multiple open platforms. Additionally, the area is home to several universities and hospitals. Notably, there are three wastewater treatment plants situated near the right ridge.

3. Materials and Methods

3.1. Field Sampling and Laboratory Analysis

A total of 31 groundwater samples were collected from wells at depths of 20–50 m across the study area. The sampling aquifer in the valley was the weathered fissure aquifer of the bedrock in red beds, while the ridge groundwater was taken from the karst aquifer of carbonate rocks. Before sampling, each well was purged by continuous pumping for 10–15 min to ensure the collection of fresh groundwater. Temperature, pH, and electrical conductivity (EC) were measured using a portable multi-parameter water quality analyzer

(Multi3 630 IDS). Groundwater samples were filtered through 0.22 μm membrane filters and stored in 1 L high-density polyethylene (HDPE) bottles pre-rinsed three times with deionized water. Immediately after filtration, concentrated nitric acid (HNO_3, 1:1) was added dropwise until the sample pH was below 2 to preserve the samples. Samples for NO_3^- dual isotopes ($\delta^{15}N–NO_3^-$ and $\delta^{18}O–NO_3^-$) were collected in 500 mL HDPE bottles, also pre-rinsed with deionized water. All samples were sealed with paraffin wax immediately after collection and stored in a portable insulated cooler at 0–4 °C. Samples were then transported to the Sichuan Geological Survey Institute for laboratory analysis.

Hydrochemical parameters were analyzed: the total hardness (TH) was detected by the EDTA volumetric method, and the total dissolved solids (TDS) was detected by the gravimetric method. Major cations (K^+, Na^+, Ca^{2+}, and Mg^{2+}) were analyzed by the inductively coupled plasma atomic emission spectrometry method (ICP-AES). Cl^- and SO_4^{2-} were detected by the ion chromatography method. HCO_3^- was measured by the HCl titration method. NO_3^- was estimated by disulfophenol spectrophotometry. F^- was determined by the ion-selective electrode method. The dual isotopes of NO_3^- ($\delta^{15}N–NO_3^-$ and $\delta^{18}O–NO_3^-$) were analyzed using isotope ratio mass spectrometry (IRMS), with an analytical precision of ±0.5‰. To ensure analytical reliability, 20% of the total samples were randomly selected for duplicate analysis as part of the quality assurance (QA) protocol. Certified reference materials (CRMs) and blank samples were included in each batch for quality control (QC).

Cation balance error (*CBE*) calculations showed that all samples had errors within ±5%, indicating acceptable ionic balance and overall data reliability (Equation (1)).

$$CBE = \frac{|cation - anion|}{cation + anion} \tag{1}$$

3.2. Entropy Water Quality Index (EWQI)

The entropy water quality index (EWQI) is a multi-parameter comprehensive evaluation method based on information entropy theory. Compared with conventional evaluation methods, the EWQI eliminates the subjectivity of manually assigned weights. As a result, it provides a more objective, systematic, and comprehensive assessment of water quality status. The calculation procedure of the EWQI involves the following steps:

(1) Scale the hydrochemical data (Equations (2) and (3)).

$$X = \begin{bmatrix} x_{11} & x_{12} & \cdots & x_{1n} \\ x_{21} & x_{22} & \cdots & x_{2n} \\ \vdots & \vdots & \ddots & \vdots \\ x_{m1} & x_{m2} & \cdots & x_{mn} \end{bmatrix} \tag{2}$$

$$Y = \begin{cases} \frac{x_{ij} - (x_{ij})_{min}}{(x_{ij})_{max} - (x_{ij})_{min}} + 0.0001 \\ \frac{(x_{ij})_{max} - x_{ij}}{(x_{ij})_{max} - (x_{ij})_{min}} + 0.0001 \end{cases} \tag{3}$$

(2) Calculate the weight of each hydrochemical parameter (Equations (4) and (5)).

$$P_{ij} = \frac{y_{ij}}{\sum_{i=1}^{m} y_{ij}} \tag{4}$$

$$e_j = -\frac{1}{lnm} \sum_{i=1}^{m} P_{ij} ln P_{ij} \tag{5}$$

$$w_j = \frac{1 - e_j}{\sum_{j=1}^{n} 1 - e_j} \tag{6}$$

(3) Determine the quantitative rating scale q (Equations (7) and (8)).

$$q_j = \frac{C_j}{S_j} \times 100 \tag{7}$$

$$q_{pH} = \begin{cases} \frac{C_{pH}-7}{8.5-7} \times 100 & C_{pH} > 7 \\ \frac{7-C_{pH}}{7-6.5} \times 100 & C_{pH} < 7 \end{cases} \tag{8}$$

(4) Calculate the EWQI value based on q and w (Equation (9)).

$$\text{EWQI} = \sum_{j=1}^{n} (w_j \times q_j) \tag{9}$$

4. Results and Discussion

4.1. Hydrochemical Characteristics of Hydrochemical Parameters

Statistical analysis and box plots of hydrochemical parameters in groundwater are summarized in Table 1 and Figure 2. The pH ranged from 6.69 to 8.20, with a mean value of 7.63, indicating that the groundwater was neutral to slightly alkaline. Total hardness (TH) varied significantly, ranging from 36.03 mg/L to 733.16 mg/L, with a mean concentration of 318.56 mg/L and a standard deviation (SD) of 118.06 mg/L. Total dissolved solids (TDS) ranged from 125.00 mg/L to 965.55 mg/L, with a mean of 461.54 mg/L and a standard deviation of 176.54 mg/L, classifying the groundwater as fresh water and slightly hard water. Piper diagrams are widely used for the classification of hydrochemical facies [60]. Most groundwater samples were categorized as Ca-HCO_3 type, while the remaining samples were classified as mixed Na·Ca-HCO_3 type and Na-HCO_3 type (Figure 3). This suggested the groundwater was undergoing a hydrochemical evolution from Ca–HCO_3 type to Na–HCO_3 type.

Table 1. Statistical analysis of hydrochemical parameters in 31 groundwater samples (units of all parameters are mg/L, except pH).

Parameters	Min	Max	Mean	SD	CV (%)	Limit	% of SEL
pH	6.69	8.20	7.63	0.36	5.00	6.5–8.5 [a]	0
TH	36.03	733.16	318.56	118.06	37.00	450 [a]	3.22
TDS	125.00	965.55	461.54	176.54	38.00	1000 [a]	0
K^+	0.79	11.73	3.59	2.40	67.00	-	-
Na^+	0.91	225.98	54.73	63.74	116.00	200 [a]	3.22
Ca^{2+}	9.85	251.55	93.47	46.46	50.00	400 [b]	0
Mg^{2+}	0.75	31.77	13.60	8.30	61.00	50 [b]	0
Cl^-	2.37	261	42.12	53.50	127.00	250 [a]	3.22
SO_4^{2-}	2.90	400.77	84.09	66.42	79.00	250 [a]	3.22
HCO_3^-	32.94	557.57	303.18	91.24	30.00	-	-
NO_3^-	0.01	66.08	13.56	16.46	121.00	20 [a]	19.35
F^-	0.08	3.69	0.61	0.73	119.00	1 [a]	16.13

[a] Standard for groundwater quality (GB/T 14848—2017). [b] Standards for drinking water quality (GB 5749—2022).

The mean concentrations of cations followed the descending order: Ca^{2+} (93.47 mg/L) > Na^+ (54.73 mg/L) > Mg^{2+} (13.60 mg/L) > K^+ (3.59 mg/L), while the anion concentrations were ranked as: HCO_3^- (303.18 mg/L) > SO_4^{2-} (84.09 mg/L) > Cl^- (42.12 mg/L) >NO_3^- (13.56 mg/L) > F^- (0.61 mg/L). According to Chinese drinking water standards [61], 3.22% of groundwater exceeded the permission limit for Na^+, SO_4^{2-}, and Cl^-, indicating minor contamination from these ions. Notably, 19.35% and 16.13% of groundwater exceeded the permission limit for NO_3^- and F^-, implying potential pollution from NO_3^- and F^-. The coefficient of variation (CV) was used to assess the degree of dispersion of hydrochemical parameters. The CV of Cl^-, NO_3^-, F^-, and Na^+ exceeded 100%, reflecting significant variability and suggesting strong anthropogenic or environmental influences on their distribution.

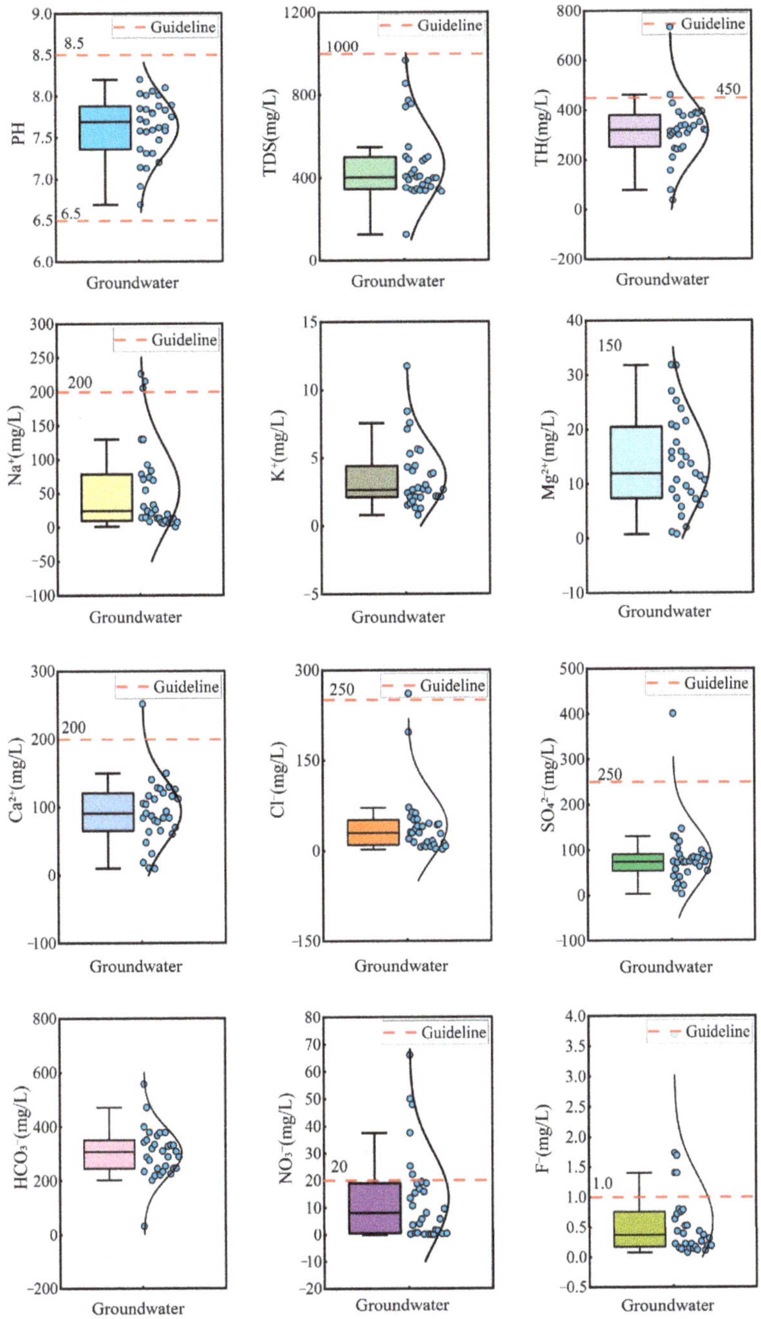

Figure 2. Box plots of hydrochemical parameters in groundwater.

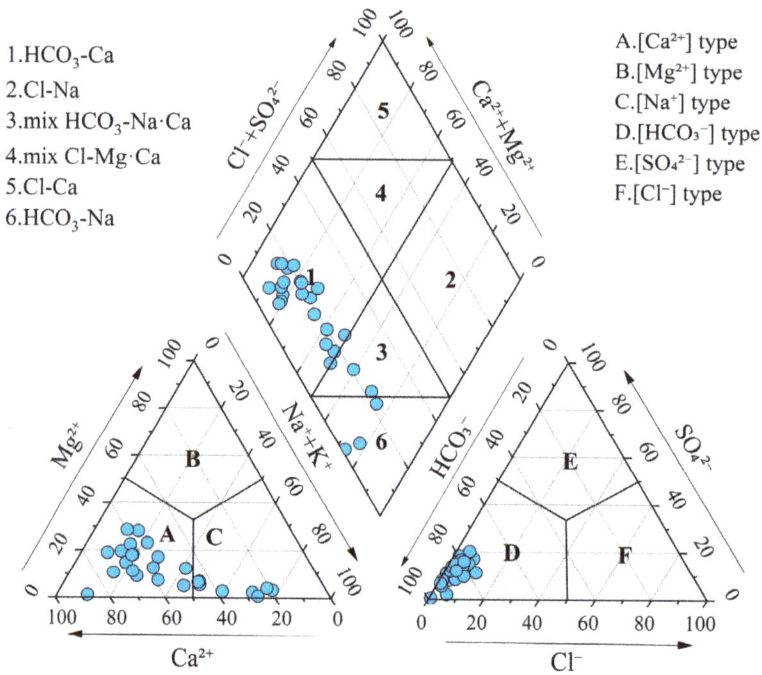

Figure 3. Piper diagram for hydrochemical types of groundwater.

4.2. Spatial Distribution Characteristics of Groundwater

The spatial distribution characteristics of hydrochemical parameters were systematically analyzed through interpolation mapping (Figure 4). Notably, Na^+ and Cl^- displayed strongly correlated distribution patterns, with their elevated concentrations predominantly clustered in the central valley area (Figure 4b,e), suggesting the same sources. In contrast, Ca^{2+}, Mg^{2+}, and $SO_4{}^{2-}$ exhibited remarkably similar spatial distributions, with their peak concentrations concentrated along the proper ridge (Figure 4c–f), indicating the same sources. The spatial distribution of $HCO_3{}^-$ showed a contrasting pattern, with elevated concentrations localized in the western left ridge region (Figure 4g). Of particular environmental concern, $NO_3{}^-$ contamination hotspots were identified in the northern sector of the proper ridge (Figure 4h). Meanwhile, elevated F^- concentrations formed a discrete anomaly in the north of the valley (Figure 4i). Notably, the spatial concentration distribution of Ca^{2+} was highly similar to that of TDS and TH, indicating that Ca^{2+} was the dominant ion controlling TDS and TH.

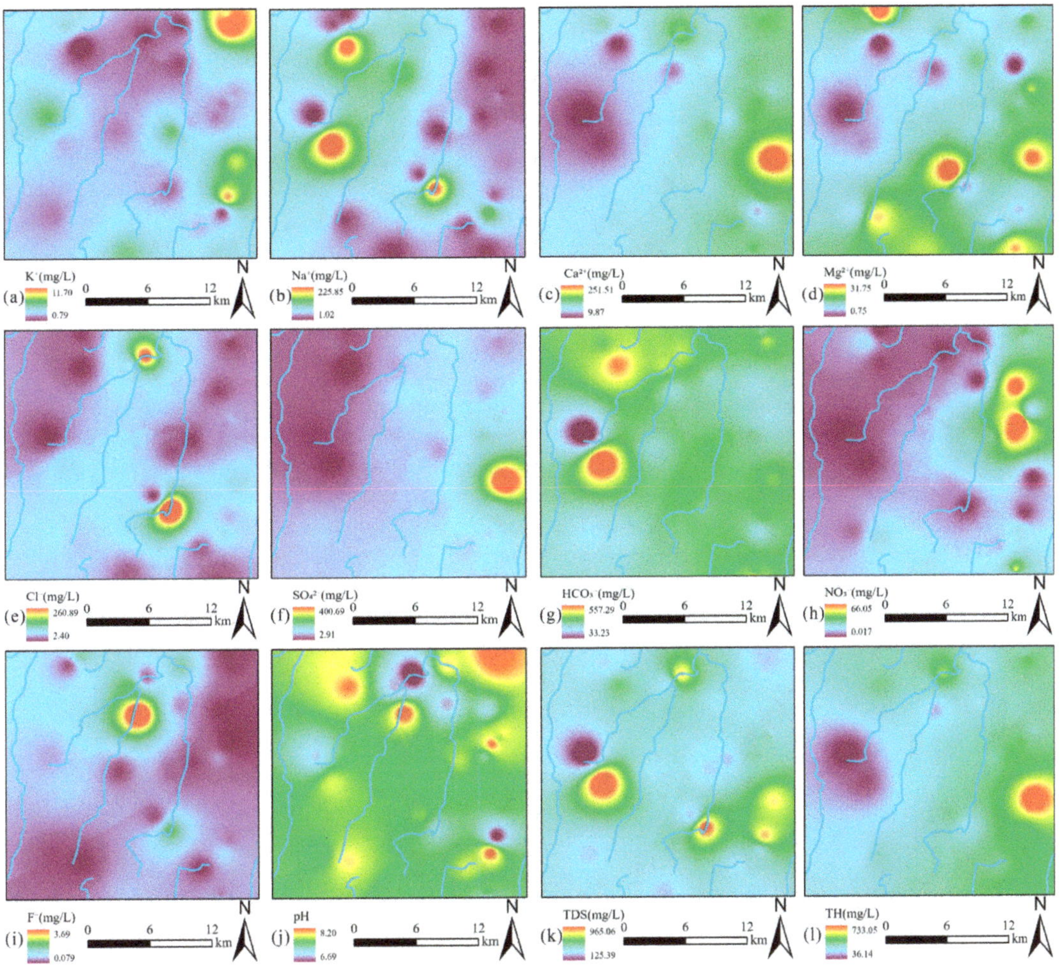

Figure 4. Spatial distribution maps of hydrochemical components in groundwater: (**a**) K^+; (**b**) Na^+; (**c**) Ca^{2+}; (**d**) Mg^{2+}; (**e**) Cl^-; (**f**) $SO_4{}^{2-}$; (**g**) $HCO_3{}^-$; (**h**) $NO_3{}^-$; (**i**) F^-; (**j**) pH; (**k**) TDS; (**l**) TH.

4.3. Natural Factors Affecting Groundwater Hydrochemistry

Groundwater chemistry is primarily governed by three natural processes: evaporation, rock weathering, and precipitation [62]. Gibbs diagrams are widely used to evaluate the dominant mechanism influencing the hydrochemical composition of groundwater [63]. All groundwater samples exhibited low TDS values (ranging from 100 to 1000 mg/L) and relatively low $Na^+/(Na^+ + Ca^{2+})$ and $Cl^-/(Cl^- + HCO_3{}^-)$ ratios (Figure 5). These patterns suggest that the hydrochemical characteristics of groundwater were predominantly controlled by water–rock interactions rather than evaporation or atmospheric precipitation inputs. Groundwater is involved in various geological processes and ecological-environmental processes. The interaction between groundwater and rock is the driving force for the evolution of the near-surface environment [64]. The migration of groundwater solutes occurring in the water–rock system greatly affects the evolution of the groundwater environment [65].

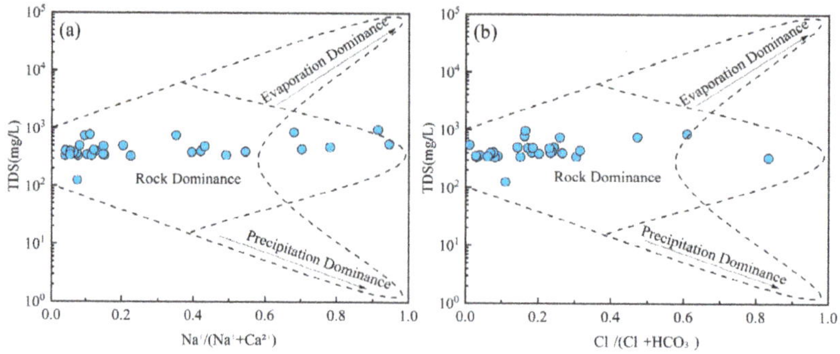

Figure 5. Gibbs diagrams of groundwater. (**a**) TDS vs. $Na^+/(Na^+ + Ca^{2+})$; (**b**) TDS vs. $Cl^-/(Cl^- + HCO_3^-)$.

To further elucidate the role of lithological influences on groundwater chemistry, Gaillardet diagrams were used to distinguish the contribution of different rock types (Figure 6). The majority of groundwater samples were situated around the carbonate end-member, indicating that carbonate rock dissolution was the primary mechanism controlling the chemical composition of groundwater in the region. A smaller number of samples aligned with the silicate weathering end-member, suggesting that silicate weathering also contributed to groundwater chemistry.

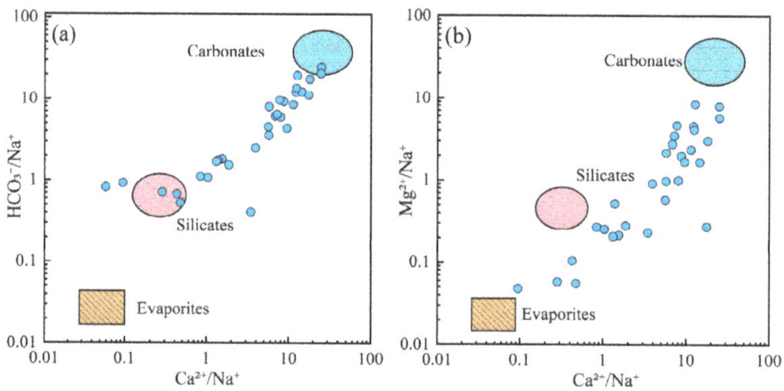

Figure 6. Gaillardet diagrams of groundwater. (**a**) HCO_3^-/Na^+ vs. Ca^{2+}/Na^+; (**b**) Mg^{2+}/Na^+ vs. Ca^{2+}/Na^+.

Previous studies have demonstrated that ionic ratios in groundwater are effective tools for identifying water–rock interactions within aquifer systems [66,67]. By analyzing various ionic ratios, it is possible to determine the sources of hydrochemical compositions, mineral weathering processes, and the influence of exogenous contamination.

The correlation between Na^+ and Cl^- is commonly used to identify the influence of halite dissolution on hydrochemical composition [68]. When Na^+ and Cl^- originate from halite dissolution, the Cl^-/Na^+ molar ratio equals 1, as represented by Equation (10). As shown in Figure 7a, some water samples exhibited Cl^-/Na^+ ratios near 1, indicating that Na^+ was likely derived from halite dissolving. Moreover, some samples were located away from this line, meaning that human activities, silicate weathering, and cation exchange processes need to be further considered.

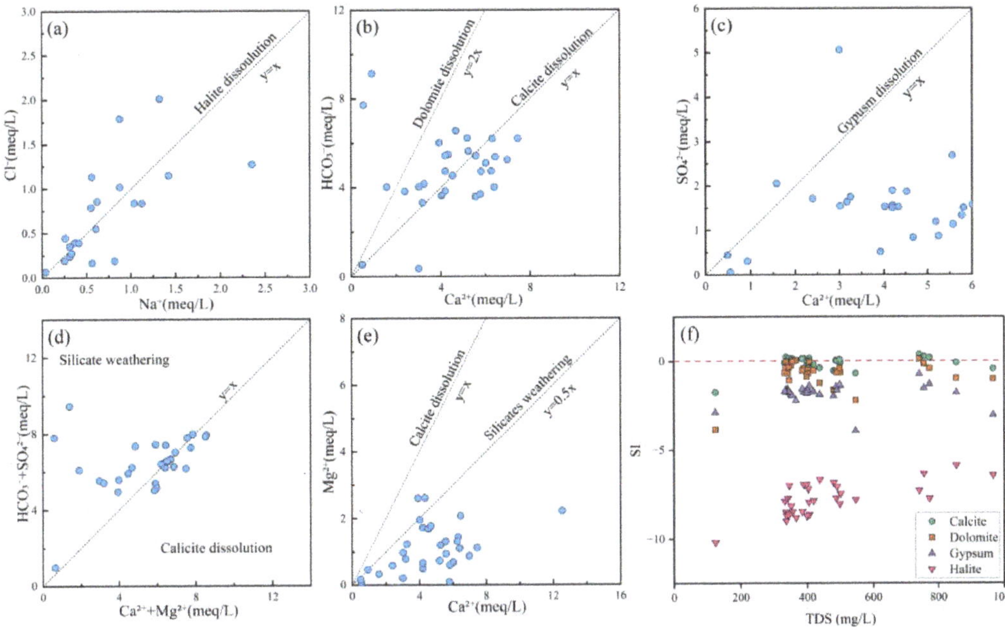

Figure 7. Ion ratio plots and SI vs. TDS plots of groundwater. (**a**) Cl^- vs. Na^+; (**b**) HCO_3^- vs. Ca^{2+}; (**c**) SO_4^{2-} vs. Ca^{2+}; (**d**) $HCO_3^- + SO_4^{2-}$ vs. $Ca^{2+} + Mg^{2+}$; (**e**) Mg^{2+} vs. Ca^{2+}; (**f**) SI vs. TDS.

The correlation between Ca^{2+} and HCO_3^- reflects dissolution processes of carbonate minerals such as calcite ($CaCO_3$) and dolomite ($CaMg(CO_3)_2$) [69], as represented by Equations (11) and (12). As illustrated in Figure 7b, most water samples were plotted on both sides of the calcite dissolution line, indicating that the hydrochemical composition of groundwater in the study area was predominantly controlled by carbonate rock dissolution. Furthermore, a few samples clustered near the dolomite dissolution line, suggesting a localized influence from dolomite dissolution in certain zones.

The Ca^{2+}/SO_4^{2-} ratio serves as an effective indicator for assessing the dissolution contribution of gypsum ($CaSO_4{\cdot}2H_2O$) [23], as expressed in Equation (13). As demonstrated in Figure 7c, most groundwater samples in the study area were plotted below the gypsum dissolution line, indicating limited hydrochemical contribution from sulfate mineral dissolution. However, a distinct subset of samples exhibited elevated SO_4^{2-} concentrations, potentially attributable to anthropogenic inputs from sulfur-containing industrial wastewater and localized weathering of sulfate-bearing minerals.

The $(HCO_3^- + SO_4^{2-})/(Ca^{2+} + Mg^{2+})$ ratio provides diagnostic insights for distinguishing between calcite dissolution and silicate weathering processes in water samples [70]. As evidenced in Figure 7d, most samples in the study area clustered along both sides of the calcite dissolution line, demonstrating that Ca^{2+} and Mg^{2+} primarily originated from carbonate mineral dissolution. Notably, a minor proportion of samples were plotted within the silicate weathering endmember zone, suggesting potential influence from weathering of silicate minerals (e.g., feldspars or pyroxenes) in localized areas.

The calcium–magnesium coefficient (Ca^{2+}/Mg^{2+} ratio) was a robust hydrogeochemical indicator for evaluating water–rock interaction intensity and discriminating ion sources between carbonate and silicate mineral origins [71]. Most samples were below the line of 1:1, indicating that carbonate and silicate mineral dissolution contributed to Ca^{2+} and Mg^{2+}.

$$NaCl \rightarrow Na^{+} + Cl^{-} \tag{10}$$

$$CaCO_3(Calcite) + H_2CO_3 \rightarrow Ca^{2+} + 2HCO_3^{-} \tag{11}$$

$$CaMg(CO_3)_2(Dolomite) + 2H_2CO_3 \rightarrow Ca^{2+} + Mg^{2+} + 4HCO_3^{-} \tag{12}$$

$$CaSO_4{\cdot}2H_2O\ (Gypsum) \rightarrow Ca^{2+} + SO_4^{2-} + 2H_2O \tag{13}$$

The Saturation Index (SI) serves as a fundamental geochemical parameter for evaluating the dissolution–precipitation equilibrium of specific minerals in groundwater and surface water systems, while simultaneously revealing potential sources of dissolved ions [72]. This study employed the PHREEQC 3.7.3 software package to systematically calculate and analyze mineral saturation indices (SI) for groundwater samples across the study area. The thermodynamic calculations for key minerals (dolomite, calcite, gypsum, and halite) were performed according to Equation (14), which rigorously accounted for aqueous speciation and mineral solubility equilibria.

$$SI = \log \frac{IAP}{K_{sp}} \tag{14}$$

where IAP denotes the ion activity product in aqueous solutions, while K_{sp} represents the temperature-specific solubility product constant of minerals.

An SI value below 0 indicated that the mineral was undersaturated, $SI = 0$ suggested a relatively equilibrium state, and an SI value above 0 signified that the mineral was in a supersaturated state [72]. In Figure 7f, the SI values of calcite ($CaCO_3$) and dolomite ($CaMg(CO_3)_2$) were near 0, indicating that the dissolution of carbonate rocks was in equilibrium. In contrast, the SI values (<0) of gypsum ($CaSO_4{\cdot}2H_2O$) and halite (NaCl) exhibited an undersaturated state. Therefore, the dissolution of sulfates and halite constituted the primary origin of the related ions, while the dissolution of carbonate rocks also contributed significantly to the relevant chemical composition.

Cation exchange often leads to variations in Na^+, K^+, Ca^{2+}, and Mg^{2+} concentrations in groundwater, causing its chemical composition to deviate from the mineral dissolution-precipitation equilibrium state. The correlation between (Na^+ + K^+)–Cl^- and (Ca^{2+} + Mg^{2+})–(HCO_3^- + SO_4^{2-}) could reflect cation exchange processes in groundwater [73]. Additionally, the Chloro-Alkaline Indices (CAI-I, Equation (7); CAI-II, Equation (8)) could determine whether cation exchange is direct or reverse, verifying the cation exchange processes occurring in the water [74]. Figure 8a,b show that most groundwater samples in the study area were distributed uniformly along the cation exchange line, indicating that the groundwater chemistry underwent direct cation exchange with clay minerals in the surrounding aquifer. This process resulted in the replacement of Ca^{2+} and Mg^{2+} by Na^+ in the groundwater.

$$CAI - I = \frac{Cl^{-} - (Na^{+} + K^{+})}{Cl^{-}} \tag{15}$$

$$CAI - II = \frac{Cl^{-} - (Na^{+} + K^{+})}{HCO_3^{-} + SO_4^{2-} + CO_3^{2-} + NO_3^{-}} \tag{16}$$

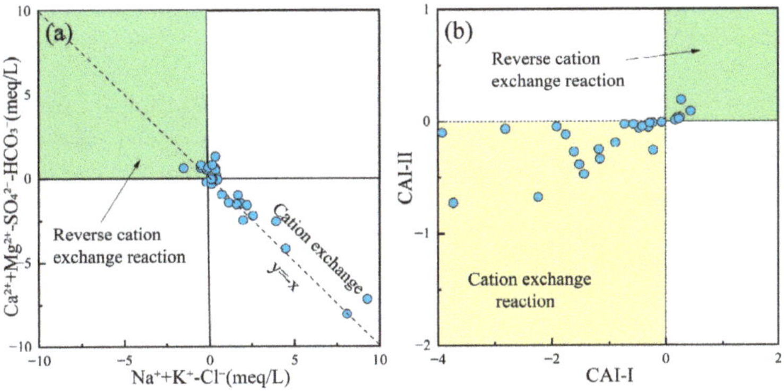

Figure 8. Cation exchange reaction plots of groundwater. (**a**) Ca^{2+} + Mg^{2+}-SO_4^{2-}-HCO_3^- vs. Na^+ + K^+-Cl^-; (**b**) CAI-II vs. CAI-I.

4.4. Anthropogenic Factors Affecting Groundwater Hydrochemistry

As discussed above, NO_3^- was identified as the primary pollutant in the groundwater of the study area. Given the relative stability and conservative behavior of Cl^- in groundwater, it is commonly used as a tracer for identifying the sources of NO_3^- contamination [75]. The relationship plot of NO_3^-/Na^+ and Cl^-/Na^+ revealed that agricultural activities and municipal sewage contributed significantly to the NO_3^- levels in groundwater (Figure 9a). The NO_3^-/Cl^- ratio typically ranges from 0.05 to 0.22 in natural water. Ratios exceeding this range generally indicate anthropogenic influences, including inputs from agricultural fertilizers, domestic wastewater, and denitrification processes. In this study, NO_3^-/Cl^- ratios varied widely, from 2.72×10^{-5} to 2.97, far exceeding the natural baseline and reflecting human impacts. The relationship between NO_3^-/Cl^- and Cl^- concentrations further supported the conclusion that domestic and municipal waste were the primary sources of NO_3^-, with additional contributions from agricultural activities (Figure 9b).

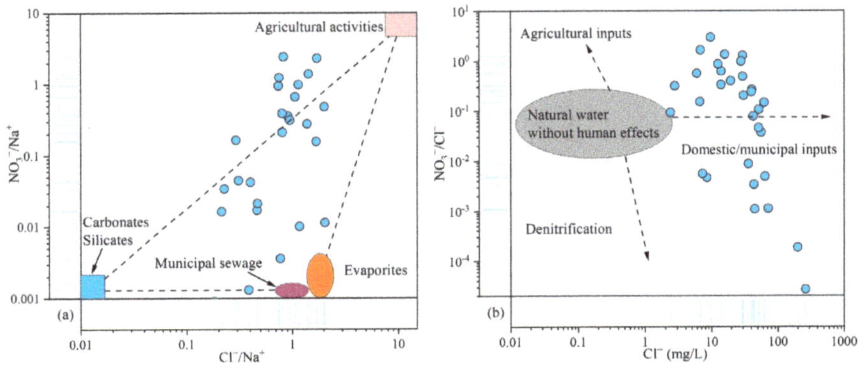

Figure 9. Relationship plots of NO_3^-, Cl^-, and Na^+ in groundwater: (**a**) NO_3^-/Na^+ vs. Cl^-/Na^+; (**b**) NO_3^-/Cl^- vs. Cl^-.

The dual isotopes of NO_3^- ($\delta^{15}N$-NO_3^- and $\delta^{18}O$-NO_3^-) are recognized as conservative tracers that provided valuable insights into the multiple sources of NO_3^- in groundwater [76]. In this study, $\delta^{15}N$-NO_3^- values ranged from −0.52‰ to 8.31‰, with a mean value of 5.07‰, while $\delta^{18}O$-NO_3^- ranged from 0.56‰ to 5.72‰, with a mean of 2.20‰. As illustrated in the $\delta^{15}N$–$\delta^{18}O$ plot (Figure 10), no groundwater samples were located in the region associated with atmospheric precipitation and NO_3^- fertilizers, suggesting that these sources had limited influence on NO_3^- levels. Two samples were plotted in the overlapping area of NH_4^+ fertilizer and soil organic nitrogen, indicating a mixed contribution from both sources. Three samples were positioned in the mixed soil organic nitrogen and sewage/manure region, suggesting combined inputs from natural and anthropogenic nitrogen sources. Furthermore, three groundwater samples fell within the mixed domain of NH_4^+ fertilizer, soil organic nitrogen, and sewage and manure, indicating that all three sources jointly contributed to NO_3^- pollution. Notably, three and two samples were located in the exclusive zones of NH_4^+ fertilizer and sewage and manure, respectively, implying that these sources were singularly responsible for the contamination in those samples. These results underscored the complex and variable origins of NO_3^- in the study area, with contributions from soil organic nitrogen, agricultural practices (e.g., fertilizer application), and anthropogenic waste (e.g., sewage discharge and manure accumulation).

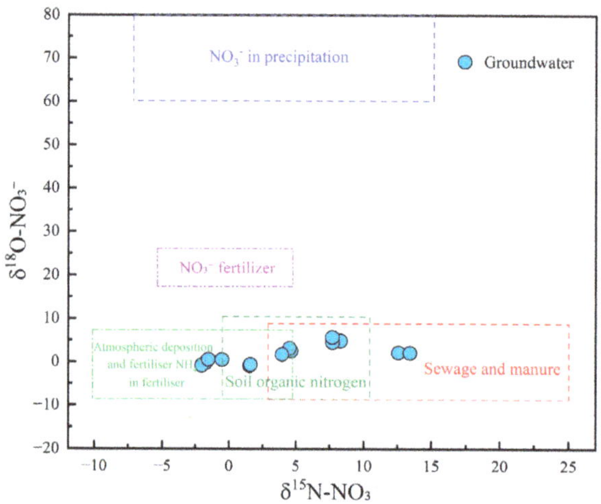

Figure 10. Cross plot of $\delta^{18}O\text{-}NO_3$ and $\delta^{15}N\text{-}NO_3$ in groundwater.

4.5. The Impact of Land Use Types on Nitrate Concentrations

Land use types exerted differential impacts on groundwater quality, particularly concerning NO_3^- concentrations [77]. As shown in Figure 11, red circles indicate groundwater sampling sites where NO_3^- concentrations exceeded the Chinese drinking water guideline (20 mg/L), while green circles represent compliant sites. Elevated NO_3^- concentrations were predominantly distributed in the northern and southern parts of the proper ridge, where land use was primarily characterized by cropland and artificial surfaces. This spatial pattern aligned well with the dual isotope evidence presented earlier. This supports the conclusion that NO_3^- in groundwater mainly originated from the mineralization of soil organic nitrogen, NH_4^+ fertilizer, and sewage and manure infiltration. In cropland areas, nitrogen fertilizers dissolve in water and are subsequently absorbed by crops in the form of NH_4^+ in the soil. Excess nitrogen not utilized by plants could be oxidized to NO_3^- and leached into the groundwater system. In rural settlements, direct discharge of domestic sewage and improper livestock manure storage could further exacerbate NO_3^- pollution through surface runoff and infiltration. Moreover, three wastewater treatment plants are situated in the study area. During heavy rainfall, substantial stormwater infiltrates the sewer system, combining with wastewater and exceeding the treatment capacity of the wastewater treatment plants. This hydraulic overload leads to frequent combined sewer overflows, indicating insufficient infrastructure development, particularly in the sewage pipe network and related facilities. In contrast, lower NO_3^- concentrations were observed in the left ridge, where grassland was the dominant land cover. Field investigations revealed extremely minimal levels of anthropogenic disturbance in grassland areas. This reflected the nitrogen fixation capacity of grassland ecosystems [78]. The spatial correlation analysis between NO_3^- concentrations and land use types revealed a transparent gradient in NO_3^- pollution potential: artificial surfaces > cropland > grassland, underscoring the importance of land management in mitigating NO_3^- contamination risks in groundwater systems.

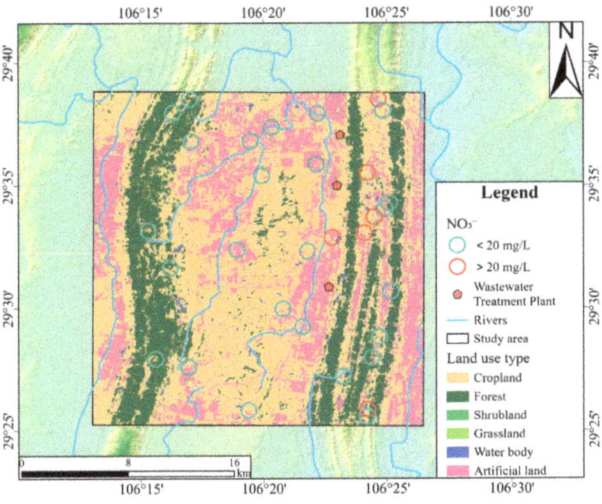

Figure 11. The land use map of the study area and the locations of groundwater samples with NO_3^- concentrations.

4.6. Drinking Water Quality Assessment

In this study, the EWQI was utilized to assess the suitability of groundwater for drinking purposes. Previous research has established that EWQI values below 100 indicate water of acceptable quality for human consumption. The groundwater quality evaluation incorporated 12 hydrochemical parameters: K^+, Na^+, Ca^{2+}, Mg^{2+}, Cl^-, SO_4^{2-}, HCO_3^-, NO_3^-, F^-, pH, TDS, and TH. Among these parameters, NO_3^- (weight = 0.17), F^- (0.17), Cl^- (0.16), and Na^+ (0.15) were assigned higher weighting coefficients, identifying them as the primary factors influencing groundwater quality for drinking purposes (Table 2). The calculated EWQI values across all samples ranged from 11.81 to 87.51, with a mean value of 38.83. Based on the EWQI classification scheme, all sampled groundwater was deemed suitable for drinking. Specifically, 77.42% of samples were classified as excellent quality, while the remaining 22.58% were considered good quality (Figure 12a). These results demonstrate that the studied groundwater sources met established drinking water standards.

Table 2. Relative weights of each hydrochemical parameter.

Parameters	K^+	Na^+	Ca^{2+}	Mg^{2+}	Cl^-	SO_4^{2-}	HCO_3^-	NO_3^-	F^-	TDS	TH	pH
w_j	0.083	0.15	0.045	0.059	0.16	0.063	0.018	0.17	0.17	0.034	0.026	0.024

A Geographic Information System (GIS) approach was employed further to analyze the geospatial distribution of drinking water quality. As illustrated in Figure 12b, elevated EWQI values were observed in the northern and eastern regions, attributable to the influence of adjacent croplands (Figure 12). Although the overall groundwater quality ranges from good to excellent, these spatial patterns suggest that anthropogenic activities, particularly agricultural practices, may contribute to localized water quality degradation. Consequently, targeted monitoring and management strategies should be implemented to mitigate potential contamination risks.

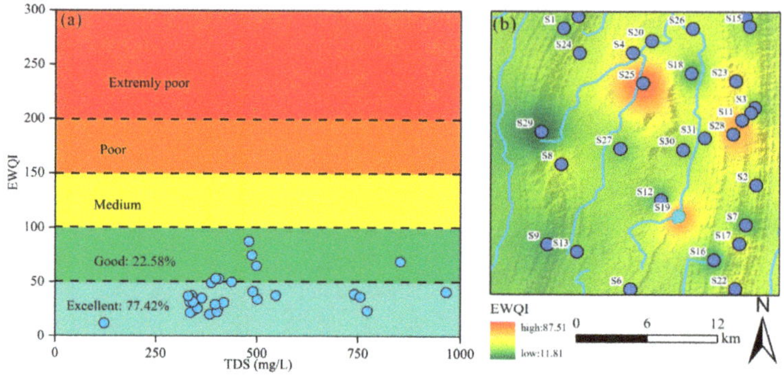

Figure 12. The drinking water quality of groundwater: (**a**) EWQI vs. TDS plot; (**b**) the spatial distribution map of EWQI in study area.

5. Conclusions

In this study, a total of 31 groundwater samples were collected from a typical urban area in southwestern China. This study elucidated the hydrochemical processes, sources of NO_3^- pollution, the impact of land use on NO_3^- levels, and drinking water safety. The key conclusions are summarized as follows:

(1) The groundwater was neutral to slightly alkaline in nature, and the groundwater type was Ca-HCO_3. The cation concentrations exhibited the following order: Ca^{2+} > Na^+ > Mg^{2+} > K^+, while the anion concentrations followed this order: HCO_3^- > SO_4^{2-} > Cl^- >NO_3^- > F^-. The Gibbs diagram, ion ratio, and mineral saturation index results indicated that the dissolution of carbonates and the weathering of silicates were the primary controlling factors for hydrochemical characteristics.

(2) The NO_3^- concentrations ranged from 0.01 mg/L to 66.08 mg/L, with 19.35% of the samples exceeding the Chinese permissible limits. The ion ratio relationship of NO_3^-, Na^+, and Cl^- showed that NO_3^- pollution originated from agricultural activities, sewage discharge, and manure. Further analysis of the dual isotopes $\delta^{15}N$ and $\delta^{18}O$ revealed that the high-concentration NO_3^- pollution in groundwater mainly resulted from NH_4^+ fertilizer, soil organic nitrogen, and domestic sewage and manure.

(3) The land use map revealed that elevated NO_3^- concentrations were found in the agricultural and artificial surface areas near the proper ridge, while low NO_3^- concentrations were found in the grassland areas. Thus, the spatial distribution of NO_3^- correlated strongly with land use types. In agricultural regions, unabsorbed NH_4^+ from nitrogen fertilizer entered the groundwater through precipitation and irrigation water infiltration. In artificial surface areas, the discharge of domestic sewage and the disposal of livestock manure contributed to NO_3^- pollution. In the grassland areas situated on the ridge, the nitrogen fixation capacity of the grassland ecosystem contributed to the low levels of NO_3^- concentration.

(4) The EWQI method indicated that 77.42% of samples were classified as excellent quality, while 22.58% were categorized as good quality. Therefore, all the groundwater was suitable for drinking. The relatively elevated EWQI values, caused by NO_3^- and F^-, were situated in the northern and eastern regions of the study area.

Author Contributions: Conceptualization, Y.Z. and S.C.; methodology, J.D.; software, S.Y.; validation, W.K.; formal analysis, Y.Y.; investigation, X.Z.; resources, Y.L.; data curation, D.F.; writing—original draft preparation, C.Y.; writing—review and editing, S.C.; visualization, Y.W.; supervision, Y.Z.; project administration, S.C.; funding acquisition, S.C. All authors have read and agreed to the published version of the manuscript.

Funding: This research was funded by the Natural Science Foundation of Chongqing (Grant number [CSTB2022NSCQMSX0881], Sichuan Science and Technology Program (Grant numbers [2025ZNSFSC0307] and [2025YFHZ0269]), the scientific program of Yibin City (Grant numbers [YB-SCXY2023020006] and [YBSCXY2023020007]), and the Open Fund of Sichuan Provincial Engineering Research Center of City Solid Waste Energy and Building Materials Conversion and Utilization Technology (Grant numbers [GF2022YB005] and [GF2024YB07]).

Institutional Review Board Statement: Not applicable.

Informed Consent Statement: Not applicable.

Data Availability Statement: The data presented in this study are available upon request from the corresponding author.

Conflicts of Interest: Author Chang Yang was employed by the company Chongqing Huadi Resources Environment Technology Co., Ltd. The remaining authors declare that the research was conducted in the absence of any commercial or financial relationships that could be construed as a potential conflict of interest.

References

1. Scanlon, B.R.; Fakhreddine, S.; Rateb, A.; De Graaf, I.; Famiglietti, J.; Gleeson, T.; Grafton, R.Q.; Jobbagy, E.; Kebede, S.; Kolusu, S.R.; et al. Author Correction: Global Water Resources and the Role of Groundwater in a Resilient Water Future. *Nat. Rev. Earth Environ.* **2023**, *4*, 351. [CrossRef]
2. Burri, N.M.; Weatherl, R.; Moeck, C.; Schirmer, M. A Review of Threats to Groundwater Quality in the Anthropocene. *Sci. Total Environ.* **2019**, *684*, 136–154. [CrossRef] [PubMed]
3. Wei, D.; Yang, S.; Zou, L.; Torres-Martínez, J.A.; Zheng, Y.; Hu, Q.; Zhang, Y. Appraisal of Potential Toxic Elements Pollution, Sources Apportionment, and Health Risks in Groundwater from a Coastal Area of SE China. *J. Environ. Manag.* **2025**, *377*, 124691. [CrossRef]
4. Sforzi, L.; Sarti, C.; Santini, S.; Martellini, T.; Cincinelli, A. Global Status, Risk Assessment, and Knowledge Gaps of Microplastics in Groundwater: A Bibliometric Analysis. *Groundw. Sustain. Dev.* **2024**, *27*, 101375. [CrossRef]
5. Wick, K.; Heumesser, C.; Schmid, E. Groundwater Nitrate Contamination: Factors and Indicators. *J. Environ. Manag.* **2012**, *111*, 178–186. [CrossRef] [PubMed]
6. Mahvi, A.H.; Nouri, J.; Babaei, A.A.; Nabizadeh, R. Agricultural Activities Impact on Groundwater Nitrate Pollution. *Int. J. Environ. Sci. Technol.* **2005**, *2*, 41–47. [CrossRef]
7. McLay, C.D.A.; Dragten, R.; Sparling, G.; Selvarajah, N. Predicting Groundwater Nitrate Concentrations in a Region of Mixed Agricultural Land Use: A Comparison of Three Approaches. *Environ. Pollut.* **2001**, *115*, 191–204. [CrossRef]
8. Selvam, S.; Nath, A.V.; Roy, P.D.; Jesuraja, K.; Muthukumar, P. Evaluation of Groundwater for Nitrate and Fluoride in Alappuzha Region from the Southwestern Coast of India and Associated Health Risks. *Environ. Res.* **2023**, *236*, 116791. [CrossRef]
9. Zhang, X.; Xu, Z.; Sun, X.; Dong, W.; Ballantine, D. Nitrate in Shallow Groundwater in Typical Agricultural and Forest Ecosystems in China, 2004–2010. *J. Environ. Sci.* **2013**, *25*, 1007–1014. [CrossRef]
10. Bijay-Singh; Craswell, E. Fertilizers and Nitrate Pollution of Surface and Ground Water: An Increasingly Pervasive Global Problem. *SN Appl. Sci.* **2021**, *3*, 518. [CrossRef]
11. Peña-Haro, S.; Llopis-Albert, C.; Pulido-Velazquez, M.; Pulido-Velazquez, D. Fertilizer Standards for Controlling Groundwater Nitrate Pollution from Agriculture: El Salobral-Los Llanos Case Study, Spain. *J. Hydrol.* **2010**, *392*, 174–187. [CrossRef]
12. Singh, B.; Singh, Y.; Sekhon, G.S. Fertilizer-N Use Efficiency and Nitrate Pollution of Groundwater in Developing Countries. *J. Contam. Hydrol.* **1995**, *20*, 167–184. [CrossRef]
13. Chaudhary, I.J.; Chauhan, R.; Kale, S.S.; Gosavi, S.; Rathore, D.; Dwivedi, V.; Singh, S.; Yadav, V.K. Groundwater Nitrate Contamination and Its Effect on Human Health: A Review. *Water Conserv. Sci. Eng.* **2025**, *10*, 33. [CrossRef]
14. Rajan, M.; Karunanidhi, D.; Jaya, J.; Preethi, B.; Subramani, T.; Aravinthasamy, P. A Comprehensive Review on Human Health Hazards Due to Groundwater Contamination: A Global Perspective. *Phys. Chem. Earth Parts A/B/C* **2024**, *135*, 103637. [CrossRef]

15. Karunanidhi, D.; Aravinthasamy, P.; Subramani, T.; Kumar, M. Human Health Risks Associated with Multipath Exposure of Groundwater Nitrate and Environmental Friendly Actions for Quality Improvement and Sustainable Management: A Case Study from Texvalley (Tiruppur Region) of India. *Chemosphere* **2021**, *265*, 129083. [CrossRef]
16. Zhang, Y.; Yan, Y.; Yao, R.; Wei, D.; Huang, X.; Luo, M.; Wei, C.; Chen, S.; Yang, C. Natural Background Levels, Source Apportionment and Health Risks of Potentially Toxic Elements in Groundwater of Highly Urbanized Area. *Sci. Total Environ.* **2024**, *935*, 173276. [CrossRef]
17. Wang, Y.; Li, R.; Wu, X.; Yan, Y.; Wei, C.; Luo, M.; Xiao, Y.; Zhang, Y. Evaluation of Groundwater Quality for Drinking and Irrigation Purposes Using GIS-Based IWQI, EWQI and HHR Model. *Water* **2023**, *15*, 2233. [CrossRef]
18. He, S.; Li, P.; Su, F.; Wang, D.; Ren, X. Identification and Apportionment of Shallow Groundwater Nitrate Pollution in Weining Plain, Northwest China, Using Hydrochemical Indices, Nitrate Stable Isotopes, and the New Bayesian Stable Isotope Mixing Model (MixSIAR). *Environ. Pollut.* **2022**, *298*, 118852. [CrossRef]
19. Liu, J.; Peng, Y.; Li, C.; Gao, Z.; Chen, S. Characterization of the Hydrochemistry of Water Resources of the Weibei Plain, Northern China, as Well as an Assessment of the Risk of High Groundwater Nitrate Levels to Human Health. *Environ. Pollut.* **2021**, *268*, 115947. [CrossRef]
20. Cao, M.; Hu, A.; Gad, M.; Adyari, B.; Qin, D.; Zhang, L.; Sun, Q.; Yu, C.-P. Domestic Wastewater Causes Nitrate Pollution in an Agricultural Watershed, China. *Sci. Total Environ.* **2022**, *823*, 153680. [CrossRef]
21. Zhang, Y.; Li, F.; Zhang, Q.; Li, J.; Liu, Q. Tracing Nitrate Pollution Sources and Transformation in Surface- and Ground-Waters Using Environmental Isotopes. *Sci. Total Environ.* **2014**, *490*, 213–222. [CrossRef]
22. Richa, A.; Touil, S.; Fizir, M. Recent Advances in the Source Identification and Remediation Techniques of Nitrate Contaminated Groundwater: A Review. *J. Environ. Manag.* **2022**, *316*, 115265. [CrossRef]
23. Yang, Q.; Wang, L.; Ma, H.; Yu, K.; Martín, J.D. Hydrochemical Characterization and Pollution Sources Identification of Groundwater in Salawusu Aquifer System of Ordos Basin, China. *Environ. Pollut.* **2016**, *216*, 340–349. [CrossRef] [PubMed]
24. Jin, Z.; Pan, Z.; Jin, M.; Li, F.; Wan, Y.; Gu, B. Determination of Nitrate Contamination Sources Using Isotopic and Chemical Indicators in an Agricultural Region in China. *Agric. Ecosyst. Environ.* **2012**, *155*, 78–86. [CrossRef]
25. Jamshidi, A.; Morovati, M.; Golbini Mofrad, M.M.; Panahandeh, M.; Soleimani, H.; Abdolahpour Alamdari, H. Water Quality Evaluation and Non-Cariogenic Risk Assessment of Exposure to Nitrate in Groundwater Resources of Kamyaran, Iran: Spatial Distribution, Monte-Carlo Simulation, and Sensitivity Analysis. *J. Environ. Health Sci. Eng.* **2021**, *19*, 1117–1131. [CrossRef] [PubMed]
26. Adimalla, N.; Qian, H. Groundwater Quality Evaluation Using Water Quality Index (WQI) for Drinking Purposes and Human Health Risk (HHR) Assessment in an Agricultural Region of Nanganur, South India. *Ecotoxicol. Environ. Saf.* **2019**, *176*, 153–161. [CrossRef]
27. Ram, A.; Tiwari, S.K.; Pandey, H.K.; Chaurasia, A.K.; Singh, S.; Singh, Y.V. Groundwater Quality Assessment Using Water Quality Index (WQI) under GIS Framework. *Appl. Water Sci.* **2021**, *11*, 46. [CrossRef]
28. Xie, Z.; Liu, W.; Chen, S.; Yao, R.; Yang, C.; Zhang, X.; Li, J.; Wang, Y.; Zhang, Y. Machine Learning Approaches to Identify Hydrochemical Processes and Predict Drinking Water Quality for Groundwater Environment in a Metropolis. *J. Hydrol. Reg. Stud.* **2025**, *58*, 102227. [CrossRef]
29. Yao, R.; Zhang, Y.; Yan, Y.; Wu, X.; Uddin, M.G.; Wei, D.; Huang, X.; Tang, L. Natural Background Level, Source Apportionment and Health Risk Assessment of Potentially Toxic Elements in Multi-Layer Aquifers of Arid Area in Northwest China. *J. Hazard. Mater.* **2024**, *479*, 135663. [CrossRef]
30. Su, K.; Wang, Q.; Li, L.; Cao, R.; Xi, Y. Water Quality Assessment of Lugu Lake Based on Nemerow Pollution Index Method. *Sci. Rep.* **2022**, *12*, 13613. [CrossRef]
31. Maskooni, E.; Naseri-Rad, M.; Berndtsson, R.; Nakagawa, K. Use of Heavy Metal Content and Modified Water Quality Index to Assess Groundwater Quality in a Semiarid Area. *Water* **2020**, *12*, 1115. [CrossRef]
32. Sadat-Noori, S.M.; Ebrahimi, K.; Liaghat, A.M. Groundwater Quality Assessment Using the Water Quality Index and GIS in Saveh-Nobaran Aquifer, Iran. *Environ. Earth Sci.* **2014**, *71*, 3827–3843. [CrossRef]
33. Boateng, T.K.; Opoku, F.; Acquaah, S.O.; Akoto, O. Groundwater Quality Assessment Using Statistical Approach and Water Quality Index in Ejisu-Juaben Municipality, Ghana. *Environ. Earth Sci.* **2016**, *75*, 489. [CrossRef]
34. Bouteraa, O.; Mebarki, A.; Bouaicha, F.; Nouaceur, Z.; Laignel, B. Groundwater Quality Assessment Using Multivariate Analysis, Geostatistical Modeling, and Water Quality Index (WQI): A Case of Study in the Boumerzoug-El Khroub Valley of Northeast Algeria. *Acta Geochim.* **2019**, *38*, 796–814. [CrossRef]
35. Kumi; Michael; Anku; Wilson, W.; Antwi; Yeboah, B.; Penny; Govender, P. Evaluation of the Suitability of Integrated Bone Char- and Biochar-Treated Groundwater for Drinking Using Single-Factor, Nemerow, and Heavy Metal Pollution Indexes. *Environ. Monit. Assess.* **2023**, *195*, 647. [CrossRef]
36. Chen, R.-H.; Li, F.-P.; Zhang, H.-P.; Jiang, Y.; Mao, L.-C.; Wu, L.-L.; Chen, L. Comparative Analysis of Water Quality and Toxicity Assessment Methods for Urban Highway Runoff. *Sci. Total Environ.* **2016**, *553*, 519–523. [CrossRef]

37. Deng, J.; Yang, G.; Yan, X.; Du, J.; Tang, Q.; Yu, C.; Pu, S. Quality Evaluation and Health Risk Assessment of Karst Groundwater in Southwest China. *Sci. Total Environ.* **2024**, *946*, 174371. [CrossRef] [PubMed]
38. Rupias, O.J.B.; Pereira, S.Y.; De Abreu, A.E.S. Hydrogeochemistry and Groundwater Quality Assessment Using the Water Quality Index and Heavy-Metal Pollution Index in the Alluvial Plain of Atibaia River- Campinas/SP, Brazil. *Groundw. Sustain. Dev.* **2021**, *15*, 100661. [CrossRef]
39. Abdessamed, D.; Jodar-Abellan, A.; Ghoneim, S.S.M.; Almaliki, A.; Hussein, E.E.; Pardo, M.Á. Groundwater Quality Assessment for Sustainable Human Consumption in Arid Areas Based on GIS and Water Quality Index in the Watershed of Ain Sefra (SW of Algeria). *Environ. Earth Sci.* **2023**, *82*, 510. [CrossRef]
40. Seifi, A.; Dehghani, M.; Singh, V.P. Uncertainty Analysis of Water Quality Index (WQI) for Groundwater Quality Evaluation: Application of Monte-Carlo Method for Weight Allocation. *Ecol. Indic.* **2020**, *117*, 106653. [CrossRef]
41. Cui, H.; Duan, L.; Pan, H.; Liu, T. Geochemical Pattern, Quality and Driving Forces of Multi-Layer Groundwater in a High-Capacity Mining Area Basin: A Comprehensive Analysis Based on the Interweaving of Multiple Factors. *J. Hydrol.* **2025**, *660*, 133376. [CrossRef]
42. Choudhary, S.; Subba Rao, N.; Chaudhary, M.; Das, R. Assessing Sources of Groundwater Quality and Health Risks Using Graphical, Multivariate, and Index Techniques from a Part of Rajasthan, India. *Groundw. Sustain. Dev.* **2024**, *27*, 101356. [CrossRef]
43. Du, J.; Jia, C.; Ding, Y.; Yang, X.; Feng, K.; Wei, M. Advancing Wetland Groundwater Pollution Zoning: A Novel Integration of Monte Carlo Health Risk Modeling and Machine Learning. *J. Hazard. Mater.* **2025**, *494*, 138412. [CrossRef]
44. Dogramaci, S.; Skrzypek, G.; Dodson, W.; Grierson, P.F. Stable Isotope and Hydrochemical Evolution of Groundwater in the Semi-Arid Hamersley Basin of Subtropical Northwest Australia. *J. Hydrol.* **2012**, *475*, 281–293. [CrossRef]
45. Ujević Bošnjak, M.; Capak, K.; Jazbec, A.; Casiot, C.; Sipos, L.; Poljak, V.; Dadić, Ž. Hydrochemical Characterization of Arsenic Contaminated Alluvial Aquifers in Eastern Croatia Using Multivariate Statistical Techniques and Arsenic Risk Assessment. *Sci. Total Environ.* **2012**, *420*, 100–110. [CrossRef] [PubMed]
46. Fu, C.; Li, X.; Ma, J.; Liu, L.; Gao, M.; Bai, Z. A Hydrochemistry and Multi-Isotopic Study of Groundwater Origin and Hydrochemical Evolution in the Middle Reaches of the Kuye River Basin. *Appl. Geochem.* **2018**, *98*, 82–93. [CrossRef]
47. Nath, A.V.; Sekar, S.; Roy, P.D.; Kamaraj, J.; Shukla, S.; Khan, R. Drinking and Irrigation Quality and Pollution Assessments of the Groundwater Resources from Alappuzha in Kerala (India) through an Integrated Approach Using WQI and GIS. *J. Geochem. Explor.* **2024**, *258*, 107391. [CrossRef]
48. Krishan, G.; Kumar, M.; Rao, M.S.; Garg, R.; Yadav, B.K.; Kansal, M.L.; Singh, S.; Bradley, A.; Muste, M.; Sharma, L.M. Integrated Approach for the Investigation of Groundwater Quality through Hydrochemistry and Water Quality Index (WQI). *Urban Clim.* **2023**, *47*, 101383. [CrossRef]
49. Yao, R.; Xu, J.; Zhou, Y.; Li, S.; Su, J.; Yan, Y.; Gan, Y.; Luo, M.; Zhang, Y. Hydrochemical Evolution and Assessment of Groundwater Quality in an Intensively Agricultural Area: Case Study of Chengdu Plain, Southwestern China. *Environ. Earth Sci.* **2025**, *84*, 211. [CrossRef]
50. Ahmed, S.; Haque, K.E.; Moniruzzaman, M.; Suborna, M.A.K.; Anonna, T.; Bhuyian, M.A.Q.; Ahsan, M.A.; Khan, A.H.A.N.; Karim, M.M.; Kasem, M.A. Hydrogeochemistry, Water Quality, and Potential Human Health Risk Assessment of Groundwater in a Drought-Prone Area, Bangladesh. *Groundw. Sustain. Dev.* **2025**, *29*, 101450. [CrossRef]
51. Vesković, J.; Deršek-Timotić, I.; Lučić, M.; Miletić, A.; Đolić, M.; Ražić, S.; Onjia, A. Entropy-Weighted Water Quality Index, Hydrogeochemistry, and Monte Carlo Simulation of Source-Specific Health Risks of Groundwater in the Morava River Plain (Serbia). *Mar. Pollut. Bull.* **2024**, *201*, 116277. [CrossRef] [PubMed]
52. Rajan, S.; Nandimandalam, J.R.; Ram, P. Hydrogeochemical Evolution of Spring Water in the Western Lower Himalayas: Seasonal Changes, Quality Assessment, and Health Risks. *Groundw. Sustain. Dev.* **2025**, *29*, 101411. [CrossRef]
53. Nisar, U.B.; Rehman, W.U.; Saleem, S.; Taufail, K.; Rehman, F.U.; Farooq, M.; Ehsan, S.A. Assessment of Water Quality Using Entropy-Weighted Quality Index, Statistical Methods and Electrical Resistivity Tomography, Moti Village, Northern Pakistan. *J. Contam. Hydrol.* **2024**, *264*, 104368. [CrossRef] [PubMed]
54. Liu, M. Response of Groundwater Chemical Characteristics to Land Use Types and Health Risk Assessment of Nitrate in Semi-Arid Areas: A Case Study of Shuangliao City, Northeast China. *Ecotoxicol. Environ. Saf.* **2022**, *236*, 113473. [CrossRef]
55. Wang, S.; Zheng, W.; Currell, M.; Yang, Y.; Zhao, H.; Lv, M. Relationship between Land-Use and Sources and Fate of Nitrate in Groundwater in a Typical Recharge Area of the North China Plain. *Sci. Total Environ.* **2017**, *609*, 607–620. [CrossRef] [PubMed]
56. Choi, W.-J.; Han, G.-H.; Lee, S.-M.; Lee, G.-T.; Yoon, K.-S.; Choi, S.-M.; Ro, H.-M. Impact of Land-Use Types on Nitrate Concentration and δ15N in Unconfined Groundwater in Rural Areas of Korea. *Agric. Ecosyst. Environ.* **2007**, *120*, 259–268. [CrossRef]
57. Iqbal, J.; Su, C.; Abbas, H.; Jiang, J.; Han, Z.; Baloch, M.Y.J.; Xie, X. Prediction of Nitrate Concentration and the Impact of Land Use Types on Groundwater in the Nansi Lake Basin. *J. Hazard. Mater.* **2025**, *487*, 137185. [CrossRef]
58. Zhang, Y.; Dai, Y.; Wang, Y.; Huang, X.; Xiao, Y.; Pei, Q. Hydrochemistry, Quality and Potential Health Risk Appraisal of Nitrate Enriched Groundwater in the Nanchong Area, Southwestern China. *Sci. Total Environ.* **2021**, *784*, 147186. [CrossRef]

59. Cui, R.; Chen, A.; Hu, W.; Fu, B.; Liu, G.; Zhang, D. Appropriate Stoichiometric Ratios of Dissolved Organic Carbon and Nitrate Can Trigger a Transition in Nitrate Removal in Groundwater around Plateau Lakes, Southwest China. *Sci. Total Environ.* **2024**, *916*, 170313. [CrossRef]
60. Piper, A.M. A Graphic Procedure in the Geochemical Interpretation of Water-Analyses. *Eos Trans. Am. Geophys. Union* **1944**, *25*, 914. [CrossRef]
61. *GB/T 14848-2017*; Standard for Groundwater Quality. AQSIQ Standards Press of China: Beijing, China, 2017.
62. Mao, M.; Wang, X.; Zhu, X. Hydrochemical Characteristics and Pollution Source Apportionment of the Groundwater in the East Foothill of the Taihang Mountains, Hebei Province. *Environ. Earth Sci.* **2021**, *80*, 14. [CrossRef]
63. Gibbs, R.J. Mechanisms Controlling World Water Chemistry. *Science* **1970**, *170*, 1088–1090. [CrossRef]
64. Samtio, M.S.; Hakro, A.A.A.D.; Jahangir, T.M.; Mastoi, A.S.; Lanjwani, M.F.; Rajper, R.H.; Lashari, R.A.; Agheem, M.H.; Noonari, M.W. Impact of Rock-Water Interaction on Hydrogeochemical Characteristics of Groundwater: Using Multivariate Statistical, Water Quality Index and Irrigation Indices of Chachro Sub-District, Thar Desert, Sindh, Pakistan. *Groundw. Sustain. Dev.* **2023**, *20*, 100878. [CrossRef]
65. Liu, Z.; Wang, X.; Wan, X.; Jia, S.; Mao, B. Evolution Origin Analysis and Health Risk Assessment of Groundwater Environment in a Typical Mining Area: Insights from Water-Rock Interaction and Anthropogenic Activities. *Environ. Res.* **2024**, *252*, 118792. [CrossRef]
66. Chowdhury, P.; Mukhopadhyay, B.P.; Bera, A. Hydrochemical Assessment of Groundwater Suitability for Irrigation in the North-Eastern Blocks of Purulia District, India Using GIS and AHP Techniques. *Phys. Chem. Earth Parts A/B/C* **2022**, *126*, 103108. [CrossRef]
67. Adjéï Kouacou, B.; Anornu, G.; Adiaffi, B.; Gibrilla, A. Hydrochemical Characteristics and Sources of Groundwater Pollution in Soubré and Gagnoa Counties, Côte d'Ivoire. *Groundw. Sustain. Dev.* **2024**, *26*, 101199. [CrossRef]
68. Yu, H.; Gui, H.; Zhao, H.; Wang, M.; Li, J.; Fang, H.; Jiang, Y.; Zhang, Y. Hydrochemical Characteristics and Water Quality Evaluation of Shallow Groundwater in Suxian Mining Area, Huaibei Coalfield, China. *Int. J. Coal Sci. Technol.* **2020**, *7*, 825–835. [CrossRef]
69. Wang, S.; Chen, J.; Zhang, S.; Zhang, X.; Chen, D.; Zhou, J. Hydrochemical Evolution Characteristics, Controlling Factors, and High Nitrate Hazards of Shallow Groundwater in a Typical Agricultural Area of Nansi Lake Basin, North China. *Environ. Res.* **2023**, *223*, 115430. [CrossRef]
70. Sheng, D.; Meng, X.; Wen, X.; Wu, J.; Yu, H.; Wu, M.; Zhou, T. Hydrochemical Characteristics, Quality and Health Risk Assessment of Nitrate Enriched Coastal Groundwater in Northern China. *J. Clean. Prod.* **2023**, *403*, 136872. [CrossRef]
71. Subba Rao, N.; Das, R.; Sahoo, H.K.; Gugulothu, S. Hydrochemical Characterization and Water Quality Perspectives for Groundwater Management for Urban Development. *Groundw. Sustain. Dev.* **2024**, *24*, 101071. [CrossRef]
72. Gao, Y.; Qian, H.; Ren, W.; Wang, H.; Liu, F.; Yang, F. Hydrogeochemical Characterization and Quality Assessment of Groundwater Based on Integrated-Weight Water Quality Index in a Concentrated Urban Area. *J. Clean. Prod.* **2020**, *260*, 121006. [CrossRef]
73. Fan, W.; Zhou, J.; Zheng, J.; Guo, Y.; Hu, L.; Shan, R. Hydrochemical Characteristics, Control Factors and Health Risk Assessment of Groundwater in Typical Arid Region Hotan Area, Chinese Xinjiang. *Environ. Pollut.* **2024**, *363*, 125301. [CrossRef] [PubMed]
74. Hao, Q.; Li, Y.; Xiao, Y.; Yang, H.; Zhang, Y.; Wang, L.; Liu, K.; Liu, G.; Wang, J.; Hu, W.; et al. Hydrogeochemical Fingerprint, Driving Forces and Spatial Availability of Groundwater in a Coastal Plain, Southeast China. *Urban Clim.* **2023**, *51*, 101611. [CrossRef]
75. Wang, D.; Li, P.; Yang, N.; Yang, C.; Zhou, Y.; Li, J. Distribution, Sources and Main Controlling Factors of Nitrate in a Typical Intensive Agricultural Region, Northwestern China: Vertical Profile Perspectives. *Environ. Res.* **2023**, *237*, 116911. [CrossRef]
76. Ding, K.; Zhang, Y.; Zhang, H.; Yu, C.; Li, X.; Zhang, M.; Zhang, Z.; Yang, Y. Tracing Nitrate Origins and Transformation Processes in Groundwater of the Hohhot Basin's Piedmont Strong Runoff Zone through Dual Isotopes and Hydro-Chemical Analysis. *Sci. Total Environ.* **2024**, *919*, 170799. [CrossRef]
77. Zhang, X.; Gao, S.; Wu, Q.; Li, F.; Wu, P.; Wang, Z.; Wu, J.; Zeng, J. Buffer Zone-Based Trace Elements Indicating the Impact of Human Activities on Karst Urban Groundwater. *Environ. Res.* **2023**, *220*, 115235. [CrossRef]
78. Chen, P.; Ma, J.; Yue, X.; Zeng, H.; Wang, C.; Huang, Q.; Zhou, Y.; Zhang, L. Nitrate Dynamics in Deep Soils of the Loess Plateau: Impact of Different Land Use Types. *Ecol. Indic.* **2025**, *175*, 113578. [CrossRef]

toxics

MDPI

Review

Arsenic Exposure and Neuropsychological Outcomes in Children: A Scoping Review

Leyre Notario-Barandiaran [1,†], Laura M. Compañ-Gabucio [2,3,4,5,*,†], Julia A. Bauer [6], Jesús Vioque [3,4,5], Margaret R. Karagas [1] and Antonio J. Signes-Pastor [3,4,5,*]

1 Department of Epidemiology, Geisel School of Medicine, Dartmouth College, Hanover, NH 03756, USA
2 Departamento de Patología y Cirugía, Universidad Miguel Hernández (UMH), 03550 Alicante, Spain
3 Instituto de Investigación Sanitaria y Biomédica de Alicante (ISABIAL), 03010 Alicante, Spain
4 Unidad de Epidemiología de la Nutrición, Departamento de Salud Pública, Historia de la Ciencia y Ginecología, Universidad Miguel Hernández (UMH), 03550 Alicante, Spain
5 Centro de Investigación Biomédica en Red de Epidemiología y Salud Pública (CIBERESP), Instituto de Salud Carlos III, 28029 Madrid, Spain
6 Division of Epidemiology and Biostatistics, School of Public Health, University of Illinois Chicago, Chicago, IL 60612-4394, USA
* Correspondence: lcompan@umh.es (L.M.C.-G.); asignes@umh.es (A.J.S.-P.); Tel.: +34-96-523-3764 (L.M.C.-G.); +34-96-591-9572 (A.J.S.-P.)
† These authors contributed equally to this work.

Abstract: A child's exposure to arsenic (As) can begin in utero through placental transfer to the fetus. There is a growing body of epidemiologic evidence suggesting an association between As exposure and neuropsychological development. Therefore, our objective was to describe the consequences of maternal and/or childhood As exposure on children's neuropsychological development. We conducted a scoping review with a systematic search of the PubMed, Scopus, EMBASE, Web of Science, and PsycINFO databases. We included studies that assessed the association between maternal and/or childhood As exposure and neuropsychological development in children up to an average of 12 years of age. A total of 77 studies were included, most of which were published between 2020 and 2024 (44.1%), conducted in the United States of America (18.2%) and Bangladesh (16.9%), and involved participants with a median age of 6.6 years. Most studies performed cross-sectional analyses (51.9%) and assessed exposure to elements other than As (64.9%). Childhood was the most frequently studied exposure window (57.2%), and urine was the most commonly used biomarker of exposure (58.4%), followed by blood or serum (32.3%). Cognition was the most frequently evaluated neuropsychological domain (94.8%), followed by psychomotor function (40.3%) and social–emotional function (29.9%). Most studies reported evidence of a negative impact of As exposure on children's neuropsychological development (73.7%), while some found no changes (27.3%) and a few suggested an improvement (1.3%). An important limitation is that most studies measured total urinary As without speciation into inorganic versus organic forms, which limits the validity of dose–response conclusions based on total arsenic concentrations. This review highlights the potential deleterious neuropsychological effects of maternal and/or childhood As exposure while also identifying areas where the evidence remains inconclusive.

Keywords: arsenic; child development; child; review; speciation techniques; neurotoxicity

1. Introduction

Arsenic (As) is a ubiquitous metalloid, and exposure to it poses a significant global health concern, impacting over 200 million people worldwide [1]. Humans are exposed to

As through diverse sources, such as water, air, soil, and food [2]. In populations without occupational exposure to As, water and food are the main sources of exposure [3]. In the environment, As can be found in both organic and inorganic (iAs) forms, with iAs being the main toxic form of As [4]. The inorganic forms of As, including As^{III} and As^{V}, are more readily absorbed by the gastrointestinal tract and have been associated with different health problems, such as cancer, cardiovascular disorders, and neurotoxicity, in the general population [5]. Exposure to As during pregnancy is particularly concerning, as this stage of life represents a uniquely vulnerable period in maternal health and fetal development [6]. Previous research shows that iAs can easily cross the placental barrier and reach fetal tissues, potentially leading to complications during pregnancy and adverse birth outcomes [7], such as spontaneous abortion, preterm birth, fetal death, and low birth weight [8]. While the precise mechanism through which iAs affects health remains unclear, some studies have suggested that iAs can interfere with physiological processes, including altering DNA and inhibiting DNA repair and cell proliferation processes [9]. Effects of prenatal As exposure on child neurological development stand out among the health concerns associated with exposure to As during pregnancy. Previous studies have shown the capacity of As to cross the blood–brain barrier [10], triggering several neurotoxic effects, such as neurotransmitter impairment, oxidative stress, brain cell apoptosis, and epigenetic modifications [11]. During late pregnancy and early childhood, the brain undergoes maximal plasticity and a series of complex events, including neurogenesis, myelination, and synaptic pruning [12,13]. Thus, minimizing iAs exposure during pregnancy is crucial to preserve optimal brain development in young children.

A previous investigation raised the possibility that As exposure during pregnancy adversely impacts child neurodevelopment in highly exposed populations [14]; however, in populations with relatively low exposure levels, studies are scarce, and results are inconsistent. For example, in a cross-sectional study carried out in Spain, a country with relatively low levels of iAs contamination, iAs exposure (average urinary concentration of 4.85 μg/L) at 4–5 years of age was inversely associated with children's neuropsychological outcomes, specifically, fine and gross motor function [15]. Another study carried out in the Health Outcomes and Measures of the Environment (HOME) study found an inverse association between urinary iAs concentrations during pregnancy and children's cognition at 3, 5, and 8 years of age [16]. However, the detrimental effect of iAs exposure on children's neuropsychological development is not well-established, as some studies find no evidence of such an effect [17,18]. Several published reviews on As exposure during pregnancy and/or childhood have been conducted to examine preterm birth [19], the effects on the central nervous system among children exposed to lower As levels [20], the health impacts of drinking As-contaminated water in children [21], and the association between As exposure, adverse pregnancy outcomes, and infant mortality [22]. One review examined the scientific evidence on neurodevelopment and behavioral disorders in children exposed to As, cadmium, and manganese [23]. One systematic review summarizes the literature on factors that modify the associations between lead, methylmercury, manganese, and As and neurodevelopment in children [24], while another review examines epidemiological studies on the effect of chronic As exposure in drinking water on children's intelligence quotient [25]. However, none of the previous reviews cover the broad scope of this scoping review, which considers various periods of As exposure and includes the literature from recent years, during which there has been a considerable increase in research. This rise in research may partly be attributed to the 2019 publication of the Healthy Babies Bright Futures (HBBF) report, which highlighted the widespread presence of toxic metals, including As, in baby foods and their potential impact on neurodevelopment, raising concerns about cognitive deficits and lower IQ in children [26,27].

Due to the disparity of results in studies on pre- and perinatal periods and infant As exposure and child neurodevelopment, as well as the existing knowledge gaps, we conducted a scoping review of epidemiological studies. The aim of this scoping review is to provide a structured overview of the existing scientific evidence addressing the following question: What are the observed neuropsychological consequences of As exposure during the prenatal and perinatal periods and early childhood?

2. Materials and Methods

We conducted a scoping review in accordance with the PRISMA Extension for Scoping Reviews (PRISMA-ScR) [28] (Table S1). We chose this approach because scoping reviews are best suited to addressing broad research questions [29,30]. Although this study is not a systematic review, we adhered to the standards outlined in the Cochrane Handbook Version 6.4, 2023 [31], to ensure content completeness, methodological transparency, and scientific rigor. We have not published a protocol of this review or registered it on PROSPERO or any other similar scientific website.

2.1. Search Strategy

Two of the authors conducted a literature search in five databases on 4 June 2024. We selected these databases based on the results of a previous study [32], which identified an optimal combination of databases for covering the broadest range of published evidence. This combination includes multidisciplinary databases, such as PubMed (MEDLINE), Scopus, EMBASE, and Web of Science, complemented by a specialized database relevant to our research question; in this case, PsycINFO.

The search strategy, developed under the supervision of a research and education librarian from the Biomedical Library at Dartmouth, utilized a combination of terms with the Boolean operators AND and OR, following the PICO structure. Specifically, we used a combination of search terms covering three main components: population (children, infants, prenatal or postnatal periods), exposure (arsenic and its inorganic forms), and outcomes (neuropsychological issues). The full search strategy applied across all five databases is detailed Table S2. The same search strategy was applied across all five databases consulted. We did not apply time restrictions or other filters in any of the databases consulted.

2.2. Eligibility Criteria

Articles had to meet the following inclusion criteria to be included in this scoping review. Language of publication: Spanish and/or English; study design: experimental and/or observational; study population: children aged, on average, $\leq$12 years without neurodevelopmental disorders (Autism Spectrum Disorder (ASD), Attention-Deficit/Hyperactivity Disorder (ADHD), intellectual disability, language disability, communication disorders, neurodevelopmental motor disorders, and specific learning disorders). This age range was selected because it represents a critical developmental phase characterized by rapid physical and psychological transitions between childhood and adolescence [33,34]. By focusing on children aged 12 or younger, we aim to address the whole of childhood [35]. Study exposure variable: exposure to As from diet, water, or industry; study outcomes: neuropsychological development. This concept is broad and has a variety of definitions. To make this scoping review more replicable, we used the Joan Forns et al. [36] definition of neuropsychological development. Based on this definition, neuropsychological development includes three different domains: psychomotor (fine and gross motor abilities), cognition (visuospatial abilities, language and/or communication, learning and memory, attention, and executive function), and socio-emotional (social competence, attachment, adaptive behavior, and emotional competence). Another inclusion criterion was that the studies had to have the full text available. No studies were excluded for this reason, as all articles were obtained through our university library services with their support. We applied and tested all inclusion criteria manually.

2.3. Study Selection and Screening

We carried out study selection and screening manually through Microsoft Excel version 16.0 (Microsoft Excel, Redmond, WA, USA). The titles found in the searches of each of the five databases were downloaded into Microsoft Excel (Redmond, WA, USA). We merged all of the titles into a single Excel sheet to eliminate duplicates. Once all duplicates were removed, we began the screening process. This process was conducted in three phases: title screening, abstract screening, and full text screening. Different Excel sheets were used for each screening phase. All three screening phases were carried out completely, carefully, and independently by two authors, discarding only those articles that we were certain did not meet the inclusion criteria. Discrepancies arising during the study selection process regarding the inclusion or exclusion of an article were resolved by a third author. The study selection flow was displayed graphically using the PRISMA diagram [37].

2.4. Data Extraction and Synthesis

Data were extracted in an Excel database, which was prepared before beginning the bibliographic search to guarantee transparency and avoid information manipulation. While this study adopts a scoping review approach, we adhered to certain guidelines outlined in the Cochrane Handbook [31] for the preparation and layout of tables to ensure comprehensive content coverage. We performed a narrative and graphical synthesis of the extracted information, using tables and figures whenever possible. We completed one table with key general information from the included articles. We elaborated different graphs and figures with R software version 4.0.4 (R Foundation for Statistical Computing, Vienna, Austria). to describe some of the variables in greater depth. On the one hand, we used Sankey plots, an increasingly utilized type of graph in the field of health sciences. These plots enable us to visualize changes over time and the evolution of specific variables in relation to a particular event [38]. This type of graph has been used in previous published review papers [39–42] to present the main characteristics of the articles in a visual and more comprehensible way. Three of the researchers were responsible for extracting data and performing the narrative and graphical synthesis of the information in the results section. The quality of the included studies was not assessed, as this is not required in scoping reviews [28].

3. Results

The initial search across selected databases was enhanced by excluding duplicate records. Subsequently, the remaining records underwent screening, identifying 221 for eligibility assessment. Finally, 77 original studies were included to conduct the current scoping review on evaluating the evidence on As exposure during the pre- and perinatal periods or childhood and its impact on various aspects of neuropsychological development (Figure 1).

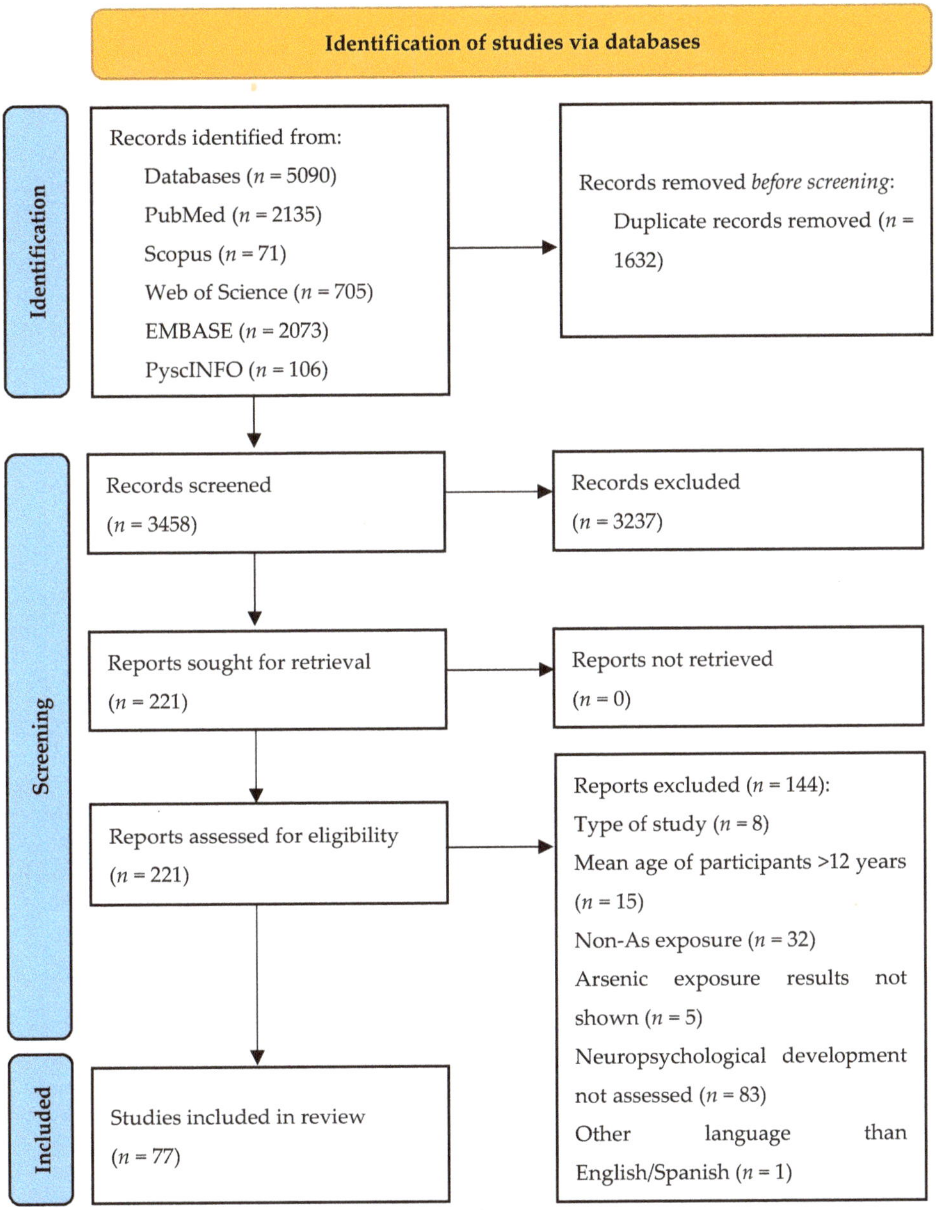

Figure 1. Flowchart of the study selection process.

3.1. Main Characteristics of the Included Studies

The included studies, spanning from 1981 to 2024, feature participants from 16 countries, with the United States of America (USA) [16,33,43–54], Bangladesh [14,17,18,55–64], China [65–74], and Mexico [75–82] being the most prevalent contributors (Figure 2).

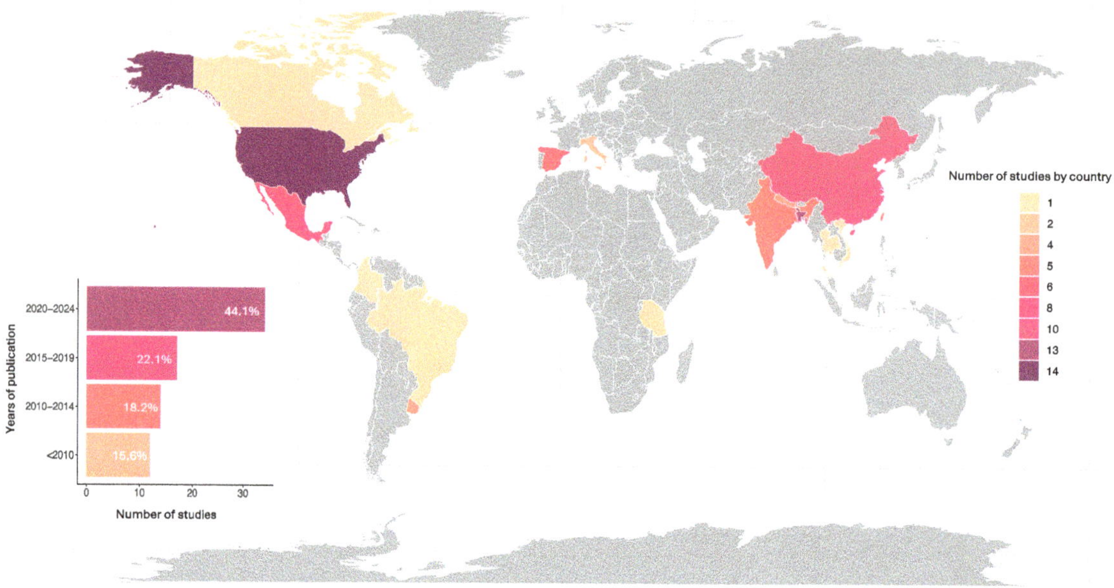

Figure 2. Geographic and temporal distribution of included studies (n = 77).

Notably, 56 (72.7%) studies explicitly focused on As exposure, as indicated by the inclusion of the word "arsenic" or its abbreviation in their stated aims. Among the selected studies, 35 (45.4%) used participants from well-defined study populations and 19 different cohorts, including cVEDA [83], the Healthy Baby Cohort Study in China [70], MABC [67], the Mining and Health prospective longitudinal study [84], the PIPA project [85], PROGRESS [75,77,78], the Taiwan Birth Panel Study (TBPS) [86], the HOME study [16], HEALS [57,59,60,64], INMA [15,87–89], MINIMat [14,17,18,63], MIREC [90], NHBCS [46,49,52], the PRISM pregnancy cohort [48,53], the Sheyang Mini Birth Cohort Study (SMBCS) [71,74], the C8 Health Project [33], the Navajo Birth Cohort Study [76], Project Viva [54], and Environmental Pollution in Young Children [73].

Among the study designs, 40 (51.9%) were cross-sectional, and 37 (48.1%) were longitudinal [14,16–18,46–49,52–57,63,67–71,74–78,84–95] (Table 1). Among longitudinal studies, 16 (43.2%) evaluated childhood exposure [14,16,17,49,52,54,56,57,63,68,71,74,75,88,91,94], 2 (5.4%) assessed perinatal exposure [86,94], and 28 (75.7%) examined prenatal exposure [14,17,18,46–49,52,53,55,63,67–70,74–78,84,85,87,89,90,92,93,95] to As. Among cross-sectional studies, 38 (95.0%) evaluated childhood exposure [15,23,33,43–45,50,51,58–62,64–66,73,79–83,96–111], 1 (2.5%) assessed perinatal exposure [112], and 2 (5.0%) examined prenatal exposure to As [72,110] (Figure 3). All studies included boys and girls, with a median final sample size of 301 participants.

Table 1. Main characteristics and findings of the included studies (n = 77).

Study Design	n (%)
Cross-sectional	40 (51.9)
Longitudinal	37 (48.1)
Number of participants	median (IQR)
	301 (148–439)

Table 1. *Cont.*

Study Design	*n* (%)
Age of participants in months	median (IQR)
Minimum age	60.0 (30.6–72.0)
Median age	78.8 (44.6–81.0)
Maximum age	96 (45.3–131.4)
Evaluated metal exposure	*n* (%)
As	27 (35.1)
As + other metals	50 (64.9)
Exposure window	*n* (%)
Perinatal/prenatal	23 (29.9)
Childhood	44 (57.2)
Both	10 (12.9)
Main source of exposure	*n* (%)
Water	20 (25.9)
Food	21 (27.3)
Food and water	8 (10.4)
Industry	14 (18.2)
Not described	14 (18.2)
Neuropsychological assessment	*n* (%)
Single test	52 (67.5)
Multiple tests	25 (32.5)
Main exposure effects	*n* (%)
Beneficial	1 (1.3)
Harmful	57 (74.0)
Null	21 (27.3)

As, arsenic; IQR, interquartile range; *n*, number.

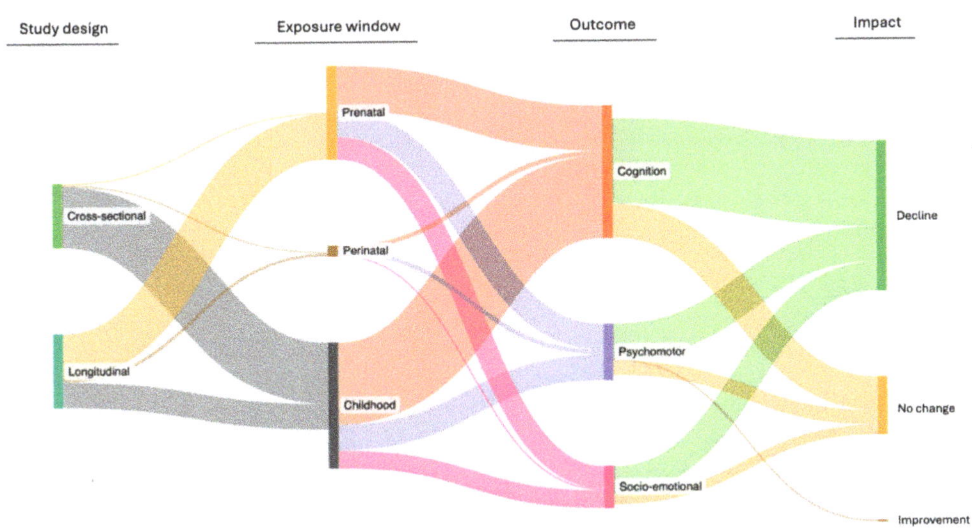

Figure 3. Sankey plot from the included studies in the scoping review ($n = 77$).

3.2. Exposure Assessment and Analytical Techniques

Among the selected studies, 50 (64.9%) studies evaluated elements beyond As, with lead assessed in 42 (54.6%) of the selected studies, cadmium in 26 (33.8%), manganese in 24 (31.7%), mercury in 16 (20.8%), and selenium in 13 (16.8%). Other elements assessed include copper, barium, beryllium, zinc, iodine, chromium, antimony, fluorine, aluminum, cesium, and lithium, each in varying proportions. The most common analytical techniques were inductively coupled plasma mass spectrometry (ICP-MS) and atomic absorption spectrometry (AAS), reported in 55 (71.4%) and 18 (23.4%) studies, respectively. Among the identified studies, 15 (19.5%) reported the use of chromatography equipment to perform speciation analysis [14–17,52,62,70,78,90,96–98,102,107,108], and 3 (20.0%) of them did not provide data on As species concentration [14,17,78].

3.3. Markers of Exposure

Urine was the most commonly used biomarker of internal exposure, appearing in 45 (58.4%) studies [14–18,23,48,52,53,57–66,70,71,73–76,79–84,87,89,90,96–98,102,103,105–109,111]. Blood and serum followed, used in 25 (32.5%) of the included studies [47,54,55,57,59,60,63–65,67–69,72,76–78,85,86,91–93,95,101,106,111]. Other biological matrices included hair [33,44,45,50,51,63,94,99–101,104,106,112] and toenails [46,49,96] or fingernails [94,99,110,112], which were also frequently employed. Several studies additionally analyzed water samples as a proxy for iAs exposure. Less frequently used biomarkers included meconium [94] and the placenta [88]. Notably, 23 (29.9%) studies utilized multiple biomarkers to capture exposure more comprehensively [49,57–66,76,82,94,96,99,101,103–106,111,112] (Figures S1–S3).

Among the studies using urine as a biomarker, 36 (80.0%) evaluated childhood exposure [14–17,23,52,57–66,71,73–75,79–83,96–98,102,103,105–109,111], 15 (33.3%) evaluated prenatal exposure [14,17,18,48,52,53,63,70,74–76,84,87,89,90], and 6 (13.3%) addressed both periods [14,17,52,63,74,75]. Regarding blood/serum biomarker studies, 12 (48.0%) evaluated childhood exposure [54,57,59,60,63–65,68,91,101,106,111], 14 (56.0%) evaluated prenatal exposure [47,55,63,67–69,72,76–78,85,92,93,95], 1 (4.0%) evaluated perinatal exposure [86], and 2 (8.0%) evaluated both prenatal and childhood exposure [63,68]. Of the 16 studies using urinary biomarkers that provided data on As speciation, the most common approach was summing iAs metabolites, with iAs median concentrations ranging from 3.63 μg/L [16] to 55.2 μg/L [79]. In studies reporting total urinary As, average concentrations ranged from 8.3 μg/L [84,106] to 181.9 μg/L [58]. For blood As, average levels ranged from 1.33 μg/L [92] to 9.46 μg/L [85]. Hair As concentrations ranged from 0.19 mg/kg [94] to 2.72 mg/kg [98], while As levels in water ranged from 5.8 μg/L [82] to 857 μg/L [105].

Of the 77 studies reviewed, 14 (18.2%) did not specify a primary As source. Of the 63 that did, 20 (25.9%) examined drinking water, 21 (27.3%) focused on dietary sources, and 8 (10.4%) assessed both pathways. In addition, 14 (18.2%) studies identified mining and industrial emissions as the principal origins of As exposure.

3.4. Neuropsychological Function and Assessment Methods

In this scoping review, we used the conceptual framework developed by Forns et al. [36] to define the different neuropsychological functional domains (cognition, psychomotor, and socio-emotional). Cognition was a major functional domain of investigation (n = 73, 94.8%) (Figures 3 and S1). Within this subset, 44 (60.3%) explored language and communication [14–17,33,43,44,47,51,53–55,57,61–64,67,68,70,71,76,80–82,84–88,91,92,94–98,100,101,103,104,109,110,112], 27 (37.0%) examined executive function [15,18,23,33,43,44,53,57,63,64,67,76–78,80,81,83,87–89,97,98,101,103,104,107,109], 26 (35.6%) investigated visuospatial abilities [14,15,33,43–45,51,54,57,61–64,68,71,80–82,87,88,97,98,101,103,107,109], 24 (32.9%) studies evaluated the subdomain of attention [23,33,44,46,49,53,60,69,72–74,79–81,88,89,93,97,98,101,104,106,109,111], and 22 (30.1%) focused on learning and memory [15,33,43,44,53,54,57,63,64,75,80,81,87,88,97,98,101,103,104,107–109].

The psychomotor functional domain was evaluated in 31 (40.2%) of the included studies (Figures 3 and S2). Within this subset, 27 (87.1%) studies evaluated the subdomain of fine motor skills [15,17,18,44,45,47,52,54,56,59,67,70,76,84–88,91,92,94–97,101,103,112], 23 (74.2%) assessed gross motor abilities [15,17,18,47,59,67,69,70,72,76,84–88,91–97,112], and

1 (3.2%) examined general motor skills [110]. It is worth noting that 20 (64.5%) studies evaluated both fine and gross skills [15,17,18,47,59,67,70,76,84–88,91,92,94–97,112].

The socio-emotional domain was the least studied among the included articles, with only 23 (29.9%) focusing on this area (Figures 3 and S3). Within this subset, 18 (78.26%) studies evaluated the subdomain social competence [33,44,46,49,60,67,73,74,76,84–86,90,106,110,111], 17 (73.9%) explored emotional competence [33,44,46,49,60,69,72–74,79,89,90,106,110,111], and 15 (65.2%) examined adaptive behavior [33,44,46,49,60,73,74,86,90,93,106,110,111]. No studies evaluated attachment.

Among the reviewed studies, a variety of tests were applied, demonstrating a diverse array of assessment methods. The Wechsler Intelligence Scale for Children (WISC) test was the most frequently used, appearing in 19 (24.7%) of the studies [14,16,33,43,44,57,61–64,68,71,80–82,90,103,106,109], followed by the Bayley Scales of Infant Development (BSID), which were used in 14 (18.2%) studies [16–18,55,56,70,91,92,94–96,100,110,112], Raven's Progressive Matrices (RPM) in 6 (7.8%) studies [58,65,66,99,103,105], and the Behavior Assessment System for Children (BASC) in 5 (6.5%) studies [33,44,46,49,90]. Only six tests evaluated all three domains simultaneously: BSID, the Comprehensive Developmental Inventory for Infants and Toddlers (CDIIT), the Malawi Developmental Assessment Tool (MDAT), the Denver Developmental Screening Test II (DDST-II), Neonatal Behavioral Neurological Assessments (NBNA), and the Ages and Stages Questionnaire Inventory (ASQ). Most of these tests are recommended for children over 5 years old, with only a few focusing on early neuropsychological assessment (BSID, ASQ, NBNA, CDIIT, and DDST-II) (Table S3). Multiple tests were applied in 25 (32.5%) studies [16–18,23,33,44,49,51,53,54,58,75,79,80,83,90,101–104,106–109,111].

3.5. *Arsenic Exposure and Neuropsychological Function*

Among the included studies, 57 (74.0%) reported neuropsychological decline, 21 (27.3%) reported no neuropsychological change, and 1 (1.3%) reported neuropsychological improvement.

In the studies evaluating cognition (Figure S1), 52 (71.2%) reported a cognitive decline, 23 (31.5%) were cohort studies, and 29 (39.73%) were cross-sectional, with most focusing on exposure during childhood. Urine was the most commonly used biomarker of As exposure. These adverse effects were predominantly reported in studies conducted in countries like Bangladesh [14,55–58,61–64], Mexico [75–82], China [66–70,72,74], and the USA [16,33,43,44,49,51,53]. Among the 16 studies that included As speciation, 15 (93.7%) evaluated cognition. Of these, 11 (73.3%) reported significant inverse associations between iAs and cognitive performance [15,16,62,70,76,79,80,87,90,96,107], and 4 (26.7%) found no statistically significant relationship [97,98,102,108]. Cognitive decline was primarily observed in the domains of language and communication (n = 33, 45.2%) [14–16,33,43,44,51,53,55,57,61–64,67,68,70,76,80–82,84,85,87,88,94,96,97,101,103,104,110,112], executive function (n = 22, 30.1%) [15,23,33,43,44,53,57,63,64,67,76,78,80,81,83,87,88,97,101,103,104,107], visuospatial skills (n = 21, 28.77%) [14,15,33,43,44,51,57,61–64,68,80–82,87,88,97,101,103,107], attention (n = 18, 24.66%), and learning and memory (n = 18, 24.7%) [15,33,43,44,53,57,63,64,75,80,81,87,88,97,101,103,104,107]. The remaining studies assessing cognition did not report statistically significant associations with As exposure.

In the studies evaluating psychomotor skills (Figure S2), 22 (70.9%) reported a decline, 9 (29.0%) observed no change, and 1 (3.2%) reported an improvement [97]. Among the 22 studies reporting a decline, 7 (31.8%) performed As speciation, confirming that the negative associations were driven specifically by iAs [15,52,70,76,87,96,97]. Of these 22 decline studies, 12 (54.6%) were cohort studies and 10 (45.5%) were cross-sectional in design and primarily focused on childhood exposure, although several also assessed prenatal and perinatal exposure. Urine and blood were the most frequently used biomarkers of As exposure. Among these 22 (70.9%) studies, 18 (81.8%) assessed fine motor abilities [15,44,52,56,59,67,70,76,84,85,87,88,94,96,97,101,103,112] and 16 (72.7%) assessed gross motor abilities [15,59,67,69,70,72,76,84,85,87,88,93,94,96,97,112]. Notably, the single study reporting a beneficial effect examined the association between the organic As species arsenobetaine and fine motor function [97].

Regarding socio-emotional development (Figure S3), 18 (78.3%) studies reported a negative impact, with 6 (26.1%) reporting no changes. Of the 18 studies reporting socio-emotional decline, 10 (55.6%) were cohorts and 8 (44.4%) cross-sectional and focused mainly on prenatal exposure, with urine and blood used as the primary biomarkers of exposure. Of the 18 studies documenting a socio-emotional decline, 3 (16.7%) included As speciation analyses and confirmed that the adverse effects were specifically linked to iAs [76,79,90]. In the studies that reported socio-emotional decline, 14 (77.8%) focused on social competence [33,44,48–50,67,74,76,84,85,90,106,110,111], 11 (61.1%) on adaptive behavior [33,44,48–50,74,90,93,106,110,111], and 13 (72.2%) on emotional competence [33, 44,48–50,69,72,74,79,90,106,110,111].

3.6. Confounding Factors

To account for socioeconomic status, various variables, such as parental education level and family income, were incorporated into 66 (85.7%) studies. Gestational age, maternal age at delivery, or child age at assessment were considered potentially confounding factors in 62 (80.5%) studies. Forty-six studies included child sex as a potential confounding factor. Maternal pre-pregnancy BMI was regarded as a potentially confounding factor in 23 (29.9%) studies. Maternal IQ was considered in 14 (18.2%) studies. Parity was considered in 14 (18.2%) studies. Home Observation for Measurement of the Environment (HOME) was accounted for in 11 (14.3%) studies. Smoking status, including passive smoking, parental smoking, and serum cotinine levels, was identified as a potentially confounding factor in 13 (16.9%) studies [16,46,49,52,65,68,70,71,74,88,98,106]. Children's head circumference was included in 9 (11.7%) studies. Additionally, diet variables, such as calorie-adjusted fish and rice intake and a healthy eating index, were recognized as potentially confounding factors in 10 (13.0%) studies [15,23,46,49,54,74,86,99,101]. We identified two studies that did not adjust for any potentially confounding factors [66,104].

4. Discussion

In this scoping review, we summarize the current state of environmental epidemiologic studies regarding As exposure and neuropsychological impact in children. Most of the evidence, 74% of the included studies, supports a negative association between As exposure and children's neuropsychological development. Except for one study suggesting a positive link [97], the rest found no association [17,18,45–47,54,60,65,71,73,74,86,89,91,92, 95,98,100,102,108,109]. The majority of the studies included in this review were published between 2019 and 2024, the most recent four years, suggesting growing scientific interest in this topic.

The USA, followed by Bangladesh, were the countries where most of the studies included in this review were conducted. For several years, there has been concern regarding exposure to As through diet and water in the USA [113,114]. The USA was included among the countries most severely affected by As contamination in groundwater, being considered one of the high-income countries with the highest As contamination levels in water [115]. However, the concentrations are typically lower than those found in Asian countries, such as Bangladesh, where contamination in drinking water is considered a massive public health issue [116], as aquifers are severely affected and millions of people consume water with high levels of As [117]. Concentrations of As measured in drinking water varied dramatically across the included studies, from 5.8 μg/L in a Mexican cohort [82] to over 857 μg/L in Bihar, India [104], more than 85 times above the World Health Organization (WHO) recommendation of 10 μg/L. In Mexico, water As at 5.8 μg/L was nonetheless linked to lower scores in performance, verbal, and full IQ scores, suggesting that harms may occur even below the 10 μg/L threshold [82]. In Bangladesh and India, where drinking water arsenic frequently exceeded 50 μg/L and in many cases surpassed 100 μg/L, multiple studies reported dose-related declines in verbal IQ, processing speed, and memory scores [58,63,99,104]. All of this evidence may explain why the USA and Bangladesh were the most studied countries in this review and why water was one of the most frequently reported sources of As exposure in the included studies.

Dietary sources of As, particularly rice-based infant cereals and toddler foods, emerged as a second critical exposure pathway. The 2019 HBBF Alliance report first quantified that baby foods with iAs levels of 100–150 μg/kg contributed substantially to an estimated loss of over 11 million IQ points across USA children aged 0–24 months, with rice products alone accounting for roughly 20 percent of total IQ reductions [26]. These concentrations exceed the USA Food and Drug Administration's action level of 100 μg/kg, yet our review found that neurodevelopmental effects were apparent at mean dietary As levels as low as 120 μg/kg. Conversely, studies focusing on arsenobetaine, a non-toxic organic As species prevalent in fish and seafood, reported no adverse cognitive outcomes and even suggested modest cognitive benefits attributable to omega-3 fatty acids [15,97,98].

Urinary total As served as the most widely used biomarker, reflecting combined exposure from water and diet [118,119]. A recent review highlights As in urine as an effective biomarker of dietary and water consumption exposure capable of detecting both inorganic and organic compounds. Consequently, the review concludes that the number of epidemiological studies utilizing urinary analysis to assess As exposure is currently escalating and poised for further expansion in the future [120]. Remarkably, in a USA birth cohort with mean urinary iAs of just 3.6 μg/L, researchers observed significant reductions in processing speed and working memory [16]. In populations with urinary As exceeding the reference value of <50 μg/L, which is considered the upper limit of acceptable exposure in non-occupationally exposed populations, the associations with decreased IQ, attention deficits, and impaired school readiness were both consistent and robust [58,79]. It should be noted that the <50 μg/L urinary As value is a commonly used reference level but not an official regulatory limit for non-occupational exposure. In contrast, cohorts with urinary arsenic under 10 μg/L generally showed no significant neuropsychological impairments, hinting at a possible threshold [60,109]; however, inter-study variability in cognitive assessments and confounding adjustment underscores that this potential cutoff requires confirmation.

Another key point discussed in the aforementioned review is the importance of As speciation in urine. As speciation is crucial due to the significant differences in toxicity among As species. Generally, inorganic As species are highly toxic, whereas organic As species are excreted unchanged in urine and therefore considered non-toxic [121]. Consequently, using urinary As for speciation analysis is regarded as a reference method for determining the specific As species to which individuals are exposed, enabling targeted interventions to reduce exposure and mitigate health effects. Nevertheless, it should be noted that few of the included studies conducted As speciation, possibly because these methods are relatively new and specific to identifying only certain arsenicals, which may not be applicable to other As species [122]. The effect of As exposure during pregnancy and/or childhood on children's neuropsychological function was negatively associated (74%) in the majority of the included studies. These results are in line with those presented in a review carried out by Ortiz-Garcia et al., whose main objective was to determine the impact of As exposure on maternal and fetal health [8]. In the previous review, maternal As exposure during pregnancy was identified as a critical window of concern, as As can easily cross the placental barrier and lead to fetal alterations, such as low birth weight and congenital anomalies [8]. Similarly, a systematic review of 92 studies conducted in Latin America highlighted the harmful effects of As exposure during pregnancy and/or childhood on several health issues, such as kidney injury, short gestational age, pulmonary illness, and decreased cognitive function in children [123]. In our review, most of the included studies analyzed the effects of maternal and/or childhood As exposure on children's cognitive function, which is frequently assessed in neuropsychological evaluations [36]. One possible reason most studies analyzed in this review focus on cognitive function is the role of As as a neurotoxic agent. As has been shown to cross the blood–brain barrier, potentially causing direct effects on the central nervous system [124–126]. Exposure to As during pregnancy may affect the central nervous system by altering neuronal plasticity, disrupting neural networks, and causing oxidative stress [125]. Nonetheless, our review also identified studies with evidence of null and/or protective effects of As exposure during pregnancy and/or infancy on neuropsychological function in children. These findings suggest that the evidence may still be inconclusive and that further research is needed. Importantly, however, the positive association found in one of the papers included in the review was derived from arsenobetaine exposure, a non-toxic organic form of As that is widely used in epidemiological studies as an indicator of fish/seafood consumption in children [15,119,127]. Fish and seafood are also foods that stand out for being rich in essential nutrients crucial for brain development, including proteins, vitamins, and highlighting omega-3 long-chain polyunsaturated fatty acids [128]. High concentrations of docosahexaenoic acid (DHA) and eicosapentaenoic acid (EPA) are found in the brain, and these are important components of neuronal cell membranes; they support synaptic plasticity and have anti-inflammatory properties [129,130]. As speciation can help tease apart the effects of beneficial nutrients in fish and seafood for neurodevelopment, whereas assessing total As could lead to misclassification of exposure, especially in populations with high fish consumption.

The present review has certain limitations. We cannot rule out selection bias, which is inherent in all studies. Our inclusion criteria, such as limiting articles to only two languages, may have increased this bias. However, one of these languages was English, which is the most commonly used language in scientific research. Additionally, using only five databases may have led to overlooking relevant articles. We should note that these databases were selected based on the recommendations of Bramer et al. [32] for an optimal combination of databases, which helps ensure comprehensive coverage of the available scientific literature. We encountered challenges in establishing our search strategy due to the broad nature of neuropsychological development and the lack of a standard definition, which complicated the selection of search terms. As a result, we may have overlooked some relevant terms during the search process, potentially introducing selection bias. To mitigate this limitation, we used the definition proposed by Joan Forns et al. [36] for neuropsychological development in epidemiological studies to guide our selection of search terms, thus enhancing the replicability of this review.

Another inherent limitation in review papers is publication bias, which can indirectly contribute to selection bias. This occurs when scientific journals do not publish studies with null results, leading to their exclusion from review articles. In our study, we categorized results as neuropsychological improvement, decline, or no change to clearly convey outcomes. However, it is important to recognize that "no change" results do not imply the absence of a neuropsychological impact; rather, they indicate that the observed effect was not statistically significant based on our study's criteria. This necessary categorization highlights primary findings but may overlook meaningful, non-significant trends. Taken together, publication bias and the limitations inherent in our reporting criteria underscore the need for caution when interpreting non-significant findings, as future research with different methodologies or larger sample sizes may provide additional insights. An additional critical limitation concerns As exposure assessment. Most included studies report only total urinary As without speciation into inorganic versus organic forms, which restricts the validity of dose–response interpretations. Finally, we acknowledge that the quality of the included articles was not assessed. Although assessing article quality is not a mandatory requirement for scoping reviews [131], the lack of such an assessment could have led us to include low-quality articles. To mitigate this limitation, we have provided figures and narrative descriptions that include information on indicators closely related to methodological quality, such as conflicts of interest, funding sources, and limitations reported by the authors of the included articles. This information is intended to help readers interpret the results presented in this scoping review.

We want to further highlight certain strengths of this scoping review. It provides a synthesis of information based on a large number of studies (n = 77) on a specific and relevant topic, namely, As exposure and its effect on neuropsychological function in childhood. In this context, it is important to note that children are considered a vulnerable study population, as disruptions during this critical stage can compromise development throughout life, increasing the scientific interest of this work. Another strength of this study is our use of a structured, detailed, and systematic methodology, ensuring the replicability of our review. Additionally, we used a combination of databases to search for articles, which is deemed optimal according to the scientific literature [32]. Finally, the greatest strength of this work and of scoping reviews in general is their ability to identify knowledge gaps. This identification helps other researchers target future investigations and complement the existing evidence. The primary knowledge gaps identified in this scoping review are the following: 1. the limited number of studies conducted in low-income countries, excluding Bangladesh, such as African and Latin American countries, and 2. the insufficient number of articles performing As speciation analysis given the critical importance of understanding exposure to different As species. Without proper speciation analysis, there is a risk of misinterpreting results and potentially underestimating or overestimating health risks, which could lead to ineffective or misguided public health interventions. Thirdly, the lack of comparability among studies due to the wide array of tests used to assess neuropsychological function makes it nearly impossible to conduct meta-analyses to evaluate the overall effects across studies.

5. Conclusions

This scoping review indicates that As exposure during prenatal, perinatal, and early-childhood stages is frequently linked to neurodevelopmental decline, particularly in cognitive function, although a minority of studies report null findings. The widespread use of urine as a biomarker of As exposure, coupled with heterogeneous speciation methods, sampling times, and neuropsychological instruments, has limited comparability and prevented clear dose–response characterization. Future work that harmonizes exposure assessment with iAs speciation, employs standardized developmental batteries across cognitive, motor, and socio-emotional domains, and follows cohorts longitudinally from gestation to school age will be essential to establish safe exposure thresholds, inform regulatory standards, and guide targeted public health interventions.

Supplementary Materials: The following supporting information can be downloaded at https://www.mdpi.com/article/10.3390/toxics13070542/s1, Table S1: Preferred Reporting Items for Systematic Reviews and Meta-Analyses Extension for Scoping Reviews (PRISMA-ScR) Checklist; Table S2: Search strategy and databases consulted; Table S3: Characteristics of the most used assessment test for neuropsychological function; Figure S1: Sankey plot from the included studies assessing the cognition functional domain; Figure S2: Sankey plot from the included studies assessing the psychomotor functional domain; Figure S3: Sankey plot from the included studies assessing the socio-emotional functional domain.

Author Contributions: Conceptualization, L.N.-B., L.M.C.-G. and A.J.S.-P.; methodology, L.N.-B., L.M.C.-G. and A.J.S.-P.; formal analysis, L.N.-B., L.M.C.-G. and A.J.S.-P.; data curation, L.N.-B., L.M.C.-G. and A.J.S.-P.; writing—original draft preparation, L.N.-B. and L.M.C.-G.; writing—review and editing, L.N.-B., L.M.C.-G., M.R.K., J.V., J.A.B. and A.J.S.-P.; supervision, A.J.S.-P. All authors have read and agreed to the published version of the manuscript.

Funding: This research was funded by Generalitat Valenciana, grant number CIDEGENT/2020/050 – ESGENT/003/2024 (Antonio J. Signes-Pastor), grant number Project AICO/2021/347 (Jesús Vioque), and by Proyectos del Fondo de Investigación Sanitaria (FIS), co-funded by the European Regional Development Fund, grant number PI20/00557; PI23/01568.

Institutional Review Board Statement: Not applicable.

Informed Consent Statement: Not applicable.

Data Availability Statement: No new data were created or analyzed in this study. The data used in this study are included in the article/Supplementary Material.

Conflicts of Interest: The authors declare no conflicts of interest. In addition, the funders had no role in the design of the study; in the collection, analyses, or interpretation of data; in the writing of the manuscript; or in the decision to publish the results.

Abbreviations

The following abbreviations are used in this manuscript:

As	Arsenic
DHA	Docosahexaenoic acid
EPA	Eicosapentaenoic acid
iAs	Inorganic arsenic
WHO	World Health Organization

References

1. Fatoki, J.O.; Badmus, J.A. Arsenic as an Environmental and Human Health Antagonist: A Review of Its Toxicity and Disease Initiation. *J. Hazard. Mater. Adv.* **2022**, *5*, 100052. [CrossRef]
2. Arslan, B.; Djamgoz, M.B.A.; Akün, E. ARSENIC: A Review on Exposure Pathways, Accumulation, Mobility and Transmission into the Human Food Chain. *Rev. Environ. Contam. Toxicol.* **2017**, *243*, 27–51. [CrossRef] [PubMed]
3. Chen, Q.Y.; Costa, M. Arsenic: A Global Environmental Challenge. *Annu. Rev. Pharmacol. Toxicol.* **2021**, *61*, 47–63. [CrossRef] [PubMed]
4. Arcella, D.; Cascio, C.; Gómez Ruiz, J. Ángel Chronic Dietary Exposure to Inorganic Arsenic. *EFSA J.* **2021**, *19*, e06380. [CrossRef]
5. Ganie, S.Y.; Javaid, D.; Hajam, Y.A.; Reshi, M.S. Arsenic Toxicity: Sources, Pathophysiology and Mechanism. *Toxicol. Res.* **2023**, *13*, tfad111. [CrossRef]
6. Miodovnik, A. Environmental Neurotoxicants and Developing Brain. *Mt. Sinai J. Med. A J. Transl. Pers. Med.* **2011**, *78*, 58–77. [CrossRef] [PubMed]
7. Punshon, T.; Davis, M.A.; Marsit, C.J.; Theiler, S.K.; Baker, E.R.; Jackson, B.P.; Conway, D.C.; Karagas, M.R. Placental Arsenic Concentrations in Relation to Both Maternal and Infant Biomarkers of Exposure in a US Cohort. *J. Expo. Sci. Environ. Epidemiol.* **2015**, *25*, 599–603. [CrossRef]
8. Ortiz-Garcia, N.Y.; Ramírez, A.I.C.; Juarez, K.; Galindo, J.B.; Briceño, G.; Martinez, E.C. Maternal Exposure to Arsenic and Its Impact on Maternal and Fetal Health: A Review. *Cureus* **2023**, *15*, e49177. [CrossRef]

9. Thomas, D.J. Arsenic Methylation-Lessons from Three Decades of Research. *Toxicology* **2021**, *457*, 152800. [CrossRef]
10. Jin, Y.; Xi, S.; Li, X.; Lu, C.; Li, G.; Xu, Y.; Qu, C.; Niu, Y.; Sun, G. Arsenic Speciation Transported Through the Placenta from Mother Mice to Their Newborn Pups. *Environ. Res.* **2006**, *101*, 349–355. [CrossRef]
11. Chakraborty, A.; Ghosh, S.; Biswas, B.; Pramanik, S.; Nriagu, J.; Bhowmick, S. Epigenetic Modifications from Arsenic Exposure: A Comprehensive Review. *Sci. Total Environ.* **2022**, *810*, 151218. [CrossRef] [PubMed]
12. Likhar, A.; Patil, M.S. Importance of Maternal Nutrition in the First 1000 Days of Life and Its Effects on Child Development: A Narrative Review. *Cureus* **2022**, *14*, e30083. [CrossRef] [PubMed]
13. Scher, M.S. "The First Thousand Days" Define a Fetal/Neonatal Neurology Program. *Front. Pediatr.* **2021**, *9*, 683138. [CrossRef] [PubMed]
14. Hamadani, J.D.; Tofail, F.; Nermell, B.; Gardner, R.; Shiraji, S.; Bottai, M.; Arifeen, S.E.; Huda, S.N.; Vahter, M. Critical Windows of Exposure for Arsenic-Associated Impairment of Cognitive Function in Pre-School Girls and Boys: A Population-Based Cohort Study. *Int. J. Epidemiol.* **2011**, *40*, 1593–1604. [CrossRef]
15. Signes-Pastor, A.J.; Vioque, J.; Navarrete-Muñoz, E.M.; Carey, M.; Garc-Villarino, M.; Fernández-Somoano, A.; Tardón, A.; Santa-Marina, L.; Irizar, A.; Casas, M.; et al. Inorganic Arsenic Exposure and Neuropsychological Development of Children of 4–5 Years of Age Living in Spain. *Environ. Res.* **2019**, *174*, 135–142. [CrossRef]
16. Signes-Pastor, A.J.; Romano, M.E.; Jackson, B.; Braun, J.M.; Yolton, K.; Chen, A.; Lanphear, B.; Karagas, M.R. Associations of Maternal Urinary Arsenic Concentrations During Pregnancy with Childhood Cognitive Abilities: The HOME Study. *Int. J. Hyg. Environ. Health* **2022**, *245*, 114009. [CrossRef]
17. Hamadani, J.D.; Grantham-McGregor, S.M.; Tofail, F.; Nermell, B.; Fängström, B.; Huda, S.N.; Yesmin, S.; Rahman, M.; Vera-Hernández, M.; Arifeen, S.E.; et al. Pre- and Postnatal Arsenic Exposure and Child Development at 18 Months of Age: A Cohort Study in Rural Bangladesh. *Int. J. Epidemiol.* **2010**, *39*, 1206–1216. [CrossRef]
18. Tofail, F.; Vahter, M.; Hamadani, J.D.; Nermell, B.; Huda, S.N.; Yunus, M.; Rahman, M.; Grantham-McGregor, S.M. Effect of Arsenic Exposure During Pregnancy on Infant Development at 7 Months in Rural Matlab, Bangladesh. *Environ. Health Perspect.* **2009**, *117*, 288–293. [CrossRef]
19. Margiana, R.; Alshahrani, S.H.; Kayumova, D.; Alawadi, A.H.R.; Hjazi, A.; Alsalamy, A.; Qasim, Q.A.; Juyal, A.; Garousi, N. Association Between Maternal Exposure to Arsenic by Drinking Water During Pregnancy and Risk of Preterm Birth: A Systematic Review and Meta-Analysis. *Int. J. Environ. Health Res.* **2023**, *34*, 2947–2956. [CrossRef]
20. Tsuji, J.S.; Garry, M.R.; Perez, V.; Chang, E.T. Low-Level Arsenic Exposure and Developmental Neurotoxicity in Children: A Systematic Review and Risk Assessment. *Toxicology* **2015**, *337*, 91–107. [CrossRef]
21. Majumdar, K.K.; Mazumder, D.N.G. Effect of Drinking Arsenic-Contaminated Water in Children. *Indian J. Public Health* **2012**, *56*, 223–226. [CrossRef]
22. Quansah, R.; Armah, F.A.; Essumang, D.K.; Luginaah, I.; Clarke, E.; Marfoh, K.; Cobbina, S.J.; Nketiah-Amponsah, E.; Namujju, P.B.; Obiri, S.; et al. Association of Arsenic with Adverse Pregnancy Outcomes/Infant Mortality: A Systematic Review and Meta-Analysis. *Environ. Health Perspect.* **2015**, *123*, 412–421. [CrossRef]
23. Rodríguez-Barranco, M.; Gil, F.; Hernández, A.F.; Alguacil, J.; Lorca, A.; Mendoza, R.; Gómez, I.; Molina-Villalba, I.; González-Alzaga, B.; Aguilar-Garduño, C.; et al. Postnatal Arsenic Exposure and Attention Impairment in School Children. *Cortex* **2016**, *74*, 370–382. [CrossRef]
24. Bauer, J.A.; Fruh, V.; Howe, C.G.; White, R.F.; Henn, B.C. Associations of Metals and Neurodevelopment: A Review of Recent Evidence on Susceptibility Factors. *Curr. Epidemiol. Rep.* **2020**, *7*, 237–262. [CrossRef]
25. Hasanvand, M.; Mohammadi, R.; Khoshnamvand, N.; Jafari, A.; Palangi, H.S.; Mokhayeri, Y. Dose-Response Meta-Analysis of Arsenic Exposure in Drinking Water and Intelligence Quotient. *J. Environ. Health Sci. Eng.* **2020**, *18*, 1691. [CrossRef]
26. Jane, H.; Charlotte, B. *What's in My Baby's Food?* Healthy Babies Bright Futures: Charlottesville, VA, USA, 2019.
27. Ryan, K.; Meghan, L. *Results of Lifetime IQ Decrement Analysis from Dietary Exposures to Lead and Inorganic Arsenic for Children 0 to <2 Years of Age*; Abt Global: Cambridge, MA, USA, 2019.
28. Tricco, A.C.; Lillie, E.; Zarin, W.; O'Brien, K.K.; Colquhoun, H.; Levac, D.; Moher, D.; Peters, M.D.J.; Horsley, T.; Weeks, L.; et al. PRISMA Extension for Scoping Reviews (PRISMA-ScR): Checklist and Explanation. *Ann. Intern. Med.* **2018**, *169*, 467–473. [CrossRef]
29. Gough, D.; Thomas, J.; Oliver, S. Clarifying Differences Between Review Designs and Methods. *Syst. Rev.* **2012**, *1*, 28. [CrossRef]
30. Munn, Z.; Peters, M.D.J.; Stern, C.; Tufanaru, C.; McArthur, A.; Aromataris, E. Systematic Review or Scoping Review? Guidance for Authors When Choosing Between a Systematic or Scoping Review Approach. *BMC Med. Res. Methodol.* **2018**, *18*, 143. [CrossRef]
31. Higgins, J.P. (Ed.) *Cochrane Handbook for Systematic Reviews of Interventions Version 6.4 (Updated August 2023)*; Cochrane: London, UK, 2023.
32. Bramer, W.M.; Rethlefsen, M.L.; Kleijnen, J.; Franco, O.H. Optimal Database Combinations for Literature Searches in Systematic Reviews: A Prospective Exploratory Study. *Syst. Rev.* **2017**, *6*, 245. [CrossRef]

33. Stein, C.R.; Wu, H.; Bellinger, D.C.; Smith, D.R.; Wolff, M.S.; Savitz, D.A. Exposure to Metal Mixtures and Neuropsychological Functioning in Middle Childhood. *Neurotoxicology* **2022**, *93*, 84–91. [CrossRef]
34. Mah, V.K.; Ford-Jones, E.L. Spotlight on Middle Childhood: Rejuvenating the 'Forgotten Years'. *Paediatr. Child Health* **2012**, *17*, 81–83. [CrossRef]
35. Voss, M.L.; Claeson, M.; Bremberg, S.; Peterson, S.S.; Alfvén, T.; Ndeezi, G. The Missing Middle of Childhood. *Glob. Health Action* **2023**, *16*, 2242196. [CrossRef]
36. Forns, J.; Aranbarri, A.; Grellier, J.; Julvez, J.; Vrijheid, M.; Sunyer, J. A Conceptual Framework in the Study of Neuropsychological Development in Epidemiological Studies. *Neuroepidemiology* **2012**, *38*, 203–208. [CrossRef]
37. Page, M.J.; McKenzie, J.E.; Bossuyt, P.M.; Boutron, I.; Hoffmann, T.C.; Mulrow, C.D.; Shamseer, L.; Tetzlaff, J.M.; Akl, E.A.; Brennan, S.E.; et al. The PRISMA 2020 Statement: An Updated Guideline for Reporting Systematic Reviews. *BMJ* **2021**, *372*, n71. [CrossRef]
38. Otto, E.; Culakova, E.; Meng, S.; Zhang, Z.; Xu, H.; Mohile, S.; Flannery, M.A. Overview of Sankey Flow Diagrams: Focusing on Symptom Trajectories in Older Adults with Advanced Cancer. *J. Geriatr. Oncol.* **2022**, *13*, 742–746. [CrossRef]
39. Gutiérrez-González, E.; García-Esquinas, E.; de Larrea-Baz, N.F.; Salcedo-Bellido, I.; Navas-Acien, A.; Lope, V.; Gómez-Ariza, J.L.; Pastor, R.; Pollán, M.; Pérez-Gómez, B. Toenails as Biomarker of Exposure to Essential Trace Metals: A Review. *Environ. Res.* **2019**, *179*, 108787. [CrossRef]
40. Ritschl, V.; Ferreira, R.J.O.; Santos, E.J.F.; Fernandes, R.; Juutila, E.; Mosor, E.; Santos-Costa, P.; Fligelstone, K.; Schraven, L.; Stummvoll, G.; et al. Suitability for e-Health of Non-Pharmacological Interventions in Connective Tissue Diseases: Scoping Review with a Descriptive Analysis. *RMD Open* **2021**, *7*, e001710. [CrossRef]
41. Wibble, T.; Pansell, T. Clinical Characteristics of Visual Motion Hypersensitivity: A Systematic Review. *Exp. Brain Res.* **2023**, *241*, 1707. [CrossRef]
42. Jiang, H.; Wang, W.; Mei, Y.; Zhao, Z.; Lin, B.; Zhang, Z. A Scoping Review of the Self-Reported Compassion Measurement Tools. *BMC Public Health* **2023**, *23*, 2323. [CrossRef]
43. Wasserman, G.A.; Liu, X.; Loiacono, N.J.; Kline, J.; Factor-Litvak, P.; van Geen, A.; Mey, J.L.; Levy, D.; Abramson, R.; Schwartz, A.; et al. A Cross-Sectional Study of Well Water Arsenic and Child IQ in Maine Schoolchildren. *Environ. Health* **2014**, *13*, 23. [CrossRef] [PubMed]
44. Wright, R.O.; Amarasiriwardena, C.; Woolf, A.D.; Jim, R.; Bellinger, D.C. Neuropsychological Correlates of Hair Arsenic, Manganese, and Cadmium Levels in School-Age Children Residing Near a Hazardous Waste Site. *Neurotoxicology* **2006**, *27*, 210–216. [CrossRef]
45. Marlowe, M.; Stellern, J.; Errera, J.; Moon, C. Main and Interaction Effects of Metal Pollutants on Visual-Motor Performance. *Arch. Environ. Health* **1985**, *40*, 221–225. [CrossRef]
46. Bauer, J.A.; Romano, M.E.; Jackson, B.P.; Bellinger, D.; Korrick, S.; Karagas, M.R. Associations of Perinatal Metal and Metalloid Exposures with Early Child Behavioral Development over Time in the New Hampshire Birth Cohort Study. *Expo. Health* **2024**, *16*, 135–148. [CrossRef]
47. Cottrell, J.; Nelson, C.; Waldron, C.; Bergeron, M.; Samson, A.; Valentovic, M. Effect of Umbilical Cord Essential and Toxic Elements, Thyroid Levels, and Vitamin d on Childhood Development. *Biomed. Pharmacother.* **2023**, *158*, 114085. [CrossRef]
48. Cowell, W.; Colicino, E.; Levin-Schwartz, Y.; Enlow, M.B.; Amarasiriwardena, C.; Andra, S.S.; Gennings, C.; Wright, R.O.; Wright, R.J. Prenatal Metal Mixtures and Sex-Specific Infant Negative Affectivity. *Environ. Epidemiol.* **2021**, *5*, e147. [CrossRef]
49. Doherty, B.T.; Romano, M.E.; Gui, J.; Punshon, T.; Jackson, B.P.; Karagas, M.R.; Korrick, S.A. Periconceptional and Prenatal Exposure to Metal Mixtures in Relation to Behavioral Development at 3 Years of Age. *Environ. Epidemiol.* **2020**, *4*, e0106. [CrossRef]
50. Marlowe, M.; Cossairt, A.; Moon, C.; Errera, J.; MacNeel, A.; Peak, R.; Ray, J.; Schroeder, C. Main and Interaction Effects of Metallic Toxins on Classroom Behavior. *J. Abnorm. Child. Psychol.* **1985**, *13*, 185–198. [CrossRef]
51. Moon, C.; Marlowe, M.; Stellern, J.; Errera, J. Main and Interaction Effects of Metallic Pollutants on Cognitive Functioning. *J. Learn. Disabil.* **1985**, *18*, 217–221. [CrossRef]
52. Butler, E.E.; Karagas, M.R.; Demidenko, E.; Bellinger, D.C.; Korrick, S.A. In Utero Arsenic Exposure and Early Childhood Motor Development in the New Hampshire Birth Cohort Study. *Front. Epidemiol.* **2023**, *3*, 1139337. [CrossRef]
53. Rosa, M.J.; Pedretti, N.F.; Goldson, B.; Mathews, N.; Merced-Nieves, F.; Xhani, N.; Enlow, M.B.; Gershon, R.; Ho, E.; Huddleston, K.; et al. Integrating Data Across Multiple Sites in the Northeastern United States to Examine Associations Between a Prenatal Metal Mixture and Child Cognition. *Am. J. Epidemiol.* **2024**, *193*, 606–616. [CrossRef] [PubMed]
54. Thilakaratne, R.; Lin, P.I.D.; Rifas-Shiman, S.L.; Landero, J.; Wright, R.O.; Bellinger, D.; Oken, E.; Cardenas, A. Cross-Sectional and Prospective Associations of Early Childhood Circulating Metals with Early and Mid-Childhood Cognition in the Project Viva Cohort. *Environ. Res.* **2024**, *246*, 118068. [CrossRef]
55. Valeri, L.; Mazumdar, M.M.; Bobb, J.F.; Henn, B.C.; Rodrigues, E.; Sharif, O.I.A.; Kile, M.L.; Quamruzzaman, Q.; Afroz, S.; Golam, M.; et al. The Joint Effect of Prenatal Exposure to Metal Mixtures on Neurodevelopmental Outcomes at 20–40 Months of Age: Evidence from Rural Bangladesh. *Environ. Health Perspect.* **2017**, *125*, 067015. [CrossRef]

56. Rodrigues, E.G.; Bellinger, D.C.; Valeri, L.; Hasan, M.O.S.I.; Quamruzzaman, Q.; Golam, M.; Kile, M.L.; Christiani, D.C.; Wright, R.O.; Mazumdar, M. Neurodevelopmental Outcomes Among 2- to 3-Year-Old Children in Bangladesh with Elevated Blood Lead and Exposure to Arsenic and Manganese in Drinking Water. *Environ. Health A Glob. Access Sci. Source* **2016**, *15*, 44. [CrossRef]
57. Wasserman, G.A.; Liu, X.; Parvez, F.; Factor-Litvak, P.; Kline, J.; Siddique, A.B.; Shahriar, H.; Uddin, M.N.; van Geen, A.; Mey, J.L.; et al. Child Intelligence and Reductions in Water Arsenic and Manganese: A Two-Year Follow-up Study in Bangladesh. *Environ. Health Perspect* **2016**, *124*, 1114–1120. [CrossRef]
58. Nahar, M.N.; Inaoka, T.; Fujimura, M. A Consecutive Study on Arsenic Exposure and Intelligence Quotient (IQ) of Children in Bangladesh. *Environ. Health Prev. Med.* **2014**, *19*, 194–199. [CrossRef]
59. Parvez, F.; Wasserman, G.A.; Factor-Litvak, P.; Liu, X.; Slavkovich, V.; Siddique, A.B.; Sultana, R.; Sultana, R.; Islam, T.; Levy, D.; et al. Arsenic Exposure and Motor Function Among Children in Bangladesh. *Environ. Health Perspect.* **2011**, *119*, 1665–1670. [CrossRef]
60. Khan, K.; Factor-Litvak, P.; Wasserman, G.A.; Liu, X.; Ahmed, E.; Parvez, F.; Slavkovich, V.; Levy, D.; Mey, J.; van Geen, A.; et al. Manganese Exposure from Drinking Water and Children's Classroom Behavior in Bangladesh. *Environ. Health Perspect.* **2011**, *119*, 1501–1506. [CrossRef]
61. Wasserman, G.A.; Liu, X.; Parvez, F.; Ahsan, H.; Factor-Litvak, P.; Kline, J.; van Geen, A.; Slavkovich, V.; Loiacono, N.J.; Levy, D.; et al. Water Arsenic Exposure and Intellectual Function in 6-Year-Old Children in Araihazar, Bangladesh. *Environ. Health Perspect.* **2007**, *115*, 285–289. [CrossRef]
62. Wasserman, G.A.; Liu, X.; Parvez, F.; Ahsan, H.; Factor-Litvak, P.; van Geen, A.; Slavkovich, V.; LoIacono, N.J.; Cheng, Z.; Hussain, I.; et al. Water Arsenic Exposure and Children's Intellectual Function in Araihazar, Bangladesh. *Environ. Health Perspect.* **2004**, *112*, 1329–1333. [CrossRef] [PubMed]
63. Vahter, M.; Skröder, H.; Rahman, S.M.; Levi, M.; Hamadani, J.D.; Kippler, M. Prenatal and Childhood Arsenic Exposure Through Drinking Water and Food and Cognitive Abilities at 10 Years of Age: A Prospective Cohort Study. *Environ. Int.* **2020**, *139*, 105723. [CrossRef] [PubMed]
64. Wasserman, G.A.; Liu, X.; Parvez, F.; Factor-Litvak, P.; Ahsan, H.; Levy, D.; Kline, J.; van Geen, A.; Mey, J.; Slavkovich, V.; et al. Arsenic and Manganese Exposure and Children's Intellectual Function. *Neurotoxicology* **2011**, *32*, 450–457. [CrossRef] [PubMed]
65. Pan, S.; Lin, L.; Zeng, F.; Zhang, J.; Dong, G.; Yang, B.; Jing, Y.; Chen, S.; Zhang, G.; Yu, Z.; et al. Effects of Lead, Cadmium, Arsenic, and Mercury Co-Exposure on Children's Intelligence Quotient in an Industrialized Area of Southern China. *Environ. Pollut.* **2018**, *235*, 47–54. [CrossRef]
66. Wang, S.-X.; Wang, Z.-H.; Cheng, X.-T.; Li, J.; Sang, Z.-P.; Zhang, X.-D.; Han, L.-L.; Qiao, X.-Y.; Wu, Z.-M.; Wang, Z.-Q. Arsenic and Fluoride Exposure in Drinking Water: Children's IQ and Growth in Shanyin County, Shanxi Province, China. *Environ. Health Perspect.* **2007**, *115*, 643–647. [CrossRef] [PubMed]
67. Liang, C.; Wu, X.; Huang, K.; Yan, S.; Li, Z.; Xia, X.; Pan, W.; Sheng, J.; Tao, R.; Tao, Y.; et al. Domain- and Sex-Specific Effects of Prenatal Exposure to Low Levels of Arsenic on Children's Development at 6 Months of Age: Findings from the Ma'anshan Birth Cohort Study in China. *Environ. Int.* **2020**, *135*, 105112. [CrossRef]
68. Wang, Y.; Wang, Y.; Yan, C. Gender Differences in Trace Element Exposures with Cognitive Abilities of School-Aged Children: A Cohort Study in Wujiang City, China. *Environ. Sci. Pollut. Res. Int.* **2022**, *29*, 64807–64821. [CrossRef]
69. Yu, X.-D.; Yan, C.-H.; Shen, X.-M.; Tian, Y.; Cao, L.-L.; Yu, X.-G.; Zhao, L.; Liu, J.-X. Prenatal Exposure to Multiple Toxic Heavy Metals and Neonatal Neurobehavioral Development in Shanghai, China. *Neurotoxicol Teratol.* **2011**, *33*, 437–443. [CrossRef] [PubMed]
70. Chen, H.; Zhang, H.; Wang, X.; Wu, Y.; Zhang, Y.; Chen, S.; Zhang, W.; Sun, X.; Zheng, T.; Xia, W.; et al. Prenatal Arsenic Exposure, Arsenic Metabolism and Neurocognitive Development of 2-Year-Old Children in Low-Arsenic Areas. *Environ. Int.* **2023**, *174*, 107918. [CrossRef]
71. Zhou, T.; Guo, J.; Zhang, J.; Xiao, H.; Qi, X.; Wu, C.; Chang, X.; Zhang, Y.; Liu, Q.; Zhou, Z. Sex-Specific Differences in Cognitive Abilities Associated with Childhood Cadmium and Manganese Exposures in School-Age Children: A Prospective Cohort Study. *Biol. Trace Elem. Res.* **2020**, *193*, 89–99. [CrossRef]
72. Wang, B.; Liu, J.; Liu, B.; Liu, X.; Yu, X. Prenatal Exposure to Arsenic and Neurobehavioral Development of Newborns in China. *Environ. Int.* **2018**, *121*, 421–427. [CrossRef]
73. Ma, J.; Geng, S.; Sun, Q.; Zhang, X.; Han, L.; Yao, X.; Zhang, B.; Zhu, L.; Wen, J. Exposure to Metal Mixtures and Young Children's Growth and Development: A Biomonitoring-Based Study in Eastern China. *Ecotoxicol. Environ. Saf.* **2023**, *268*, 115726. [CrossRef]
74. Dai, Y.; Lu, H.; Zhang, J.; Ding, J.; Wang, Z.; Zhang, B.; Qi, X.; Chang, X.; Wu, C.; Zhou, Z. Sex-Specific Associations of Maternal and Childhood Urinary Arsenic Levels with Emotional Problems Among 6-Year-Age Children: Evidence from a Longitudinal Cohort Study in China. *Ecotoxicol. Environ. Saf.* **2023**, *267*, 115658. [CrossRef] [PubMed]
75. Merced-Nieves, F.M.; Chelonis, J.; Pantic, I.; Schnass, L.; Téllez-Rojo, M.M.; Braun, J.M.; Paule, M.G.; Wright, R.J.; Wright, R.O.; Curtin, P. Prenatal Trace Elements Mixture Is Associated with Learning Deficits on a Behavioral Acquisition Task Among Young Children. *New Dir. Child. Adolesc. Dev.* **2022**, *2022*, 53–66. [CrossRef]

76. Nozadi, S.S.; Li, L.; Luo, L.; Mackenzie, D.; Erdei, E.; Du, R.; Roman, C.W.; Hoover, J.; O'donald, E.; Burnette, C.; et al. Prenatal Metal Exposures and Infants' Developmental Outcomes in a Navajo Population. *Int. J. Environ. Res. Public Health* **2021**, *19*, 425. [CrossRef] [PubMed]
77. de Water, E.; Curtin, P.; Gennings, C.; Chelonis, J.J.; Paule, M.; Bixby, M.; McRae, N.; Svensson, K.; Schnaas, L.; Pantic, I.; et al. Prenatal Metal Mixture Concentrations and Reward Motivation in Children. *Neurotoxicology* **2022**, *88*, 124–133. [CrossRef]
78. Levin-Schwartz, Y.; Gennings, C.; Schnaas, L.; Chávez, M.D.C.H.; Bellinger, D.C.; Téllez-Rojo, M.M.; Baccarelli, A.A.; Wright, R.O. Time-Varying Associations Between Prenatal Metal Mixtures and Rapid Visual Processing in Children. *Environ. Health* **2019**, *18*, 92. [CrossRef] [PubMed]
79. Roy, A.; Kordas, K.; Lopez, P.; Rosado, J.L.; Cebrian, M.E.; Vargas, G.G.; Ronquillo, D.; Stoltzfus, R.J. Association Between Arsenic Exposure and Behavior Among First-Graders from Torreón, Mexico. *Environ. Res.* **2011**, *111*, 670–676. [CrossRef]
80. Rosado, J.L.; Ronquillo, D.; Kordas, K.; Rojas, O.; Alatorre, J.; Lopez, P.; Garcia-Vargas, G.; Caamaño, M.C.; Cebrián, M.E.; Stoltzfus, R.J. Arsenic Exposure and Cognitive Performance in Mexican Schoolchildren. *Environ. Health Perspect.* **2007**, *115*, 1371–1375. [CrossRef]
81. Calderón, J.; Navarro, M.E.; Jimenez-Capdeville, M.E.; Santos-Diaz, M.A.; Golden, A.; Rodriguez-Leyva, I.; Borja-Aburto, V.; Díaz-Barriga, F. Exposure to Arsenic and Lead and Neuropsychological Development in Mexican Children. *Environ. Res.* **2001**, *85*, 69–76. [CrossRef]
82. Rocha-Amador, D.; Navarro, M.E.; Carrizales, L.; Morales, R.; Calderón, J. Decreased Intelligence in Children and Exposure to Fluoride and Arsenic in Drinking Water. *Cad. Saude Publica* **2007**, *23* (Suppl. S4), S579–S587. [CrossRef]
83. Vaidya, N.; Holla, B.; Heron, J.; Sharma, E.; Zhang, Y.; Fernandes, G.; Iyengar, U.; Spiers, A.; Yadav, A.; Das, S.; et al. Neurocognitive Analysis of Low-Level Arsenic Exposure and Executive Function Mediated by Brain Anomalies Among Children, Adolescents, and Young Adults in India. *JAMA Netw. Open* **2023**, *6*, E2312810. [CrossRef]
84. Nyanza, E.C.; Bernier, F.P.; Martin, J.W.; Manyama, M.; Hatfield, J.; Dewey, D. Effects of Prenatal Exposure and Co-Exposure to Metallic or Metalloid Elements on Early Infant Neurodevelopmental Outcomes in Areas with Small-Scale Gold Mining Activities in Northern Tanzania. *Environ. Int.* **2021**, *149*, 106104. [CrossRef]
85. Araujo, M.S.D.A.; de Figueiredo, N.D.; Froes-Asmus, C.I.R. Prenatal Exposure to Metals and Neurodevelopment in Infants at Six Months: Rio Birth Cohort Study of Environmental Exposure and Childhood Development (PIPA Project). *ISEE Conf. Abstr.* **2022**, *2022*, 4295. [CrossRef]
86. Lin, C.-C.; Chen, Y.-C.; Su, F.-C.; Lin, C.-M.; Liao, H.-F.; Hwang, Y.-H.; Hsieh, W.-S.; Jeng, S.-F.; Su, Y.-N.; Chen, P.-C. In Utero Exposure to Environmental Lead and Manganese and Neurodevelopment at 2 Years of Age. *Environ. Res.* **2013**, *123*, 52–57. [CrossRef]
87. Soler-Blasco, R.; Murcia, M.; Lozano, M.; Sarzo, B.; Esplugues, A.; Riutort-Mayol, G.; Vioque, J.; Lertxundi, N.; Marina, L.S.; Lertxundi, A.; et al. Prenatal Arsenic Exposure, Arsenic Methylation Efficiency, and Neuropsychological Development Among Preschool Children in a Spanish Birth Cohort. *Environ. Res.* **2022**, *207*, 112208. [CrossRef]
88. Freire, C.; Amaya, E.; Gil, F.; Fernández, M.F.; Murcia, M.; Llop, S.; Andiarena, A.; Aurrekoetxea, J.; Bustamante, M.; Guxens, M.; et al. Prenatal Co-Exposure to Neurotoxic Metals and Neurodevelopment in Preschool Children: The Environment and Childhood (INMA) Project. *Sci. Total Environ.* **2018**, *621*, 340–351. [CrossRef]
89. Forns, J.; Fort, M.; Casas, M.; Cáceres, A.; Guxens, M.; Gascon, M.; Garcia-Esteban, R.; Julvez, J.; Grimalt, J.O.; Sunyer, J. Exposure to Metals During Pregnancy and Neuropsychological Development at the Age of 4 Years. *Neurotoxicology* **2014**, *40*, 16–22. [CrossRef]
90. Patti, M.A.; Kelsey, K.T.; MacFarlane, A.J.; Papandonatos, G.D.; Arbuckle, T.E.; Ashley-Martin, J.; Fisher, M.; Fraser, W.D.; Lanphear, B.P.; Muckle, G.; et al. Maternal Folate Status and the Relation Between Gestational Arsenic Exposure and Child Health Outcomes. *Int. J. Environ. Res. Public Health* **2022**, *19*, 11332. [CrossRef]
91. Parajuli, R.P.; Fujiwara, T.; Umezaki, M.; Watanabe, C. Home Environment and Cord Blood Levels of Lead, Arsenic, and Zinc on Neurodevelopment of 24 Months Children Living in Chitwan Valley, Nepal. *J. Trace Elem. Med. Biol.* **2015**, *29*, 315–320. [CrossRef]
92. Parajuli, R.P.; Fujiwara, T.; Umezaki, M.; Furusawa, H.; Watanabe, C. Home Environment and Prenatal Exposure to Lead, Arsenic and Zinc on the Neurodevelopment of Six-Month-Old Infants Living in Chitwan Valley, Nepal. *Neurotoxicol. Teratol.* **2014**, *41*, 89–95. [CrossRef]
93. Parajuli, R.P.; Fujiwara, T.; Umezaki, M.; Watanabe, C. Association of Cord Blood Levels of Lead, Arsenic, and Zinc with Neurodevelopmental Indicators in Newborns: A Birth Cohort Study in Chitwan Valley, Nepal. *Environ. Res.* **2013**, *121*, 45–51. [CrossRef]
94. Jiang, C.B.; Kao, C.S.; Chien, L.C.; Chen, Y.J.; Liao, K.W. Associations Among Prenatal and Postnatal Arsenic, Lead, and Cadmium Exposures and Motor Development in 3-Year-Old Children: A Longitudinal Birth Cohort Study in Taiwan. *Environ. Sci. Pollut. Res. Int.* **2022**, *29*, 43191–43200. [CrossRef]

95. Parajuli, R.P.; Umezaki, M.; Fujiwara, T.; Watanabe, C. Association of Cord Blood Levels of Lead, Arsenic, and Zinc and Home Environment with Children Neurodevelopment at 36 Months Living in Chitwan Valley, Nepal. *PLoS ONE* **2015**, *10*, e0120992. [CrossRef]
96. Jiang, C.B.; Hsueh, Y.M.; Kuo, G.L.; Hsu, C.H.; Chang, J.H.; Chien, L.C. Preliminary Study of Urinary Arsenic Concentration and Arsenic Methylation Capacity Effects on Neurodevelopment in Very Low Birth Weight Preterm Children Under 24 Months of Corrected Age. *Medicine* **2018**, *97*, e12800. [CrossRef]
97. Notario-Barandiaran, L.; Díaz-Coto, S.; Jimenez-Redondo, N.; Guxens, M.; Vrijheid, M.; Andiarena, A.; Irizar, A.; Riaño-Galan, I.; Fernández-Somoano, A.; Llop, S.; et al. Latent Childhood Exposure to Mixtures of Metals and Neurodevelopmental Outcomes in 4–5-Year-Old Children Living in Spain. *Expo. Health* **2023**, *16*, 1053–1066. [CrossRef]
98. Desai, G.; Barg, G.; Queirolo, E.I.; Vahter, M.; Peregalli, F.; Mañay, N.; Kordas, K. A Cross-Sectional Study of General Cognitive Abilities Among Uruguayan School Children with Low-Level Arsenic Exposure, Potential Effect Modification by Methylation Capacity and Dietary Folate. *Environ. Res.* **2018**, *164*, 124–131. [CrossRef]
99. Manju, R.; Hegde, A.M.; Parlees, P.; Keshan, A. Environmental Arsenic Contamination and Its Effect on Intelligence Quotient of School Children in a Historic Gold Mining Area Hutti, North Karnataka, India: A Pilot Study. *J. Neurosci. Rural Pract.* **2017**, *8*, 364–367. [CrossRef]
100. Kordas, K.; Ardoino, G.; Coffman, D.L.; Queirolo, E.I.; Ciccariello, D.; Mañay, N.; Ettinger, A.S. Patterns of Exposure to Multiple Metals and Associations with Neurodevelopment of Preschool Children from Montevideo, Uruguay. *J. Environ. Public Health* **2015**, *2015*, 493471. [CrossRef]
101. la Ossa, C.A.D.; Ramírez-Giraldo, A.F.; Arroyo-Alvis, K.; Marrugo-Negrete, J.; Díez, S. Neuropsychological Effects and Cognitive Deficits Associated with Exposure to Mercury and Arsenic in Children and Adolescents of the Mojana Region, Colombia. *Environ. Res.* **2023**, *216*, 114467. [CrossRef]
102. Kunwittaya, S.; Ruksee, N.; Khamnong, T.; Jiawiwatkul, A.; Kleebpung, N.; Chumchua, V.; Plitponkarnpim, A.; Nopparat, C.; Permpoonputtana, K. Inorganic Arsenic Contamination and the Health of Children Living Near an Inactive Mining Site: Northern Thailand. *EXCLI J.* **2022**, *21*, 1007–1014. [CrossRef] [PubMed]
103. von Ehrenstein, O.S.; Poddar, S.; Yuan, Y.; Mazumder, D.G.; Eskenazi, B.; Basu, A.; Hira-Smith, M.; Ghosh, N.; Lahiri, S.; Haque, R.; et al. Children's Intellectual Function in Relation to Arsenic Exposure. *Epidemiology* **2007**, *18*, 44–51. [CrossRef] [PubMed]
104. Kumar, A.; Rahman, M.S.; Kumar, R.; Ali, M.; Niraj, P.K.; Srivastava, A.; Singh, S.K.; Ghosh, A.K. Arsenic Contamination in Groundwater Causing Impaired Memory and Intelligence in School Children of Simri Village of Buxar District of Bihar. *J. Ment. Health Hum. Behav.* **2019**, *24*, 132–138. [CrossRef]
105. Ghosh, S.B.; Chakraborty, D.; Mondal, N.K. Effect of Arsenic and Manganese Exposure on Intellectual Function of Children in Arsenic Stress Area of Purbasthali, Burdwan, West Bengal. *Expo. Health* **2017**, *9*, 1–11. [CrossRef]
106. Renzetti, S.; Cagna, G.; Calza, S.; Conversano, M.; Fedrighi, C.; Forte, G.; Giorgino, A.; Guazzetti, S.; Majorani, C.; Oppini, M.; et al. The Effects of the Exposure to Neurotoxic Elements on Italian Schoolchildren Behavior. *Sci. Rep.* **2021**, *11*, 9898. [CrossRef]
107. Desai, G.; Barg, G.; Vahter, M.; Queirolo, E.I.; Peregalli, F.; Mañay, N.; Millen, A.E.; Yu, J.; Kordas, K. Executive Functions in School Children from Montevideo, Uruguay and Their Associations with Concurrent Low-Level Arsenic Exposure. *Environ. Int.* **2020**, *142*, 105883. [CrossRef]
108. Desai, G.; Barg, G.; Vahter, M.; Queirolo, E.I.; Peregalli, F.; Mañay, N.; Millen, A.E.; Yu, J.; Browne, R.W.; Kordas, K. Low Level Arsenic Exposure, b-Vitamins, and Achievement Among Uruguayan School Children. *Int. J. Hyg. Environ. Health* **2020**, *223*, 124–131. [CrossRef]
109. Lucchini, R.G.; Guazzetti, S.; Renzetti, S.; Conversano, M.; Cagna, G.; Fedrighi, C.; Giorgino, A.; Peli, M.; Placidi, D.; Zoni, S.; et al. Neurocognitive Impact of Metal Exposure and Social Stressors Among Schoolchildren in Taranto, Italy. *Environ. Health A Glob. Access Sci. Source* **2019**, *18*, 67. [CrossRef]
110. Egwunye, J.; Cardoso, B.R.; Braat, S.; Ha, T.; Hanieh, S.; Hare, D.; Duan, A.X.; Doronila, A.; Tran, T.; Tuan, T.; et al. The Role of Fingernail Selenium in the Association Between Arsenic, Lead and Mercury and Child Development in Rural Vietnam: A Cross-Sectional Analysis. *Br. J. Nutr.* **2022**, *129*, 1589–1597. [CrossRef]
111. Huang, C.; Pan, S.; Chin, W.; Hsu, J.; Guo, Y.L. Urinary Heavy Metals and Attention-Deficit/Hyperactivity Symptoms of Preschool Children: A Mixed-Exposure Analysis. *Ecotoxicol. Environ. Saf.* **2023**, *268*, 115714. [CrossRef]
112. Kao, C.S.; Fan, Y.T.; Chien, L.C.; Liao, K.W.; Chang, J.H.; Hsu, C.H.; Chen, Y.J.; Jiang, C.B. Effects of Preterm Birth and Postnatal Exposure to Metal Mixtures on Neurodevelopment in Children at 24 Months of Age. *Environ. Sci. Pollut. Res. Int.* **2023**, *30*, 86856–86865. [CrossRef]
113. Wilson, D. Arsenic Consumption in the United States. *J. Environ. Health* **2015**, *78*, 8–44.
114. Nachman, K.E.; Ginsberg, G.L.; Miller, M.D.; Murray, C.J.; Nigra, A.E.; Pendergrast, C.B. Mitigating Dietary Arsenic Exposure: Current Status in the United States and Recommendations for an Improved Path Forward. *Sci. Total Environ.* **2017**, *581–582*, 221. [CrossRef]

115. Shaji, E.; Santosh, M.; Sarath, K.V.; Prakash, P.; Deepchand, V.; Divya, B.V. Arsenic Contamination of Groundwater: A Global Synopsis with Focus on the Indian Peninsula. *Geosci. Front.* **2021**, *12*, 101079. [CrossRef]
116. Rahman, M.A.; Rahman, A.; Khan, M.Z.K.; Renzaho, A.M.N. Human Health Risks and Socio-Economic Perspectives of Arsenic Exposure in Bangladesh: A Scoping Review. *Ecotoxicol. Environ. Saf.* **2018**, *150*, 335–343. [CrossRef] [PubMed]
117. Rahaman, M.S.; Mise, N.; Ichihara, S. Arsenic Contamination in Food Chain in Bangladesh: A Review on Health Hazards, Socioeconomic Impacts and Implications. *Hyg. Environ. Health Adv.* **2022**, *2*, 100004. [CrossRef]
118. Wongsasuluk, P.; Chotpantarat, S.; Siriwong, W.; Robson, M. Human Biomarkers Associated with Low Concentrations of Arsenic (as) and Lead (Pb) in Groundwater in Agricultural Areas of Thailand. *Scientific Reports* **2021**, *11*, 13896. [CrossRef]
119. Martinez-Morata, I.; Sobel, M.; Tellez-Plaza, M.; Navas-Acien, A.; Howe, C.G.; Sanchez, T.R. A State-of-the-Science Review on Metal Biomarkers. *Curr. Environ. Health Rep.* **2023**, *10*, 215–249. [CrossRef]
120. Yoshinaga, J. Urinary Arsenic as a Biomarker: Speciation Analysis for the Assessment of Dietary Exposure. In *Biomarkers in Disease: Methods, Discoveries and Applications*; Springer: Berlin/Heidelberg, Germany, 2022; pp. 173–193. [CrossRef]
121. Jomova, K.; Jenisova, Z.; Feszterova, M.; Baros, S.; Liska, J.; Hudecova, D.; Rhodes, C.J.; Valko, M. Arsenic: Toxicity, Oxidative Stress and Human Disease. *J. Appl. Toxicol.* **2011**, *31*, 95–107. [CrossRef]
122. Virk, R.K.; Garla, R.; Kaushal, N.; Bansal, M.P.; Garg, M.L.; Mohanty, B.P. The Relevance of Arsenic Speciation Analysis in Health & Medicine. *Chemosphere* **2023**, *316*, 137735. [CrossRef]
123. Khan, K.M.; Chakraborty, R.; Bundschuh, J.; Bhattacharya, P.; Parvez, F. Health Effects of Arsenic Exposure in Latin America: An Overview of the Past Eight Years of Research. *Sci. Total Environ.* **2020**, *710*, 136071. [CrossRef]
124. Concha, G.; Vogler, G.; Lezcano, D.; Nermell, B.; Vahter, M. Exposure to Inorganic Arsenic Metabolites During Early Human Development. *Toxicol. Sci. Off. J. Soc. Toxicol.* **1998**, *44*, 185–190. [CrossRef]
125. Tolins, M.; Ruchirawat, M.; Landrigan, P. The Developmental Neurotoxicity of Arsenic: Cognitive and Behavioral Consequences of Early Life Exposure. *Ann. Glob. Health* **2014**, *80*, 303–314. [CrossRef]
126. Cervantes, G.I.V.; Esquivel, D.F.G.; Ortega, D.R.; Ayala, T.B.; Chávez, L.A.R.; López-López, H.E.; Salazar, A.; Flores, I.; Pineda, B.; Gómez-Manzo, S.; et al. Mechanisms Associated with Cognitive and Behavioral Impairment Induced by Arsenic Exposure. *Cells* **2023**, *12*, 2537. [CrossRef]
127. Jones, M.R.; Tellez-Plaza, M.; Vaidya, D.; Grau, M.; Francesconi, K.A.; Goessler, W.; Guallar, E.; Post, W.S.; Kaufman, J.D.; Navas-Acien, A. Estimation of Inorganic Arsenic Exposure in Populations with Frequent Seafood Intake: Evidence from MESA and NHANES. *Am. J. Epidemiol.* **2016**, *184*, 590–602. [CrossRef]
128. Nevins, J.E.H.; Donovan, S.M.; Snetselaar, L.; Dewey, K.G.; Novotny, R.; Stang, J.; Taveras, E.M.; Kleinman, R.E.; Bailey, R.L.; Raghavan, R.; et al. Omega-3 Fatty Acid Dietary Supplements Consumed During Pregnancy and Lactation and Child Neurodevelopment: A Systematic Review. *J. Nutr.* **2021**, *151*, 3483–3494. [CrossRef]
129. Sherzai, D.; Moness, R.; Sherzai, S.; Sherzai, A. A Systematic Review of Omega-3 Fatty Acid Consumption and Cognitive Outcomes in Neurodevelopment. *Am. J. Lifestyle Med.* **2022**, *17*, 649–685. [CrossRef]
130. Basak, S.; Mallick, R.; Duttaroy, A.K. Maternal Docosahexaenoic Acid Status During Pregnancy and Its Impact on Infant Neurodevelopment. *Nutrients* **2020**, *12*, 3615. [CrossRef] [PubMed]
131. Peters, M.D.J.; Marnie, C.; Tricco, A.C.; Pollock, D.; Munn, Z.; Alexander, L.; McInerney, P.; Godfrey, C.M.; Khalil, H. Updated Methodological Guidance for the Conduct of Scoping Reviews. *JBI Evid. Synth.* **2020**, *18*, 2119–2126. [CrossRef] [PubMed]

toxics

MDPI

Article

Leachability and Health Risk Assessment of Cadmium and Other Heavy Metals in Agricultural Soils from the Mae Tao Watershed, Northern Thailand

Nipada Santha [1], Thanan Watcharamai [1], Rungroj Benjakul [1] and Schradh Saenton [1,2,3,*]

[1] Department of Geological Sciences, Faculty of Science, Chiang Mai University, Chiang Mai 50200, Thailand; nipada.santha@cmu.ac.th (N.S.); t.watcharamai@gmail.com (T.W.); rungroj.b@cmu.ac.th (R.B.)

[2] Environmental Science Research Center, Faculty of Science, Chiang Mai University, Chiang Mai 50200, Thailand

[3] Advanced Research Center for Computational Simulation, Faculty of Science, Chiang Mai University, Chiang Mai 50200, Thailand

[*] Correspondence: schradh.saenton@cmu.ac.th; Tel.: +66-5394-3302

Abstract: Decades of unregulated zinc mining activities in the Mae Tao watershed, located in Mae Sot District, Tak Province, northern Thailand, have resulted in the pervasive contamination of agricultural soils with heavy metals, particularly cadmium (Cd), zinc (Zn), lead (Pb), and manganese (Mn). This legacy pollution has significantly impacted multiple environmental compartments—including surface water, groundwater, and sediments—and poses chronic health risks to local populations. This study investigates the key geochemical and physicochemical factors governing the leachability and mobility of these metals from contaminated soils and evaluates the associated human health risks. Controlled leaching experiments demonstrated that ionic strength exerts a more pronounced influence on metal mobilization than pH or other tested variables, suggesting that the electrolyte composition of pore water plays a dominant role in heavy metal transport. Despite elevated total concentrations of Cd, Zn, Pb, and Mn in the soils, hazard quotient (HQ) calculations indicated no significant non-carcinogenic risk under typical exposure scenarios. However, Cd exhibited a carcinogenic risk above the acceptable threshold at both average and peak soil concentrations, underscoring its potential to adversely affect human health. These findings enhance the understanding of heavy metal behavior in contaminated agroecosystems and provide a scientific basis for targeted risk management and long-term monitoring strategies in the Mae Sot region.

Keywords: heavy metals; agricultural soils; health risk assessment; Mae Tao watershed; leachability

1. Introduction

Heavy metal contamination in soils has been one of the major environmental problems leading to several detrimental concerns regarding the environment and human health [1–3]. Heavy metals are released into the environment from the dissolution of primary sources such as minerals and rocks, or they may be leached from soils and sediments from sorbed metal ions, which are called a secondary source. The released metal ions are toxic and may pose a human health risk after prolonged exposure. Health risks arising from heavy metal exposure can be intensified in areas with industrial operations [4], dumpsites [4], and mining activities, including mining operations, smelting activity, mine waste, and tailings disposal [5–7]. Heavy metal pollution in agricultural soils has been widely reported across various regions. In African contexts, particularly within the six geopolitical zones of Nigeria, the literature indicates that the primary sources of heavy metals such as Cd, Cu, Ni, Pb, Zn, Co, Cr, Fe, and As include urban and industrial effluents, deteriorating sewage infrastructure, sewage sludge, fertilizers, and pesticides [4]. The concentrations of these metals span a wide range, with maximum levels frequently exceeding the WHO's permissible limits [4]. Their presence in agricultural soils poses serious environmental and public health concerns, as evidenced by their accumulation in crops, cow's milk, and animal organs [8–11]. Similarly to Kafr El-Zayat—a major agricultural and industrial city in Egypt—which faces significant soil contamination, particularly with aluminum, arsenic, cadmium, chromium, and nickel [12], these metals are known carcinogens and can cause damage to the brain, kidneys, liver, and other organs, primarily by inducing oxidative stress through free radical generation [12–14]. Furthermore, a review of heavy metal contamination in agricultural soils resulting from mining activities identified 72 affected mining areas in the southern and eastern regions of China [15]. The results indicated potential non-carcinogenic risks from the ingestion and dermal absorption of As, Cd, Cr, Cu, Ni, Pb, Zn, and Hg, as well as potential carcinogenic risks from As [15]. It was observed that non-carcinogenic risks were especially associated with lead–zinc, manganese, and tungsten mines, more so than with copper, gold, and iron mines [15]. These findings highlight that heavy metals in agricultural soils contribute to increasing toxicity levels and elevated public health risks, as evidenced by bioaccumulation and biomagnification.

The Pa Daeng zinc mine, located in Mae Tao watershed, Mae Sot district, Tak province, northern Thailand, is the largest zinc mine in Southeast Asia, which has operated for almost four decades. The mining activity continuously feeds cadmium, zinc, and lead contamination into nearby surface water, groundwater, and soils [16,17]. The Pa Daeng zinc deposit is characterized as a secondary mineral deposit [18], which was formed via the infiltration of zinc solution as a result of oxidized primary zinc sulfide and subsequent groundwater transport before it was redeposited along the fault planes, fractures, and pore spaces in the rock [19]. The redeposited ore was weathered by rain, surface water, and groundwater, thus generating a continuing flux of dissolved heavy metals. The Mae Tao Creek and its tributaries are the major transport route of the sediments and have contributed to the distribution of heavy metals in soils and sediments downstream [20]. Moreover, the Mae Tao Creek system is the main recharge into the Mae Tao watershed, making it a chief water supply for local people for drinking, consuming, and agriculture [17]. The Mae Tao watershed, which is close to a mining area and human communities, could cause environmental and health issues.

Over almost two decades, there have been numerous studies on health-related and environmental problems in the Mae Tao watershed. It has been revealed that this area has a wide range contaminants, such as cadmium, zinc, lead, and manganese in soil, water, and plants. The most dangerous situation is that the high mobility of cadmium and zinc in soil–plant systems permits their easy entry into food chains [21], where they may stimulate both human diseases [22] and toxic effects on animals, microorganisms, and plants [23]. The harmful health impacts on cadmium-exposed populations in the Mae Sot district have been reported [24–28]. Contamination has led to health risk assessment, where it has been reported that Mae Sot residents have irregularly higher cadmium contamination in their blood, bones, and urine than normal Thai residents, resulting in a 10–86% increase in disability-adjusted life years [29]. Recently, it was found that Mae Tao's rice can take up heavy metals, which accumulate in the grain [30–32], and that rice consumption causes health impacts. Almost 20% of 159 samples collected in six villages of the Mae Tao sub-district, Mae Sot district, Tak province, were contaminated with a high content of Cd [30]. Sticky rice contained, on average, 0.1–0.7 mg/kg of Cd, with 2.60 mg/kg as the maximum value, while white rice showed 0.1–0.3 mg/kg, with the maximum at 1.94 mg/kg [30]. A health risk assessment of rice consumption was conducted and the result identified was that both types were regarded as having a significant hazard quotient (>1) at the maximum Cd concentration; however, only sticky rice poses a health risk when considering the average Cd content [30]. The stream sediments of the Mae Tao Creek showed hazardous levels of cadmium and zinc, with maxima of 37.11 mg/kg and 1231 mg/kg, and the highest values of suspended solids were found to be 18.27 mg/kg of Cd and 17,767 mg/kg of Zn [17]. Intense concentrations of Cd and Zn in the soil of the agricultural area receiving water from the Mae Tao Creek were found—5.93 to 30.4 mg/kg and 286 to 594 mg/kg, respectively [32]. Heavy metal contamination in the Mae Tao watershed has led to a particular focus on health risk assessments of rice consumption, but the persistence of solid concentrations in soil and stream sediments that function as the main heavy metal source for plant uptake, to which humans are directly exposed, have been considered only in concentration and distribution. Therefore, potential health risk assessments of heavy metals in the agricultural soil of the Mae Tao watershed require further study.

Heavy metal contamination can be determined from its intensity and its leaching potential, as well as chemical speciation, since the toxicity of heavy metals depends on the concentrations of species [33,34]. Heavy metals will be more harmful when they are leaching and being distributed into biota (bioavailability). John and Leventhal [35] gave the definition of bioavailability as the ratio of total element concentrations that are available for integration into biota or bioaccumulation [35]. They also explained that the plant root can take up the heavy metals crossing the membrane of epidermal cells into plants' cells (bioaccumulation) from the contaminated soil; afterwards, the metal accumulated in plants is consumed by herbivores and humans [35]. In this case, the trace elements enter through the food chain and accumulate in the higher-order consumer, which causes significant health risks across multiple physiological systems [36]. For instance, cadmium exposure is extensively linked to a range of adverse outcomes, including respiratory issues like lung diseases, hypertension, and various malignancies such as prostate, bladder, pancreatic, kidney, and breast cancers [37]. Furthermore, Cd has been implicated in neurological disorders, notably Alzheimer's and Parkinson's diseases [38–40]. Its ability to cross the placental barrier also raises concerns about its potential impact on fetal development [41]. Similarly, lead exposure is known to disrupt a wide array of bodily functions, affecting the neurological, skeletal, reproductive, hematopoietic, renal, and cardiovascular systems [42]. While essential in moderate amounts, excessive zinc intake can lead to toxicity, manifesting as adverse effects on the neuronal, gastrointestinal, or respiratory systems [43]. Likewise, chronic overexposure to manganese can result in progressive and permanent neurotoxicity, often accompanied by toxicity to the lungs, heart, liver, and reproductive system [44,45]. Therefore, the bioavailability of heavy metals that have been identified as having a significant impact on human health will be used for potential health risk assessments.

The primary objectives of this study are to characterize the geochemical properties of native soils and quantify the total concentrations of cadmium (Cd), zinc (Zn), lead (Pb), and manganese (Mn). Soil samples exhibiting elevated levels of these metals will be further evaluated for their leachability under controlled conditions, with a particular focus on pH and ionic strength—two key factors influencing metal bioavailability [35]. This assessment aims to elucidate the potential for metal mobilization within the environment. In addition, a comprehensive human health risk assessment will be conducted by considering all relevant exposure pathways associated with the use of contaminated agricultural soils.

2. Materials and Methods

2.1. Study Area

Soil sampling was conducted within the Mae Tao watershed, an agricultural region located in Mae Sot District, Tak Province, northern Thailand. From an environmental geology perspective, this area has been identified as a hotspot for heavy metal contamination, particularly along the Mae Tao Creek, which flows downstream from the Pa Daeng zinc mine [17] (see Figure 1). The creek also serves as a primary source of irrigation water for local agriculture. This study aimed not only to assess the total metal concentrations, but also to evaluate the bioavailability of heavy metals in agricultural soils. The sampling design was primarily based on land use classification, with soil collected specifically from paddy fields currently used by local farmers for cultivating rice and corn.

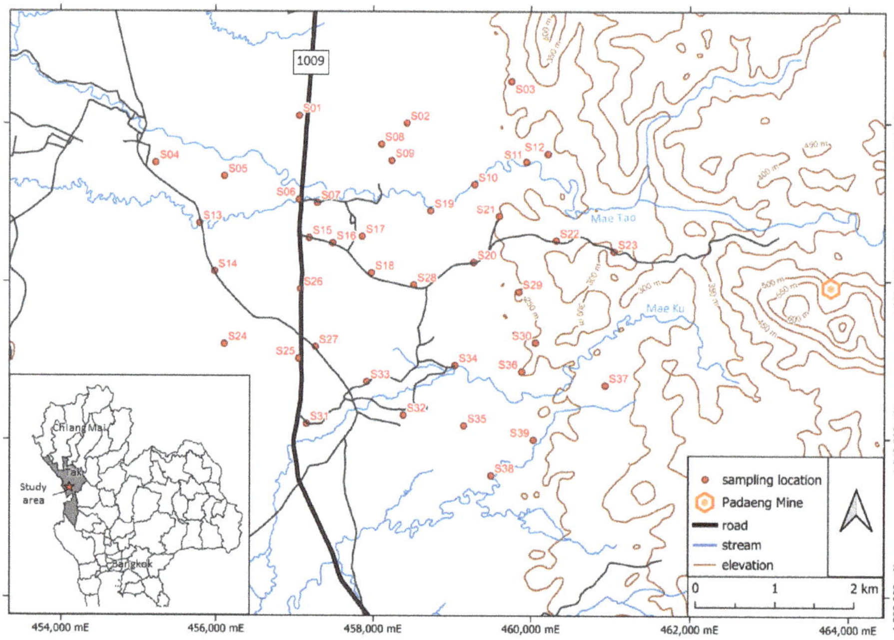

Figure 1. The locations of sampling points in the study area and Pa Daeng zinc mine, Mae Tao watershed (modified from Royal Thai Survey Department, 1999 [46]).

2.2. Field Sampling

A total of 39 agricultural soil samples were collected across an area of approximately 25 km^2. Sampling locations were selected randomly from active farmland currently used by local residents for cultivating corn and rice. To minimize spatial autocorrelation, each sampling point was spaced at least 200 m apart, as illustrated in Figure 1. The sampling sites were georeferenced based on the topographic map of Mae Sot District (Sheet 4742 III) [46]. Soil samples were collected from a depth of 30 cm below the ground surface, which covered the plowing depth in the study area, using a spade. Approximately 1 kg of soil was placed into clean, transparent polyethylene zipper bags while maintaining field-moist conditions. The bags were sealed immediately after sampling to preserve the natural moisture and chemical properties of the soil, as required for subsequent geochemical and physicochemical analyses in the laboratory.

2.3. Chemical Analysis

2.3.1. Basic Soil Properties Analysis

The basic soil properties were determined using a 1:2 soil-to-water suspension method [47]. For each sample, 50 g of soil was placed into a 250 mL flask, followed by the addition of 100 mL of distilled water. The flasks were sealed with Parafilm® (Bemis Company, Inc., Neenah, WI, USA) to prevent evaporation and contamination. Measurements of key soil parameters—temperature (°C), pH (hydrogen ion activity), electrical conductivity (EC, in μS/cm), and oxidation-reduction potential (ORP, in mV)—were conducted at multiple time intervals: 0.5, 1.5, 4, 12, 24, 36, and 48 h after mixing. The pH and oxidation-reduction potential (ORP) were measured using a portable pH meter (HI9125, Hanna Instruments, Cluj Napoca, Romania), while electrical conductivity (EC) was determined with a multi-range conductivity meter (HI8733, Hanna Instruments, Amorim, Portugal). These measurements provided insights into the temporal stability and interaction of soil's chemical properties in an aqueous suspension.

2.3.2. Chemical Analysis for Total Element Concentration

Prior to chemical analysis for the total elemental concentrations of four heavy metals—cadmium (Cd), zinc (Zn), lead (Pb), and manganese (Mn)—approximately 500 g of each soil sample was oven-dried at 70 °C for 72 h. Once dried, samples were manually cleaned to remove extraneous materials such as rock fragments, plant residues, and other debris. The cleaned soil was then finely ground using a porcelain mortar and pestle to obtain a homogenous powder. Each powdered sample was divided into two subsamples: one sieved through an 80-mesh nylon sieve (<177 μm) and the other through a 200-mesh nylon sieve (<74 μm).

Subsequently, approximately 500 mg of each sieved subsample was subjected to acid digestion using a tri-acid mixture comprising 20 mL of concentrated nitric acid (HNO_3), 20 mL of hydrofluoric acid (HF), and 2 mL of perchloric acid ($HClO_4$) in a ratio of 10:10:1 [48]. For sample digestion and other analytical procedures, all acids (including concentrated HNO_3 and HCl) and reagents used were of analytical grade and supplied by MERCK (Darmstadt, Germany). Digestion was conducted under a laboratory chemical fume hood with controlled heating until the solution reached boiling and was maintained for 30 min. After cooling to an ambient laboratory temperature, the digested mixture was filtered to remove undissolved residues, and the filtrate was stored in clean 150 mL polyethylene bottles as stock solutions for subsequent analysis.

The total concentrations of Cd, Zn, Pb, and Mn were quantified using atomic absorption spectrometry (AAS) [49]. The measurements were performed with a PerkinElmer Analyst 200 instrument,(PerkinElmer, Shelton, CT, USA) operated using AA WinLab32 software version 6.5. Flame AAS (FAAS) was employed for atomization, utilizing a segmented solid-state detector with a double-beam optical system to ensure the high accuracy and reproducibility of metal quantification. As part of the quality assurance and quality control protocol, all calibration curves were rigorously evaluated for linearity, with R^2 values required to exceed 0.995 to ensure a strong linear response across the working concentration range, with deionized water and standard solutions used for each calibration curve, along with the supplier (MERCK, Germany). Moreover, the limits of quantification (LOQs) were determined by analyzing multiple replicates of samples (3 times). Analysis of the samples alternates between the samples and the standard solution (known concentration) in order to ensure acceptable accuracy and precision.

2.3.3. Determination of Heavy Metal Leachability

Three soil samples exhibiting significantly high concentrations of cadmium (Cd), lead (Pb), zinc (Zn), and manganese (Mn) were selected for heavy metal leachability testing. Each selected sample was subdivided into six sets (and each set has three replicates), with approximately 20 g of finely powdered soil (both 80-mesh and 200-mesh fractions) allocated per set. Each set was used to evaluate leachability under a distinct solution condition, simulating variations in natural environmental parameters.

The 20 g soil subsamples were placed into clean 150 mL polyethylene bottles, to which 120 mL of a test solution was added. These test solutions varied in pH and ionic strength to represent plausible environmental conditions that influence metal mobility. pH adjustments were made using diluted hydrochloric acid (HCl) and sodium hydroxide (NaOH), while ionic strength was controlled using sodium perchlorate ($NaClO_4$), a non-reactive salt that does not interfere with other chemical species in solution. The six experimental conditions, combining different levels of pH and ionic strength, are summarized in Table 1. Sodium perchlorate was chosen as the electrolyte due to its chemical inertness and minimal complexation with metals, thereby minimizing interference with metal speciation during the experiment [50,51]. The selected ionic strengths and pH values reflect the soil solution conditions typically observed in contaminated agricultural environments and allow us to systematically assess their influence on metal mobility. These parameters are commonly employed in similar leaching studies to evaluate potential environmental risk and metal bioavailability [52,53].

Table 1. The conditions of solution in leaching test.

Condition	pH	Ionic Strength (mol/L)
1	4.0	0.01
2	7.0	0.01
3	10.0	0.01
4	4.0	0.1
5	7.0	0.1
6	10.0	0.1

Each soil–solution mixture was thoroughly agitated to ensure complete interaction between the soil particles and the test solution, facilitating the accurate assessment of metal leaching behavior under the specified conditions.

For each duration time, the resulting solution would be separated from 20 mL of solution using a centrifuge and by pipetting only clear solution. The centrifuge was used in each mixture for 3500 rounds per minute (rpm) for ten minutes. The heavy metal leachabilities in this study were sampled for five duration times, including 1 day, 3 days, 7 days, 14 days, and 28 days. After each duration of sampling, every mixture needed to be shaken for mixing again. The separated clear solutions for each duration were measured for the leached amount of cadmium, zinc, lead, and manganese using an atomic absorption spectrometer to determine the heavy metals' leachability in the various solution conditions. Finally, the raw data was analyzed using a two-way ANOVA statistical method to identify whether the pH or ionic strength was a significant factor in heavy metal leaching and to determine the relationship between two variances.

2.4. Mineralogical Analysis Using X-Ray Diffractometer

In this step, the mineral composition of the powder soil samples in 200-mesh sieves, which are the same samples as in the heavy metal leaching test, were analyzed using BrukerD8 X-ray diffractometer (Bruker Corporation, Billerica, MA, USA), equipped with copper anode, at the Department of Geological Sciences, Faculty of Science, Chiang Mai University. X-ray diffraction (XRD) was performed on powder samples using an X-ray wavelength of 1.540598 nm. The identification and quantification of samples were assisted by EVA (search-match program) in the database of the International Centre for Diffraction Data (ICDD). The conditions of use were as follows: voltage 40 kV, stop 2θ 2 degrees, step size 0.04 degrees, time/step divergence 0.5 degrees, and anti-scattering slit.

2.5. Health Risk Assessment Method

In general, health risk assessment for metal-contaminated soils involves evaluating exposure through three primary pathways: ingestion, inhalation, and dermal absorption [54]. In this study, the assessment focused on the ingestion and dermal contact routes, as the inhalation pathway is considered to contribute minimally to heavy metal exposure in agricultural settings. The average daily dose (ADD) for both ingestion and dermal absorption was calculated following the guidelines provided by the United States Environmental Protection Agency (USEPA) [55]. The calculations were performed using the standard equations outlined by the USEPA, as shown in Equation (1) for ingestion and Equation (2) for dermal absorption, respectively.

$$ADD_{\text{ing}} = \frac{C \times I \times EF \times D}{A \times W}, \quad (1)$$

where C is the concentration of heavy metals in soil (mg/kg) (data from laboratory analysis), I is the ingestion rate (100 mg/d for adults and 200 mg/d for children [56]), EF is the exposure frequency (350 days per year [56]), D is the exposure duration (70 years for adults and 6 years for children), W is the body weight (56.7 kg for adults [57] and 16.2 kg for children [54]), and A is average lifespan (70 × 350 d for adults and 6 × 350 d for children [56]).

$$ADD_{\text{derm}} = \frac{C \times ESA \times AF \times ABS \times EF \times D}{A \times W} \quad (2)$$

ESA is the exposed skin surface area (5700 cm^2 for adults 2800 cm^2 for children [58]), AF is the adherence factor (0.07 for adults 0.2 for children mg cm^{-2} [58]), and ABS is the dermal absorption factor (0.001 for all metals) [59,60].

It is important to note that soil exposure may occur through ingestion, dermal absorption, or a combination of both pathways. When heavy metal contamination was considered a non-carcinogenic health risk, the hazard quotient (HQ) was also evaluated using Equation (3) for ingestion and Equation (4) for dermal absorption [55]:

$$HQ_{\text{oral}} = \frac{ADD_{\text{ing}}}{RfD_{\text{o}}}, \quad (3)$$

$$HQ_{\text{abs}} = \frac{ADD_{\text{derm}}}{RfD_{\text{abs}}}, \quad (4)$$

where RfD_o is an oral references dose (1.00 × 10^{-3} mg/kg·d for Cd, 1.40 × 10^{-3} mg/kg·d for Pb, 3.00 × 10^{-1} mg/kg·d [61,62] for Zn, and 1.40 × 10^{-1} mg/kg·d for Mn [62]), RfD_{abs} is a dermal adsorption dose (1.00 × 10^{-5} mg/kg·d for Cd and 4.20 × 10^{-4} mg/kg·d for Pb [56,58,61,62]). To interpret the HQ, a value of the HQ greater than 1 indicates human health risk. The maximum acceptable carcinogenic risk is 1 × 10^{-4}, which is considered an acceptable hazard [55].

It should be noted that carcinogenic health risk assessment only considered Cd and Pb as probable human carcinogens [56], not Zn or Mn, because the USEPA defined the Zn and

Mn groups as not classifiable regarding human carcinogenicity [63,64]. Lifetime Cancer Risk (LCR) (Equations (5) and (6)) was analyzed for potential carcinogenic health risks [65].

$$LCR_{\text{oral}} = ADD_{\text{ing}} \times CSF_{\text{oral}}, \quad (5)$$

$$LCR_{\text{derm}} = ADD_{\text{derm}} \times CSF_{\text{derm}}, \quad (6)$$

where CSF_{oral} is the oral carcinogenic slope factor (6.1 mg/kg·d for Cd [66–68], 0.0085 mg/kg·d for Pb [66,69]). CSF_{derm} is the dermal carcinogenic slope factor (6.1 mg/kg·d for Cd [66], 0.0085 mg/kg·d for Pb [66,70]).

In addition, the total cancer risk from lifetime exposure (LCR) for Cd and Pb was estimated based on the accumulative values of all exposure pathways using Equation (7) [69]:

$$LCR_{\text{cancer}} = LCR_{\text{ing}} + LCR_{\text{derm}} \quad (7)$$

If the carcinogenic risk (LCR) value is less than 1×10^{-6}, it is considered to pose no significant cancer risk. A LCR value between 1×10^{-6} and 1×10^{-4} is generally regarded as an acceptable or tolerable risk range for humans. However, if the LCR value exceeds 1×10^{-4}, it indicates a high risk of cancer development and may warrant regulatory concern or remediation.

3. Results

3.1. Basic Soil Properties

Basic soil properties—including the temperature, hydrogen ion activity (pH), electrical conductivity (EC), and oxidation-reduction potential (ORP or Eh)—were measured at multiple time intervals: 0.5, 1.5, 4.0, 12, 24, 36, and 48 h. Table 2 presents the descriptive statistics (minimum, maximum, mean, median, standard deviation, and 95% confidence intervals) for these parameters across the 39 soil samples.

Table 2. The values of pH, EC and ORP of soils measured at different times.

Variable		Min	Max	Mean	Median	SD	95% IC
pH	0.5 h	7.17	8.86	7.96	7.79	0.53	0.17
	1.5 h	6.94	8.56	7.89	7.93	0.39	0.12
	4 h	6.61	8.47	7.79	7.79	0.36	0.11
	12 h	5.96	7.63	6.98	7.09	0.44	0.14
	24 h	6.06	7.60	6.95	6.95	0.41	0.13
	36 h	5.75	7.29	6.71	6.79	0.37	0.12
EC (μS/cm)	0.5 h	2.60	194	21.4	9.7	36	11.4
	1.5 h	3.90	268	34.0	17.3	55.5	17.6
	4 h	5.40	320	44.7	28.0	66.4	21.1
	12 h	6.50	588	78.8	43.5	125	39.8
	24 h	11.10	755	110	76.2	158	50.1
	36 h	11.80	817	126	83.1	183	58.3
	48 h	13.00	932	139	90.2	204	64.7
ORP (mV)	0.5 h	123	299	217	223	40.0	12.6
	1.5 h	138	329	223	224	43.1	13.6
	4 h	160	337	241	241	33.8	10.7
	12 h	168	319	214	210	36.1	11.4
	24 h	206	366	244	237	35.3	11.3
	36 h	209	306	247	243	24.5	7.73
	48 h	214	342	262	262	25.3	7.99

The pH values ranged from 5.75 to 8.86, indicating that the soil varied from weakly acidic to weakly alkaline conditions over the course of the two-day observation period. A general decreasing trend in pH was observed with increasing time, suggesting progressive stabilization. Specifically, the pH values at each time interval were as follows: 7.17–8.86 (0.5 h), 6.94–8.56 (1.5 h), 6.61–8.47 (4.0 h), 5.96–7.63 (12 h), 6.06–7.60 (24 h), 5.75–7.29 (36 h), and 6.09–7.20 (48 h). At equilibrium, soil samples tended toward a neutral pH, with both mean and median values falling within the range of 6.91 to 7.91.

Electrical conductivity (EC) exhibited a wide range across samples, from 2.60 to 932 μS/cm, with mean and median values ranging from 9.7 to 139 μS/cm. In contrast, the oxidation-reduction potential (ORP) values showed relatively narrow variation, ranging between 214 and 366 mV, reflecting stable redox conditions across the samples.

3.2. Metal Contents, Leachability Test and Mineralogical Contants

3.2.1. Metal Concentrations

The total metal concentrations were determined using an acid digestion method, in which concentrated acids were applied to extract metals from soil samples prepared in both 80-mesh and 200-mesh fractions. The concentrations of four target heavy metals—cadmium (Cd), zinc (Zn), lead (Pb), and manganese (Mn)—were analyzed in soil samples collected from the Mae Tao watershed. A statistical summary of the metal concentrations, along with the relevant regulatory limits, was compiled based on data from 39 samples. Table 3 presents the descriptive statistics for each heavy metal, including the minimum, maximum, mean, median, and standard deviation.

Table 3. Heavy metal contents in soils.

Metals	Size	Min	Max	Mean	Median	Regulated Standard
Cd (mg/kg)	80-mesh	6.59	135	23.90	17.40	3 *
	200-mesh	7.40	251	30.00	17.80	3 *
Zn (mg/kg)	80-mesh	45.60	2756	338	128	300 *
	200-mesh	51.70	3269	406	147	300 *
Pb (mg/kg)	80-mesh	5.40	144	38.90	31.40	300 *
	200-mesh	7.20	235	45.40	33.40	300 *
Mn (mg/kg)	80-mesh	139	2917	892	792	1800 **
	200-mesh	166	3255	936	810	1800 **

Sources: European Union (EU) maximum permissible [71] *, the Pollution Control Department (PCD) maximum permissible [72] **.

The cadmium contents of 80-mesh sieved soil samples ranged from 6.59 to 135 mg/kg, and the average was 23.90 mg/kg. All the values had a lower cadmium content in 200-mesh sieved soil samples, ranging from 7.40 to 251 mg/kg, and the mean was 30.00 mg/kg. In the same way, the amounts of zinc, lead, and manganese in 80-mesh sieved samples are smaller than those in 200-mesh soil samples. The zinc contents of 80-mesh sieved soil samples ranged from 45.6 to 2756 mg/kg, and the average was 338 mg/kg, lower than in 200-mesh samples, whose contents ranged from 51.6 to 3269 mg/kg, with a mean value of 406 mg/kg. The lead contents of 80-mesh sieved soil samples ranged from 5.40 to 144 mg/kg, with an average of 38.9 mg/kg. In the case of 200-mesh-sized soil samples, the lead content ranged from 7.20 to 235 mg/kg, and the mean was 45.4 mg/kg. Moreover, in 80-mesh samples, the amount of manganese was found to range from approximately 139 to 2917 mg/kg, with a mean value of 892 mg/kg. The manganese contents ranged from 166 to 3255 mg/kg, with an average of 936 mg/kg in 200-mesh samples. This study showed that the larger soil samples (80-mesh) contained lower contents than the small, sieved soil samples (200-mesh) of all the types of heavy metals.

In addition, the comparison of the measured heavy metals with the standard values from the European Union (EU) [71] and the Pollution Control Department (PCD) maximum permissible [72] was considered, in order to carry out a preliminary risk assessment. All the statistical values of cadmium content were significantly greater than the maximum allowance, despite minimum values of Cd (6.59 mg/kg for 80-mesh and 7.40 mg/kg for 200-mesh) that exceeded approximately two times the guideline values (3 mg/kg), but all values of Pb were lower than the standard values (300 mg/kg). Considering Zn and Mn, the average values of Zn (338 mg/kg for 80-mesh and 406 mg/kg for 200-mesh) were over the European Union maximum permissible, which is 300 mg/kg, but Mn (892 mg/kg for 80-mesh and 936 mg/kg for 200-mesh) had acceptable values; however, the maximum values of Zn (2756 mg/kg for 80-mesh and 3269 mg/kg for 200-mesh) and Mn (2917 mg/kg for 80-mesh and 3255 mg/kg for 200-mesh) were above the EU maximum permissible by 10 and 1.6 times, respectively. Notably, the total concentrations of Cd, Zn, Pb, and Mn were 5952 mg/kg for 80-mesh and 7010 mg/kg for 200-mesh.

3.2.2. Leachability of Metals

Three soil samples (S04, S12, and S19), exhibiting the highest total concentrations of cadmium (Cd), zinc (Zn), lead (Pb), and manganese (Mn), were selected for the leaching experiment. For each site, both 80-mesh and 200-mesh sieved fractions were tested, yielding a total of six experimental soil samples. The leachability of heavy metals was assessed under controlled laboratory conditions, by varying two key factors: ionic strength (0.01 M and 0.10 M $NaClO_4$) and pH (adjusted to 4, 7, and 10). The leaching tests were conducted over a 28-day period, with leachate samples collected on days 1, 3, 7, 14, and 28. The concentrations of leached metals were quantified using atomic absorption spectrometry (AAS).

Overall, the leached concentrations were substantially lower than the corresponding total metal contents, and the leaching behavior was strongly influenced by both the soil particle size and ionic strength (see Tables S1–S4). The data also showed variability over time. To assess whether steady-state leaching was achieved, leached concentrations after 14 days were analyzed using linear regression. Following the criterion that a steady state is reached when the slope of the concentration-versus-time curve is not statistically different from zero [73], regression analysis (performed in Microsoft Excel®) confirmed that the leaching rates had stabilized by day 14 ($p > 0.05$); see Figures S1–S4.

The leaching percentages for Cd, Zn, Pb, and Mn were calculated as the ratio of metal leached at day 14 to the total metal content in each sample. These percentages were further analyzed using two-way ANOVA, with replications to evaluate the significance of pH and ionic strength effects. The results are summarized in Table 4. A statistically significant effect was indicated when the calculated *F*-value exceeded the *F*-critical value [74].

The analysis revealed that ionic strength had a significant impact on Cd leaching in both 80-mesh and 200-mesh samples, and on Pb leaching in the 80-mesh fraction. In contrast, Zn and Mn leaching were not significantly affected by ionic strength. As shown in Figure 2, soils subjected to the higher ionic strength (0.10 M) exhibited markedly greater leaching of Cd and Pb compared to the 0.01 M treatment. This indicates enhanced metal mobility under higher ionic strength conditions.

Regarding pH, no statistically significant differences in leaching were observed across the pH levels tested. Furthermore, the metal leaching trends varied: Cd, Zn, and Pb showed relatively stable leaching profiles, while Mn (particularly in the 200-mesh samples) exhibited fluctuating concentrations over time (Figure 2), indicating an unstable leaching pattern.

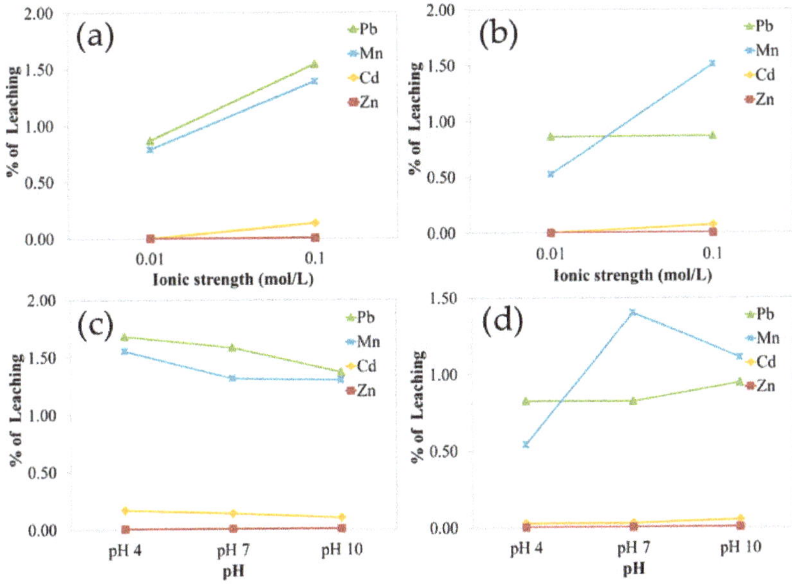

Figure 2. Graph summarizing two-way ANOVA results of the percentages of leaching of lead, manganese, cadmium, and zinc for ionic strength; (**a**) ionic strength factor of 80-mesh-sized samples, (**b**) ionic strength factor of 200-mesh-sized samples; (**c**) pH factor of 80-mesh-sized samples; and (**d**) pH factor of 200-mesh-sized samples.

Table 4. Two-way ANOVA results of cadmium, zinc, lead, and manganese in both 80- and 200-mesh-sized samples.

Cd		**80-Mesh**					**200-Mesh**			
Variation	**Sum of Squares**	**Mean Square**	**F**	***p*-Value**	**F-Crit**	**Sum of Squares**	**Mean Square**	**F**	***p*-Value**	**F-Crit**
Ionic strength	0.083	0.084	71.710	0.000	4.747	0.023	0.023	21.592	0.001	4.747
pH	0.004	0.002	1.525	0.257	3.885	0.002	0.001	1.026	0.388	3.885
Interaction	0.004	0.002	1.525	0.257	3.885	0.002	0.001	1.026	0.388	3.885
Within	0.014	0.001				0.013	0.001			
Total	0.105					0.040				
Zn		80-mesh					200-mesh			
Ionic strength	1.27×10^{-6}	1.27×10^{-6}	0.048	0.830	4.747	4.67×10^{-8}	4.67×10^{-8}	0.0073	0.933	4.747
pH	2.52×10^{-5}	1.26×10^{-5}	0.475	0.633	3.885	2.73×10^{-5}	1.37×10^{-5}	2.1251	0.162	3.885
Interaction	2.57×10^{-5}	1.28×10^{-5}	0.485	0.627	3.885	1.42×10^{-5}	7.11×10^{-6}	1.1053	0.363	3.885
Within	3.18×10^{-4}	2.65×10^{-5}				7.72×10^{-5}	6.43×10^{-6}			
Total	3.70×10^{-4}					1.19×10^{-4}				
Pb		80-mesh					200-mesh			
Ionic strength	2.028	2.028	4.966	0.046	4.747	0.000	0.000	0.000	0.985	4.747
pH	0.409	0.205	0.501	0.618	3.885	0.058	0.029	0.119	0.889	3.885
Interaction	0.094	0.047	0.115	0.892	3.885	0.328	0.164	0.680	0.525	3.885
Within	4.902	0.409				2.891	0.241			
Total	7.434					3.276				
Mn		80-mesh					200-mesh			
Ionic strength	1.604	1.604	1.383	0.262	4.747	4.380	4.380	3.425	0.089	4.747
pH	0.019	0.009	0.008	0.992	3.885	2.286	1.143	0.894	0.435	3.885
Interaction	0.287	0.143	0.124	0.885	3.885	0.946	0.473	0.370	0.699	3.885
Within	13.917	1.160				15.350	1.279			
Total	15.826					22.962				

3.2.3. Mineralogical Composition of Soils

An X-ray diffraction technique was used to analyze the mineral compositions and carry out the semi-quantitative analysis, in which the results were applied in the stimulation part. This procedure used three 200-mesh sieved representatives. The XRD results are shown in Figure S3 and the semi-quantitative analysis of mineral compositions can be tentatively carried out using an X-ray diffractometer (XRD), using the percentage of all mineral compositions in the soil samples, as summarized in Table 5. The mineral compositions of samples mainly consisted of four minerals, namely quartz (76.65–87.91%), magnesium calcite (2.84–4.82%), dolomite (5.45–11.91%), and clay mineral (0–7.87%). Clay minerals disappeared from one out of three samples. The clay mineral contents of the two samples were from different categories. One sample found only illite (3.79%), and the other contained both illite (3.30%) and chlorite (4.57%). It should be noted that X-ray diffraction (XRD) was employed to identify the major mineralogical composition of soil samples. While XRD is effective for detecting dominant, well-crystallized phases, its utility is limited for trace-level or poorly crystalline constituents. The detection threshold for XRD typically exceeds 1–2 wt%, meaning that low-abundance Cd-bearing phases may not be observable. Furthermore, XRD does not provide information on the chemical speciation or binding state of trace metals such as cadmium. Therefore, the absence of identifiable Cd minerals in our analysis does not preclude the presence of Cd in labile or adsorbed forms. This limitation was considered in the interpretation of results and highlights the need for complementary techniques in future studies.

Table 5. Results of semi-quantitative analysis of soil samples using XRD.

Sample No. (Name)	Composition				
	Quartz (%)	Magnesium Calcite (%)	Dolomite (%)	Clay Minerals	
				Illite (%)	Chlorite (%)
1 (S4)	84.62	3.47	11.91	-	-
2 (S12)	87.91	2.84	5.45	3.79	-
3 (S19)	76.65	4.82	10.66	3.30	4.57

3.3. *Health Risk Assessment Calculation*

Non-carcinogenic risk was assessed through the calculation of the average daily dose (ADD, in mg/kg × d) and total hazard quotient (HQ) for four heavy metals: cadmium (Cd), lead (Pb), zinc (Zn), and manganese (Mn). These calculations were based on the mean, minimum, and maximum concentrations of each metal in two particle size fractions (80-mesh and 200-mesh) and considered two exposure pathways: ingestion and dermal contact. The ADD and HQ values were evaluated separately for adults and children, as summarized in Tables 6 and 7, respectively.

The results indicated that the ADD values for ingestion (ADD_{ing}) were significantly higher than those for dermal absorption (ADD_{derm}) across all metals and age groups. The ADD values were comparable between adults and children, but ingestion remained the dominant exposure route. A positive correlation was observed between the metal concentration and ADD, with the ranking of average intake levels as follows: Mn > Zn > Cd > Pb.

In contrast, the hazard quotient (HQ) values followed a different trend: Cd > Pb > Zn > Mn. Notably, the HQ values for children were approximately one to two times higher than those for adults, reflecting their increased vulnerability. Additionally, the HQs were higher for the smaller particle size (200-mesh), indicating the greater bioavailability of metals in finer soils.

Table 6. Average daily dose (*ADD*) and hazard quotient (*HQ*) of heavy metals via ingestion and dermal adsorption from agricultural soils for adults.

Metals	Value	Size	Concentration (mg/kg)	ADD_{ing}	ADD_{derm}	HQ_{ing}	HQ_{derm}
Cd	Mean	80-mesh	23.90	4.04×10^{-5}	1.61×10^{-7}	4.04×10^{-2}	1.61×10^{-2}
		200-mesh	30.00	5.07×10^{-5}	2.02×10^{-7}	5.07×10^{-2}	2.02×10^{-2}
	Min	80-mesh	6.59	1.11×10^{-5}	4.45×10^{-8}	1.11×10^{-2}	4.45×10^{-3}
		200-mesh	7.40	1.25×10^{-5}	4.99×10^{-8}	1.25×10^{-2}	4.99×10^{-3}
	Max	80-mesh	135.00	2.28×10^{-4}	9.11×10^{-7}	2.28×10^{-1}	9.11×10^{-2}
		200-mesh	251.00	4.24×10^{-4}	1.69×10^{-6}	4.24×10^{-1}	1.69×10^{-1}
Pb	Mean	80-mesh	38.90	6.58×10^{-5}	2.62×10^{-7}	4.70×10^{-2}	6.25×10^{-4}
		200-mesh	45.40	7.68×10^{-5}	3.06×10^{-7}	5.48×10^{-2}	7.29×10^{-4}
	Min	80-mesh	5.40	9.13×10^{-6}	3.64×10^{-8}	6.52×10^{-3}	8.68×10^{-5}
		200-mesh	7.20	1.22×10^{-5}	4.86×10^{-8}	8.70×10^{-3}	1.16×10^{-4}
	Max	80-mesh	144.00	2.44×10^{-4}	9.72×10^{-7}	1.74×10^{-1}	2.31×10^{-3}
		200-mesh	235.00	3.97×10^{-4}	1.59×10^{-6}	2.84×10^{-1}	3.78×10^{-3}
Zn	Mean	80-mesh	338.00	5.72×10^{-4}	2.28×10^{-6}	1.91×10^{-3}	-
		200-mesh	406.00	6.87×10^{-4}	2.74×10^{-6}	2.29×10^{-3}	-
	Min	80-mesh	45.60	7.71×10^{-5}	3.08×10^{-7}	2.57×10^{-4}	-
		200-mesh	51.70	8.74×10^{-5}	3.49×10^{-7}	2.91×10^{-4}	-
	Max	80-mesh	2756.00	4.66×10^{-3}	1.86×10^{-5}	1.55×10^{-2}	-
		200-mesh	3269.00	5.53×10^{-3}	2.21×10^{-5}	1.84×10^{-2}	-
Mn	Mean	80-mesh	892.00	1.51×10^{-3}	6.02×10^{-6}	1.08×10^{-2}	-
		200-mesh	936.00	1.58×10^{-3}	6.32×10^{-6}	1.13×10^{-2}	-
	Min	80-mesh	139.00	2.35×10^{-4}	9.38×10^{-7}	1.68×10^{-3}	-
		200-mesh	166.00	2.81×10^{-4}	1.12×10^{-6}	2.01×10^{-3}	-
	Max	80-mesh	2917.00	4.93×10^{-3}	1.97×10^{-5}	3.52×10^{-2}	-
		200-mesh	3255.00	5.50×10^{-3}	2.20×10^{-5}	3.93×10^{-2}	-

Table 7. Average daily dose (*ADD*) and hazard quotient (*HQ*) of heavy metals via ingestion and dermal adsorption from agricultural soils for children.

Metals	Value	Size	Concentration (mg/kg)	ADD_{ing}	ADD_{derm}	HQ_{ing}	HQ_{derm}
Cd	Mean	80-mesh	23.90	2.43×10^{-5}	6.79×10^{-8}	2.43×10^{-2}	6.79×10^{-3}
		200-mesh	30.00	3.04×10^{-5}	8.52×10^{-8}	3.04×10^{-2}	8.52×10^{-3}
	Min	80-mesh	6.59	6.69×10^{-6}	1.87×10^{-8}	6.69×10^{-3}	1.87×10^{-3}
		200-mesh	7.40	7.51×10^{-6}	2.10×10^{-8}	7.51×10^{-3}	2.10×10^{-3}
	Max	80-mesh	135.00	1.37×10^{-4}	3.84×10^{-7}	1.37×10^{-1}	3.84×10^{-2}
		200-mesh	251.00	2.55×10^{-4}	7.13×10^{-7}	2.55×10^{-1}	7.13×10^{-2}
Pb	Mean	80-mesh	38.90	3.95×10^{-5}	1.11×10^{-7}	2.82×10^{-2}	2.63×10^{-4}
		200-mesh	45.40	4.61×10^{-5}	1.29×10^{-7}	3.29×10^{-2}	3.07×10^{-4}
	Min	80-mesh	5.40	5.48×10^{-6}	1.53×10^{-8}	3.91×10^{-3}	3.65×10^{-5}
		200-mesh	7.20	7.31×10^{-6}	2.05×10^{-8}	5.22×10^{-3}	4.87×10^{-5}
	Max	80-mesh	144.00	1.46×10^{-4}	4.09×10^{-7}	1.04×10^{-1}	9.74×10^{-4}
		200-mesh	235.00	2.38×10^{-4}	6.68×10^{-7}	1.70×10^{-1}	1.59×10^{-3}
Zn	Mean	80-mesh	338.00	3.43×10^{-4}	9.60×10^{-7}	1.14×10^{-3}	-
		200-mesh	406.00	4.12×10^{-4}	1.15×10^{-6}	1.37×10^{-3}	-
	Min	80-mesh	45.60	4.63×10^{-5}	1.30×10^{-7}	1.54×10^{-4}	-
		200-mesh	51.70	5.25×10^{-5}	1.47×10^{-7}	1.75×10^{-4}	-
	Max	80-mesh	2756.00	2.80×10^{-3}	7.83×10^{-6}	9.32×10^{-3}	-
		200-mesh	3269.00	3.32×10^{-3}	9.29×10^{-6}	1.11×10^{-2}	-
Mn	Mean	80-mesh	892.00	9.05×10^{-4}	2.53×10^{-6}	6.47×10^{-3}	-
		200-mesh	936.00	9.50×10^{-4}	2.66×10^{-6}	6.78×10^{-3}	-
	Min	80-mesh	139.00	1.41×10^{-4}	3.95×10^{-7}	1.01×10^{-3}	-
		200-mesh	166.00	1.68×10^{-4}	4.72×10^{-7}	1.20×10^{-3}	-
	Max	80-mesh	2917.00	2.96×10^{-3}	8.29×10^{-6}	2.11×10^{-2}	-
		200-mesh	3255.00	3.30×10^{-3}	9.25×10^{-6}	2.36×10^{-2}	-

Despite these differences, all HQ values for both adults and children remained below the threshold of 1, suggesting no significant non-carcinogenic health risk. However, it is important to note that both adults and children are sensitive to the ingestion of Cd and Pb, while adults represent a more sensitive population group for the dermal adsorption of Cd. These results may require special attention in long-term exposure scenarios.

For the carcinogenic risk assessment (Table 8), the calculated cancer risk (CR) values for cadmium (Cd) indicate that exposure via ingestion at both the maximum and average values of concentrations in both 80-mesh and 200-mesh soil exceeds the acceptable threshold ($CR > 1 \times 10^{-4}$) for both adults and children, suggesting a potentially significant cancer risk from this pathway. In the case of dermal adsorption, the concentrations of cadmium and lead in both grain sizes resulted in a lifetime cancer risk (LCR) value of less than 1×10^{-4}, indicating no carcinogenic concern via dermal exposure.

Table 8. Cancer risk of assessed metals via ingestion and dermal adsorption from agricultural soils.

Metals	Value	Age	Size	Concentration (mg/kg)	LCR_{ing}	LCR_{derm}	LCR_{cancer}
Cd	Mean	Adults	80-mesh	23.90	2.47×10^{-4}	9.84×10^{-7}	2.48×10^{-4}
			200-mesh	30.00	3.09×10^{-4}	1.23×10^{-6}	3.11×10^{-4}
		Children	80-mesh	23.90	1.48×10^{-4}	4.14×10^{-7}	1.48×10^{-4}
			200-mesh	30.00	1.86×10^{-4}	5.20×10^{-7}	1.86×10^{-4}
	Min	Adults	80-mesh	6.59	6.80×10^{-5}	2.71×10^{-7}	6.83×10^{-5}
			200-mesh	7.40	7.63×10^{-5}	3.05×10^{-7}	7.66×10^{-5}
		Children	80-mesh	6.59	4.08×10^{-5}	1.14×10^{-7}	4.09×10^{-5}
			200-mesh	7.40	4.58×10^{-5}	1.28×10^{-7}	4.59×10^{-5}
	Max	Adults	80-mesh	135.00	1.39×10^{-3}	5.56×10^{-6}	1.40×10^{-3}
			200-mesh	251.00	2.59×10^{-3}	1.03×10^{-5}	2.60×10^{-3}
		Children	80-mesh	135.00	8.36×10^{-4}	2.34×10^{-6}	8.38×10^{-4}
			200-mesh	251.00	1.55×10^{-3}	4.35×10^{-6}	1.56×10^{-3}
Pb	Mean	Adults	80-mesh	38.90	5.59×10^{-7}	2.23×10^{-9}	5.61×10^{-7}
			200-mesh	45.40	6.53×10^{-7}	2.60×10^{-9}	6.55×10^{-7}
		Children	80-mesh	38.90	3.36×10^{-7}	9.39×10^{-10}	3.36×10^{-7}
			200-mesh	45.40	3.92×10^{-7}	1.10×10^{-9}	3.93×10^{-7}
	Min	Adults	80-mesh	5.40	7.76×10^{-8}	3.10×10^{-10}	7.79×10^{-8}
			200-mesh	7.20	1.04×10^{-7}	4.13×10^{-10}	1.04×10^{-7}
		Children	80-mesh	5.40	4.66×10^{-8}	1.30×10^{-10}	4.67×10^{-8}
			200-mesh	7.20	6.21×10^{-8}	1.74×10^{-10}	6.23×10^{-8}
	Max	Adults	80-mesh	144.00	2.07×10^{-6}	8.26×10^{-9}	2.08×10^{-6}
			200-mesh	235.00	3.38×10^{-6}	1.35×10^{-8}	3.39×10^{-6}
		Children	80-mesh	144.00	1.24×10^{-6}	3.48×10^{-9}	1.25×10^{-6}
			200-mesh	235.00	2.03×10^{-6}	5.68×10^{-9}	2.03×10^{-6}

These findings suggest that elevated Cd concentrations in fine soil particles (200-mesh) pose a notable carcinogenic risk through ingestion for both adults and children, and through the ingestion of the average and maximum concentration levels. In contrast, the LCR values for lead (Pb) through both the oral and dermal exposure pathways ranged from 9.39×10^{-10} (dermal) to 1.24×10^{-6} (oral). These values fall within or below the generally accepted risk range of 1×10^{-6} to 1×10^{-4}, with values below 1×10^{-6} considered to indicate no significant carcinogenic risk. Therefore, the carcinogenic risk associated with Pb exposure in this study is considered negligible to acceptable under the current exposure conditions.

4. Discussion

4.1. Basic Properties of Soils

The analysis of basic soil properties across samples collected from the Mae Tao watershed revealed dynamic changes in soil chemistry over time. In particular, the pH values of soil–water suspensions demonstrated a notable shift from weakly alkaline to neutral or slightly acidic conditions over a 48 h observation period. The initial alkaline conditions gradually declined with increasing time, suggesting the onset of chemical equilibration. This trend can be attributed to the activity of hydrogen (H^+) and aluminum (Al^{3+}) ions, which are key contributors to soil acidity. Aluminum ions undergo hydrolysis in the aqueous phase, forming $AlOH^{2+}$ and releasing protons (H^+), which in turn reduce pH levels over time [75,76]. Each Al^{3+} ion can produce up to three H^+ ions, amplifying the acidifying effect in the soil solution during equilibration. Despite this acidification, the mean and median pH values across all samples remained within the neutral range of 6.91–7.91, indicating that the soils tend to stabilize around a neutral pH under natural conditions.

The electrical conductivity (EC) values varied widely among the samples, ranging from 2.60 to 932 μS/cm. This wide distribution indicates heterogeneity in the soluble salt content, likely influenced by agricultural practices such as the use of chemical fertilizers and irrigation water sourced from contaminated streams, particularly the Mae Tao Creek [77]. Over time, the EC values exhibited a consistent increase, driven by the dissolution of salts and ions present in the soil into the initially deionized water used for extraction. As equilibrium was approached, ion release stabilized, reflecting the chemical dynamics of soil–water interactions.

Oxidation-reduction potential (Eh) measurements showed moderate oxidation conditions across all samples, with values ranging between 214 and 366 mV. None of the samples exhibited reducing conditions (Eh < 100 mV). These values suggest that the soils are aerated and not waterlogged, which is consistent with agricultural usage. The combined data for pH, EC, and Eh further suggest that the soils exhibit considerable variability in physicochemical properties, likely due to the differential impacts of anthropogenic inputs, including fertilizers, pesticides, irrigation, and industrial discharges. This heterogeneity has important implications for the behavior, mobility, and bioavailability of heavy metals in the soil matrix.

4.2. Total Heavy Metal Concentrations and Soil Particle Size Effects

The total concentrations of cadmium (Cd), zinc (Zn), lead (Pb), and manganese (Mn) were determined for both 80-mesh and 200-mesh sieved soil fractions. Notably, the finer 200-mesh particles consistently exhibited higher concentrations of all four metals. For instance, the Cd levels in 200-mesh samples ranged up to 251 mg/kg (mean: 30.00 mg/kg), significantly higher than the 135 mg/kg maximum observed in 80-mesh samples (mean: 23.90 mg/kg). Similar trends were observed for Zn, Pb, and Mn. This size-dependent distribution reflects the greater specific surface area of finer particles, which enhances the sorption capacity and metal retention via surface complexation and ion exchange [76].

These findings confirm that particle size plays a critical role in determining the total metal concentrations in soil, and that finer particles present a greater risk to metal mobilization and bioavailability. The increased surface area in smaller particles provides more reactive sites for the adsorption of metals, making them more relevant when assessing environmental risk and remediation potential.

4.3. Comparison with Regulatory Standards

When compared with the regulatory limits established by the European Union (EU) [71] and Thailand's Pollution Control Department (PCD) [72], the concentrations of Cd in all samples exceeded the permissible thresholds, with even the minimum values surpassing the guideline levels by a factor of two or more. The Zn and Mn levels were also elevated; the average Zn concentration exceeded the EU limits, and the maximum concentrations exceeded the permissible levels by 10-fold (Zn) and 1.6-fold (Mn). In contrast, the Pb concentrations remained below the regulatory thresholds in all cases.

These results indicate that the Mae Tao watershed is significantly contaminated with Cd, moderately contaminated with Zn and Mn, and not notably impacted by Pb. The elevated Cd levels raise concern due to its high toxicity, persistence, and potential for bioaccumulation in agricultural systems. The excessive Cd concentrations are consistent with previous studies linking contamination in this area to long-term zinc mining and industrial activity upstream of the Mae Tao Creek.

4.4. Heavy Metal Leachability

Leaching experiments were conducted on the three most contaminated soil samples (S04, S12, and S19) under varying pH and ionic strength conditions. Results showed that metal leachability was strongly influenced by ionic strength, especially for Cd and Pb. At a higher ionic strength (0.10 M), the electrical double layer surrounding soil particles is compressed, reducing electrostatic repulsion and increasing metal sorption, thus decreasing leachability [78–80]. Conversely, lower ionic strength (0.01 M) expands the double layer, allowing for the greater mobility and leaching of Cd and Pb into solution.

In contrast, the leachability of Zn and Mn was not significantly influenced by ionic strength, suggesting that their retention in soil is governed by other factors, such as specific adsorption to mineral surfaces or incorporation into secondary minerals. The influence of pH on leachability was not statistically significant in this study, although its role in surface protonation and the deprotonation of soil minerals is known to affect metal mobility under different conditions [81,82]. These results support the conclusion that ionic strength is a dominant factor in Cd and Pb leaching in the study soils, while pH plays a more complex and secondary role.

4.5. Mineralogical Evidence and Metal Speciation

X-ray diffraction (XRD) analysis was conducted on representative samples to evaluate mineralogical compositions. Major phases included quartz (76.65–87.91%), dolomite (5.45–11.91%), magnesium calcite (2.84–4.82%), and minor clay minerals (0–7.87%). Notably, no crystalline phases of cadmium-bearing minerals such as otavite ($CdCO_3$) or greenockite (CdS) were detected, suggesting that Cd exists in the soil primarily in adsorbed or amorphous forms rather than as discrete mineral phases. This supports the interpretation that the observed Cd in leachates originates from loosely bound or exchangeable fractions rather than from the dissolution of cadmium minerals.

4.6. Health Risk Assessment

The human health risk assessment included an evaluation of both non-carcinogenic and carcinogenic risks via the ingestion and dermal exposure pathways. Non-carcinogenic risks were quantified using the hazard quotient (HQ) based on the average daily dose (ADD). The HQ values for all metals (Cd, Pb, Zn, and Mn) in both adults and children were below the threshold of 1.0, indicating no significant health risk under the current exposure conditions. Ingestion was the dominant exposure route, with the ADD_{ing} consistently higher than the ADD_{derm} across all metals. This aligns with the established exposure patterns for soil contaminants.

Despite earlier studies reporting adverse health outcomes in the local population and concerns about Cd exposure from rice consumption [27,31], the present assessment based on soil exposure pathways (ingestion and dermal contact) suggests that non-carcinogenic risk is minimal. One possible explanation for this is that the sampled soils, collected at a depth of 30 cm, may represent a leached zone with lower bioavailable concentrations of contaminants compared to surface soils, where human exposure is more direct.

However, the carcinogenic risk assessment revealed more concerning results. The lifetime cancer risk (LCR) values for Cd exceeded the critical threshold of 1×10^{-4} in some scenarios, particularly in ingestion pathways for both adults and children, using both 80-mesh and 200-mesh soil samples. This suggests that Cd in fine soil particles poses a potential carcinogenic risk, especially through the oral pathway. The dermal-based exposure LCR for Cd exceeded the acceptable limits all of concentrations in both 80-mesh and 200-mesh samples [55]. In contrast, Pb's LCR values were within or below the acceptable range (1×10^{-6} to 1×10^{-4} and less than 1×10^{-6}), indicating negligible carcinogenic risk.

Adults exhibited slightly higher vulnerability compared to children, with estimated hazard quotient (HQ) and lifetime cancer risk (LCR) values approximately one to two times greater across all exposure pathways. This increased susceptibility is primarily attributed to the longer lifespan of adults, which extends the duration of exposure and thereby amplifies both non-carcinogenic and carcinogenic risk accumulations over time, particularly from early life stages onward.

4.7. Implications and Recommendations

The results of this study highlight the complex interplay between soil properties, heavy metal contamination, particle size, leachability, and potential health risks. The Mae Tao watershed remains heavily impacted by cadmium contamination, with measurable risks for long-term human exposure, particularly through the oral pathway with both coarse and fine soil particles. While non-carcinogenic risk appears low under the current exposure assumptions, the carcinogenic potential of Cd warrants continued monitoring, the stricter regulation of agricultural land use, and targeted remediation efforts.

Future work should prioritize surface soil sampling (<15 cm depth), seasonal variations, and food chain transfer, particularly in rice and vegetable crops. Additional studies on metal speciation, bioavailability, and long-term soil–plant–human interactions are essential for developing a comprehensive risk management strategy for the region.

5. Conclusions

This study provides a comprehensive evaluation of heavy metal contamination, leachability behavior, and the associated human health risks in agricultural soils of the Mae Tao watershed, Mae Sot District, Tak Province, Thailand. The findings revealed elevated concentrations of cadmium (Cd), zinc (Zn), lead (Pb), and manganese (Mn) in both surface and subsurface soils, with Cd levels significantly exceeding the national and international regulatory thresholds [72]. While Zn and Mn were present at moderately elevated levels, Pb concentrations generally remained within the acceptable limits.

Leachability experiments demonstrated that ionic strength is a key factor influencing the mobility of heavy metals, particularly Cd and Pb, in the soil environment. Higher ionic strength was associated with increased metal desorption and mobility, thereby enhancing the potential for bioavailability and environmental transport. In contrast, pH showed no consistent or statistically significant effect on leaching behavior, likely due to its complex interactions with soil mineral surfaces and competing ions. These findings suggest that conventional assessments based solely on total metal concentrations may overestimate or underestimate actual environmental risk unless the influence of soil chemistry is properly accounted for.

Health risk assessments indicated that, under the current exposure scenarios via ingestion and dermal contact, none of the evaluated heavy metals posed a significant non-carcinogenic risk to either adults or children. However, carcinogenic risk from cadmium, particularly through ingestion exposure to both coarse (80-mesh) and fine soil particles (200-mesh), was found to exceed the acceptable threshold (1×10^{-4}), signaling potential long-term health concerns. Lead also presented a minor carcinogenic risk, but remained within the acceptable limits.

Overall, this study underscores the importance of understanding the geochemical behavior of heavy metals in agricultural soils. Factors such as particle size, ionic strength, and mineralogical composition play critical roles in determining metal mobility and risk potential. The complex and dynamic nature of soil systems necessitates an integrated approach for effective environmental monitoring and risk management in contaminated agricultural landscapes.

Supplementary Materials: The following supporting information can be downloaded at: https://www.mdpi.com/article/10.3390/toxics13080687/s1. Table S1: Cadmium leachability test results; Table S2: Zinc leachability test results; Table S3: Lead leachability test results; Table S4: Manganese leachability test results; Figure S1: Percent leached concentrations of cadmium from soil samples number S04, S12, and S19 in 80-mesh and 200-mesh size in test time of 28 days; Figure S2: Percent leached concentrations of zinc from soil samples number S04, S12, and S19 in 80-mesh and 200-mesh size in test time of 28 days; Figure S3: Percent leached concentrations of lead from soil samples number S04, S12, and S19 in 80-mesh and 200-mesh size in test time of 28 days; and Figure S4: Percent leached concentrations of manganese from soil samples number S04, S12, and S19 in 80-mesh and 200-mesh size in test time of 28 day.

Author Contributions: N.S., conceptualization; methodology; formal analysis; writing—original draft; and writing—review and editing; T.W., experimental investigation; data assimilation; data curation; methodology; and formal analysis, R.B., formal analysis; writing—original draft; and writing—review and editing, S.S., conceptualization; resources; supervision; validation; writing—review and editing; and funding acquisition. All authors have read and agreed to the published version of the manuscript.

Funding: This research work was partially supported by Chiang Mai University and the Fundamental Fund 2025, Chiang Mai University. There was no additional external funding received for this project. The funders had no role in the study design, data collection, or analysis; the decision to publish the results; or the preparation of the manuscript.

Institutional Review Board Statement: Not applicable.

Informed Consent Statement: Not applicable.

Data Availability Statement: The groundwater flow model simulation input/output files, groundwater level measurements, and stable isotopic data are available upon request.

Acknowledgments: This research was partially supported by Chiang Mai University and the Fundamental Fund 2025, Chiang Mai University. The Development and Promotion of Science and Technology Talent Project (DPST)'s scholarship is also acknowledged for Thanan Watcharamai's financial support.

Conflicts of Interest: The authors declare no conflicts of interest. The funders had no role in the design of the study; in the collection, analyses, or interpretation of data; in the writing of the manuscript; or in the decision to publish the results.

References

1. Pouyat, R.V.; McDonnell, M.J. Heavy metal accumulations in forest soils along an urban-rural gradient in southeastern New York, USA. *Water Air Soil Pollut.* **1991**, *57*, 797–807. [CrossRef]
2. Rainbow, P.S. Trace metal bioaccumulation: Models, metabolic availability and toxicity. *Environ. Int.* **2007**, *33*, 576–582. [CrossRef] [PubMed]
3. Järup, L. Hazards of heavy metal contamination. *Br. Med. Bull.* **2003**, *68*, 167–182. [CrossRef] [PubMed]
4. Oloruntoba, A.; Omoniyi, A.O.; Shittu, Z.A.; Ajala, R.O.; Kolawole, S.A. Heavy metal contamination in soils, water, and food in nigeria from 2000–2019: A systematic review on methods, pollution level and policy implications. *Water Air Soil Pollut.* **2024**, *235*, 586. [CrossRef]
5. Kemper, T.; Sommer, S. Estimate of heavy metal contamination in soils after a mining accident using reflectance spectroscopy. *Environ. Sci. Technol.* **2002**, *36*, 2742–2747. [CrossRef] [PubMed]
6. Bünemann, E.K.; Bongiorno, G.; Bai, Z.; Creamer, R.E.; De Deyn, G.; De Goede, R.; Fleskens, L.; Geissen, V.; Kuyper, T.W.; Mäder, P. Soil quality–A critical review. *Soil Biol. Biochem.* **2018**, *120*, 105–125. [CrossRef]
7. Down, C.G. *Stocks, The Environmental Impact of Mining*; Applied Science Publisher Ltd.: London, UK, 1977.
8. Okoye, C.O.B.; Aneke, A.U.; Ibeto, C.N.; Ihedioha, I.J.N. Heavy metals analysis of local and exotic poultry meat. *Int. J. Appl. Environ. Sci.* **2011**, *6*, 49–55.
9. Adekiya, A.O.; Oloruntoba, A.P.; Ojeniyi, S.O.; Ewulo, B.S. Heavy metal composition of maize and tomato grown on contaminated soils. *Open Agric.* **2018**, *3*, 414–426. [CrossRef]
10. Nnadozie, C.U.; Birnin-Yauri, U.A.; Muhammad, C.; Umar, A. Assessment of some diary products sold in Sokoto Metropolis, Nigeria. *Int. J. Adv. Res. Chem. Sci.* **2014**, *1*, 31–37.
11. Taiwo, A.M.; Oyebode, A.O.; Salami, F.O.; Okewole, I.; Gbogboade, A.S.; Agim, C.; Oladele, T.O.; Kamoru, T.A.; Abdullahi, K.L.; Davidson, N. Carcinogenic and non-carcinogenic evaluations of heavy metals in protein foods from southwestern Nigeria. *J. Food Compos. Anal.* **2018**, *73*, 60–66. [CrossRef]
12. Shaheen, M.E.; Tawfik, W.; Mankola, A.F.; Gagnon, J.E.; Fryer, B.J.; El-Mekawy, F.M. Assessment of contamination levels of heavy metals in the agricultural soils using ICP-OES. *Soil Sediment Contam. Int. J.* **2023**, *32*, 665–691. [CrossRef]
13. Ali, H.; Khan, E.; Ilahi, I. Environmental chemistry and ecotoxicology of hazardous heavy metals: Environmental persistence, toxicity, and bioaccumulation. *J. Chem.* **2019**, *2019*, 6730305. [CrossRef]
14. Briffa, J.; Sinagra, E.; Blundell, R. Heavy metal pollution in the environment and their toxicological effects on humans. *Heliyon* **2020**, *6*, e04691. [CrossRef] [PubMed]
15. Li, Z.; Ma, Z.; van der Kuijp, T.J.; Yuan, Z.; Huang, L. A review of soil heavy metal pollution from mines in China: Pollution and health risk assessment. *Sci. Total Environ.* **2014**, *468–469*, 843–853. [CrossRef]
16. Khaokaew, S.; Landrot, G. A field-scale study of cadmium phytoremediation in a contaminated agricultural soil at Mae Sot District, Tak Province, Thailand:(1) Determination of Cd-hyperaccumulating plants. *Chemosphere* **2015**, *138*, 883–887. [CrossRef]
17. Akkajit, P. Review of the current situation of Cd contamination in agricultural field in the Mae Sot district, Tak province, northwestern Thailand. *Appl. Environ. Res.* **2015**, *37*, 71–82. [CrossRef]
18. Totirakul, V.; Sooksamiti, P. Report Environmental Quality of Zinc Mine in Pa Daeng Deposit, Mae Sot District, Tak Province (2003–2004). Department of Primary Industries and Mines of Thailand: Bangkok, Thailand, 2004.
19. Kaowichakorn, P. GIS Application for Mineral Resource and Environmental Management: Case Study of Zinc Deposit, Mae Sot District, Tak Province. Master's Thesis, Chulalongkorn University, Bangkok, Thailand, 2006.
20. Moonthongnoi, C.; Arunlertaree, C. Water Quality and Contamination of Heavy Metal in Surface Water of Huai Mae Tao, Tak Province. *Environ. Nat. Resour. J.* **2008**, *6*, 103–112.
21. Ryan, J.A.; Pahren, H.R.; Lucas, J.B. Controlling cadmium in the human food chain: A review and rationale based on health effects. *Environ. Res.* **1982**, *28*, 251–302. [CrossRef]
22. Nogawa, K.; Honda, R.; Kido, T.; Tsuritani, I.; Yamada, Y. Limits to Protect People Eating Cadmium in Rice, Based on Epidemiological Studies. In Proceedings of the University of Missouri's 21st Annual Conference on Trace Substances in Environmental Health, St. Louis, MO, USA, 25–28 May 1987.
23. Vassilev, A.; Vangronsveld, J.; Yordanov, I. Cadmium phytoextraction: Present state, biological backgrounds and research needs. *Bulg. J. Plant Physiol.* **2002**, *28*, 68–95.
24. Limpatanachote, P.; Swaddiwudhipong, W.; Mahasakpan, P.; Krintratun, S. Cadmium-exposed population in Mae Sot District, Tak Province: 2. Prevalence of renal dysfunction in the adults. *Med. J. Med. Assoc. Thail.* **2009**, *92*, 1345.
25. Limpatanachote, P.; Swaddiwudhipong, W.; Nishijo, M.; Honda, R.; Mahasakpan, P.; Nambunmee, K.; Ruangyuttikarn, W. Cadmium-exposed population in Mae Sot District, Tak Province: 4 bone mineral density in persons with high cadmium exposure. *Med. J. Med. Assoc. Thail.* **2010**, *93*, 1451.

26. Witaya Swaddiwudhipong, M.D.; Limpatanachote, P.; Krintratun, S.; Chantana Padungtod, M.D. Cadmium-exposed population in Mae Sot District, Tak Province: 1. Prevalence of high urinary cadmium levels in the adults. *J. Med. Assoc. Thai.* **2007**, *90*, 143–148.
27. Swaddiwudhipong, W.; Limpatanachote, P.; Nishijo, M.; Honda, R.; Mahasakpan, P.; Krintratun, S. Cadmium-exposed population in Mae Sot district, Tak province: 3. Associations between urinary cadmium and renal dysfunction, hypertension, diabetes, and urinary stones. *J. Med. Assoc. Thail.* **2010**, *93*, 231–238.
28. Teeyakasem, W.; Nishijo, M.; Honda, R.; Satarug, S.; Swaddiwudhipong, W.; Ruangyuttikarn, W. Monitoring of cadmium toxicity in a Thai population with high-level environmental exposure. *Toxicol. Lett.* **2007**, *169*, 185–195. [CrossRef] [PubMed]
29. Songprasert, N.; Sukaew, T.; Kusreesakul, K.; Swaddiwudhipong, W.; Padungtod, C.; Bundhamcharoen, K. Additional burden of diseases associated with cadmium exposure: A case study of cadmium contaminated rice fields in Mae Sot District, Tak Province, Thailand. *Int. J. Environ. Res. Public Health* **2015**, *12*, 9199–9217. [CrossRef]
30. Suwatvitayakorn, P.; Ko, M.-S.; Kim, K.-W.; Chanpiwat, P. Human health risk assessment of cadmium exposure through rice consumption in cadmium-contaminated areas of the Mae Tao sub-district, Tak, Thailand. *Environ. Geochem. Health* **2020**, *42*, 2331–2344. [CrossRef]
31. Kosolsaksakul, P.; Oliver, I.W.; Graham, M.C. Evaluating cadmium bioavailability in contaminated rice paddy soils and assessing potential for contaminant immobilisation with biochar. *J. Environ. Manag.* **2018**, *215*, 49–56. [CrossRef]
32. Sriprachote, A.; Kanyawongha, P.; Ochiai, K.; Matoh, T. Current situation of cadmium-polluted paddy soil, rice and soybean in the Mae Sot District, Tak Province, Thailand. *Soil Sci. Plant Nutr.* **2012**, *58*, 349–359. [CrossRef]
33. Ahn, Y.; Yun, H.-S.; Pandi, K.; Park, S.; Ji, M.; Choi, J. Heavy metal speciation with prediction model for heavy metal mobility and risk assessment in mine-affected soils. *Environ. Sci. Pollut. Res.* **2020**, *27*, 3213–3223. [CrossRef]
34. Tchounwou, P.B.; Yedjou, C.G.; Patlolla, A.K.; Sutton, D.J. Heavy metal toxicity and the environment. *Mol. Clin. Environ. Toxicol.* **2012**, *101*, 133–164.
35. John, D.A.; Leventhal, J.S. Bioavailability of metals. In *Preliminary Compilation of Descriptive Geoenvironmental Mineral Deposit Models*; U.S. Geological Survey: Reston, VA, USA, 1995; pp. 10–18.
36. Chi, Q.-q.; Zhu, G.-w.; Langdon, A. Bioaccumulation of heavy metals in fishes from Taihu Lake, China. *J. Environ. Sci.* **2007**, *19*, 1500–1504. [CrossRef]
37. Rasin, P.; A V, A.; Basheer, S.M.; Haribabu, J.; Santibanez, J.F.; Garrote, C.A.; Arulraj, A.; Mangalaraja, R.V. Exposure to cadmium and its impacts on human health: A short review. *J. Hazard. Mater. Adv.* **2025**, *17*, 100608. [CrossRef]
38. Lin, H.-C.; Hao, W.-M.; Chu, P.-H. Cadmium and cardiovascular disease: An overview of pathophysiology, epidemiology, therapy, and predictive value. *Rev. Port. De Cardiol.* **2021**, *40*, 611–617. [CrossRef]
39. Niture, S.; Lin, M.; Qi, Q.; Moore, J.T.; Levine, K.E.; Fernando, R.A.; Kumar, D. Role of autophagy in cadmium-induced hepatotoxicity and liver diseases. *J. Toxicol.* **2021**, *2021*, 9564297. [CrossRef]
40. Parida, L.; Patel, T.N. Systemic impact of heavy metals and their role in cancer development: A review. *Environ. Monit. Assess.* **2023**, *195*, 766. [CrossRef]
41. Geng, H.-X.; Wang, L. Cadmium: Toxic effects on placental and embryonic development. *Environ. Toxicol. Pharmacol.* **2019**, *67*, 102–107. [CrossRef] [PubMed]
42. Collin, M.S.; Venkatraman, S.K.; Vijayakumar, N.; Kanimozhi, V.; Arbaaz, S.M.; Stacey, R.G.S.; Anusha, J.; Choudhary, R.; Lvov, V.; Tovar, G.I.; et al. Bioaccumulation of lead (Pb) and its effects on human: A review. *J. Hazard. Mater. Adv.* **2022**, *7*, 100094. [CrossRef]
43. Van, H.-T.; Hoang, V.H.; Nga, L.T.Q.; Nguyen, V.Q. Effects of Zn pollution on soil: Pollution sources, impacts and solutions. *Results Surf. Interfaces* **2024**, *17*, 100360. [CrossRef]
44. Crossgrove, J.; Zheng, W. Manganese toxicity upon overexposure. *NMR Biomed.* **2004**, *17*, 544–553. [CrossRef] [PubMed]
45. Santamaria, A.B. Manganese exposure, essentiality & toxicity. *Indian J. Med. Res.* **2008**, *128*, 484–500.
46. *Amphoe Mae Sot Topographic Map Sheet 4742 III, Series L7018, Scale 1:50000*; Royal Thai Survey Department: Bangkok, Thailand, 1999.
47. Gleason, S.M.; Ewel, K.C.; Hue, N. Soil redox conditions and plant–soil relationships in a micronesian mangrove forest. *Estuar. Coast. Shelf Sci.* **2003**, *56*, 1065–1074. [CrossRef]
48. Dijkstra, J.J.; Meeussen, J.C.L.; Comans, R.N.J. Leaching of heavy metals from contaminated soils: An experimental and modeling study. *Environ. Sci. Technol.* **2004**, *38*, 4390–4395. [CrossRef] [PubMed]
49. Hill, S.J.; Fisher, A.S. Atomic absorption, methods and instrumentation. In *Encyclopedia of Spectroscopy and Spectrometry*, 2nd ed.; Elsevier: Amsterdam, The Netherlands, 2010; pp. 46–53.
50. Sparks, D.L.; Singh, B.; Siebecker, M.G. *Environmental Soil Chemistry*; Elsevier: Amsterdam, The Netherlands, 2022.
51. Violante, A.; Cozzolino, V.; Perelomov, L.; Caporale, A.G.; Pigna, M. Mobility and bioavailability of heavy metals and metalloids in soil environments. *J. Soil Sci. Plant Nutr.* **2010**, *10*, 268–292. [CrossRef]
52. McLean, J.E.; Bledsoe, B.E. *Behavior of Metals in Soils: Ground Water Issue, EPA/540/S-92/018*; U.S. Environmental Protection Agency: Washington, DC, USA, 1992.

53. Kabata-Pendias, A. *Trace Elements in Soils and Plants*; CRC Press: Boca Raton, FL, USA, 2000.
54. Šljivić Husejnović, M.; Janković, S.; Nikolić, D.; Antonijević, B. Human health risk assessment of lead, cadmium, and mercury co-exposure from agricultural soils in the Tuzla Canton (Bosnia and Herzegovina). *Arh. Za Hig. Rada I Toksikol.* **2021**, *72*, 268–279. [CrossRef] [PubMed]
55. U.S. Environmental Protection Agency. *Guidelines for Carcinogen Risk Assessment US Environmental Protection Agency*; EPA/630/P-03; U.S. Environmental Protection Agency: Washington, DC, USA, 2005.
56. U.S. Environmental Protection Agency. *Supplemental Guidance for Developing Soil Screening Levels for Superfund Sites*; U.S. Environmental Protection Agency: Washington, DC, USA, 2002.
57. Zhang, K.; Li, X.; Song, Z.; Yan, J.; Chen, M.; Yin, J. Human health risk distribution and safety threshold of cadmium in soil of coal chemical industry area. *Minerals* **2021**, *11*, 678. [CrossRef]
58. U.S. Environmental Protection Agency. *Appendix A Generics SSLs for the Residential and Commercial/Industrstrial Scenarios*; U.S. Environmental Protection Agency: Washington, DC, USA, 2002.
59. U.S. Environmental Protection Agency. *Risk Assessment Guidance for Superfund (RAGS): Part E*; U.S. Environmental Protection Agency: Washington, DC, USA, 2004.
60. U.S. Environmental Protection Agency. *Exposure Factors Handbook*, 2011st ed.; U.S. Environmental Protection Agency: Washington, DC, USA, 2011.
61. Nguyen, K.T.; Nguyen, H.M.; Truong, C.K.; Ahmed, M.B.; Huang, Y.; Zhou, J.L. Chemical and microbiological risk assessment of urban river water quality in Vietnam. *Environ. Geochem. Health* **2019**, *41*, 2559–2575. [CrossRef]
62. Nag, R.; Cummins, E. Human health risk assessment of lead (Pb) through the environmental-food pathway. *Sci. Total Environ.* **2022**, *810*, 151168. [CrossRef]
63. Integrated Risk Information System (IRIS); U.S. Environmental Protection Agency. *Manganese*; CASRN 7439-96-5; U.S. Environmental Protection Agency: Washington, DC, USA, 1995.
64. Integrated Risk Information System (IRIS); U.S. Environmental Protection Agency. *Zinc and Compounds*; CASRN 7440-66-6; U.S. Environmental Protection Agency: Washington, DC, USA, 2005.
65. Wu, H.; Liao, Q.; Chillrud, S.N.; Yang, Q.; Huang, L.; Bi, J.; Yan, B. Environmental exposure to cadmium: Health risk assessment and its associations with hypertension and impaired kidney function. *Sci. Rep.* **2016**, *6*, 29989. [CrossRef]
66. Gui, H.; Yang, Q.; Lu, X.; Wang, H.; Gu, Q.; Martín, J.D. Spatial distribution, contamination characteristics and ecological-health risk assessment of toxic heavy metals in soils near a smelting area. *Environ. Res.* **2023**, *222*, 115328. [CrossRef]
67. Yu, L.; Liu, H.; Liu, W.; Qin, P.; Yu, J.; Zhou, B.; Zhang, F.; Chen, Z.; Zhao, Y.; Shi, Z. Spatial distribution, pollution characteristics, and health risk assessment of heavy metals in soils from a typical agricultural county, East China. *Agriculture* **2022**, *12*, 1565. [CrossRef]
68. Miletić, A.; Lučić, M.; Onjia, A. Exposure Factors in Health Risk Assessment of Heavy Metal(loid)s in Soil and Sediment. *Metals* **2023**, *13*, 1266. [CrossRef]
69. Ahmad, W.; Alharthy, R.D.; Zubair, M.; Ahmed, M.; Hameed, A.; Rafique, S. Toxic and heavy metals contamination assessment in soil and water to evaluate human health risk. *Sci Rep* **2021**, *11*, 17006. [CrossRef]
70. Kusin, F.M.; Azani, N.N.M.; Hasan, S.N.M.S.; Sulong, N.A. Distribution of heavy metals and metalloid in surface sediments of heavily-mined area for bauxite ore in Pengerang, Malaysia and associated risk assessment. *Catena* **2018**, *165*, 454–464. [CrossRef]
71. Council of the European Union. *Council Directive 86/278/EEC of 12 June 1986 on the Protection of the Environment, and in Particular of the Soil, When Sewage Sludge Is Used in Agriculture*; Council of the European Union: Brussels, Belgium, 1986; pp. 6–12.
72. Pollution Control Department. *Notification of the National Environmental Board on Soil Quality Standard*; Royal Gazette: Bangkok, Thailand, 2021.
73. Alvarez, P.J.J. Field and Laboratory Methods to Determine Parameters for Modeling Contamination Fate and Transport in Groundwater. *Bioremediation Nat. Attenuation Process Fundam. Math. Models* **2005**, *27*, 283.
74. Kao, L.S.; Green, C.E. Analysis of variance: Is there a difference in means and what does it mean? *J. Surg. Res.* **2008**, *144*, 158–170. [CrossRef] [PubMed]
75. Sparks, D.L. 5—Sorption Phenomena on Soils. In *Environmental Soil Chemistry*, 2nd ed.; Sparks, D.L., Ed.; Academic Press: Burlington, VT, USA, 2003; pp. 133–186.
76. Sparks, D.L. 2—Inorganic Soil Components. In *Environmental Soil Chemistry*, 2nd ed.; Sparks, D.L., Ed.; Academic Press: Burlington, VT, USA, 2003; pp. 43–73.
77. Hanlon, E.A. *Soil pH and Electrical Conductivity: A County Extension Soil Laboratory Manual*; Economic Development Innovations Singapore: Singapore, 2009.
78. Garcia-Miragaya, J.; Page, A.L. Influence of ionic strength and inorganic complex formation on the sorption of trace amounts of Cd by montmorillonite. *Soil Sci. Soc. Am. J.* **1976**, *40*, 658–663. [CrossRef]
79. Homann, P.S.; Zasoski, R.J. *Solution Composition Effects on Cadmium Sorption by Forest Soil Profiles*; Wiley Online Library: Hoboken, NJ, USA, 1987.

80. Naidu, R.; Bolan, N.S.; Kookana, R.S.; Tiller, K.G. Ionic-strength and pH effects on the sorption of cadmium and the surface charge of soils. *Eur. J. Soil Sci.* **1994**, *45*, 419–429. [CrossRef]
81. Gupta, V.; Miller, J.D. Surface force measurements at the basal planes of ordered kaolinite particles. *J. Colloid Interface Sci.* **2010**, *344*, 362–371. [CrossRef]
82. Liu, J.; Miller, J.D.; Yin, X.; Gupta, V.; Wang, X. Influence of ionic strength on the surface charge and interaction of layered silicate particles. *J. Colloid Interface Sci.* **2014**, *432*, 270–277. [CrossRef]

toxics

MDPI

Article

Soil Quality and Trace Element Risk in Urban and Rural Kitchen Gardens: A Comparative Analysis

Diego Arán [1,2,*], Osvaldo Santos [2,3], Rodrigo Feteira-Santos [2,3], Yacine Benhalima [1] and Erika S. Santos [1,2]

1 LEAF—Linking Landscape, Environment, Agriculture and Food Research Center, Instituto Superior de Agronomia, Universidade de Lisboa, Tapada da Ajuda, 1349-017 Lisbon, Portugal; yacinebenhalima@isa.ulisboa.pt (Y.B.); erikasantos@isa.ulisboa.pt (E.S.S.)

2 Associate Laboratory TERRA, Instituto Superior de Agronomia, Universidade de Lisboa, Tapada da Ajuda, 1349-017 Lisbon, Portugal; osantos@medicina.ulisboa.pt (O.S.)

3 Institute of Environmental Health, Lisbon School of Medicine, Universidade de Lisboa, Av. Prof. Egas Moniz, 1649-028 Lisbon, Portugal

* Correspondence: diegoaran@isa.ulisboa.pt

Abstract: The development and use of urban spaces for food production is increasing in response to the search for healthier foods and contact with nature. These spaces can be created or built on materials of various types, which might contain potentially toxic elements (PTEs). This study focuses on the evaluation of soil fertility and contamination levels in urban and rural kitchen gardens in Lisbon, Portugal. Soils of twenty kitchen gardens (n_{urban} = 15; n_{rural} = 5) were sampled, and their physicochemical characteristics and the contents of PTEs in the total and available fractions were analyzed. The results were used to calculate contamination indices and associated ecological risk. The soils of the urban and rural kitchen gardens had a neutral pH, with the presence of carbonate forms, and moderate-to-high organic matter contents, although with a clear nutritional imbalance. Regarding PTEs, both urban and rural kitchen gardens soils showed elevated levels of certain elements (e.g., Cr, Ni, Cu), exceeding the maximum allowable values established by Portuguese regulations. However, the available fraction of these elements was generally low. Contamination indices ranged from mild to considerable in isolated cases, with no general multi-element contamination or ecological risk. This suggests that associated environmental and health risks are minimal, although periodic monitoring of kitchen gardens' soil quality is necessary to ensure and maximize the health benefits.

Keywords: pollution index; enrichment factor; PLI; potential ecological risk; community gardens; allotment gardens; urban soils

1. Introduction

Urbanization is a growing, worldwide phenomenon. The United Nations (UN) has estimated that the urban population will increase from 30% in 1950 to 68% of total population in 2050, a proportion even higher in more developed countries (86%) [1]. Movements and concentration of population in urban areas have positively affected societies' wellbeing and public health, by improving their residents' living standards [2,3], associated with more opportunities or access to healthcare, for instance by providing more opportunities and better access to healthcare.

However, urban living and exposure to urban environment-associated risk factors can also be harmful to health and mortality [3,4], posing health challenges to individuals related to the urban environment and lifestyle. Estimations of mortality attributable to

https://doi.org/10.3390/toxics13080697

urban environmental risks highlight a negative impact, for instance, from living exposed to excessive air pollution [5,6], extreme heat [7] and insufficient green spaces [8,9].

Green infrastructure in cities and urban gardening is particularly important, preserving health and wellbeing of urban populations and offering recreational opportunities, physical activity and social cohesion [10–14]. The positive effects on health and wellbeing, especially mental health, of being in contact with natural environments within the urban setting have also been documented [15,16]. Despite heterogeneity across study designs and populations, findings seem to agree on the fact that exposure to green spaces has benefits for mental health outcomes [17–20]. Human–nature interactions in cities also provide opportunities for people to develop pro-environmental attitudes toward the environment, helping to increase their awareness of environmental issues [13,21].

Reconnecting urban populations with nature and agriculture has also been highlighted in the dimension of food systems serving cities. With many people already living in metropolitan areas and many more coming from rural areas, 80% of all food produced is expected to be consumed in cities by 2050 [22]. However, food supply chains for cities are becoming increasingly longer, often starting far from cities in order to deliver enough nutritious and safe food to urban populations [23]. In some cases, urbanization has led to phenomena known as food deserts and food swamps, which are geographic areas characterized by either limited access to diverse, fresh or nutritious foods or to an overabundance of high-energy and low-nutritional density foods, respectively [23]. As a result, some population subgroups are at higher risk of poor health outcomes, since limited access to nutritious, safe and affordable food is limited. Many times, paradoxically, malnutrition can manifest in multiple forms, and an association has been observed between food insecurity and obesity [24–26].

In accordance with the Food and Agriculture Organization framework for the Urban Food Agenda, urban food policy should focus on addressing food insecurity and building resilience in urban food systems, for instance, by promoting rural–urban linkages and leveraging sustainable food production in or near cities [27]. Different strategies have thus been proposed to shrink the gap between where food is produced and where it is consumed, namely urban agriculture and in-city farming, with benefits in different domains, such as promoting local supply chains, improving wellbeing and access to food and hence food security, as well as other positive impacts for biodiversity and climate resilience [28]. Moreover, urban gardening can have multiple positive outcomes, including physical, psychological and social benefits, which have been often described as being beneficial for human health in urban areas [11,12].

Nevertheless, urban gardens also have environmental risks which can harm human health. Soil contamination by potentially toxic elements (e.g., Pb, Cd, Cr, Ni, Zn) is one of the main concerns since these often accumulate in urban soils, due to long-term deposition from vehicle traffic, industrial emissions, and multiple other sources. Human exposure (with several possible transmission pathways) to these soil contaminants can occur through ingestion of home-grown produce, inhalation of resuspended soil particles, or direct contact with soil during gardening. Several studies have found PTE concentrations in urban garden soils often at toxic levels, posing potential risks for human health, namely for more vulnerable groups [29]. To address this issue, we evaluated soil fertility and PTE contamination in kitchen gardens (also designated as community gardens or allotment gardens) across urban Lisbon, Portugal, using rural kitchen gardens as a baseline for comparison. Our study addresses a critical knowledge gap regarding PTE contamination in Lisbon's urban kitchen gardens. It contributes to clarifying the extent of PTE enrichment in these soils and associated exposure risks for local gardeners, providing evidence to inform public health guidelines, urban food policies, and environmental planning decisions for safer and more sustainable urban gardening.

2. Materials and Methods

2.1. Characterization of Study Area and Sampling Sites

The Lisbon Metropolitan Area is one of the most dynamic and urbanized regions in Portugal. The northern zone includes nine municipalities located north of Lisbon and the Tagus River. These municipalities combine urban and suburban areas with high population density, heavy traffic and industrial centers.

The climate is mild Mediterranean, classified as Csb/Csa under the Köppen system [30], is characterized by warm–hot and dry summers, while winters are mild and rainy. The proximity of the Atlantic Ocean attenuates the temperatures, particularly in coastal areas. The average annual rainfall is between 600 and 1000 mm, particularly between October and January (climatological normal for the Lisbon station, 1981–2010 and 1991–2020; [30]). The average annual temperature ranges from 11 to 24 °C (climatological normal for the Lisbon station, 1981–2010 and 1991–2020; [30]).

A total of 20 kitchen gardens were selected for the study: 15 in the northern zone of the Lisbon Metropolitan Area (N-LMA) and five in the adjacent rural zone, which served as a control group. The sampling areas were selected based on the presence of crops intended for human consumption, high levels of automotive traffic, and proximity (or not) to other urban sources of contamination. A homogeneous distribution was also ensured across the study area (Figure 1). In general, most of the urban kitchen gardens have a total area of 0.3 ha, divided into small plots between 10 and 20 m^2. They are surrounded by households and intensive traffic flow.

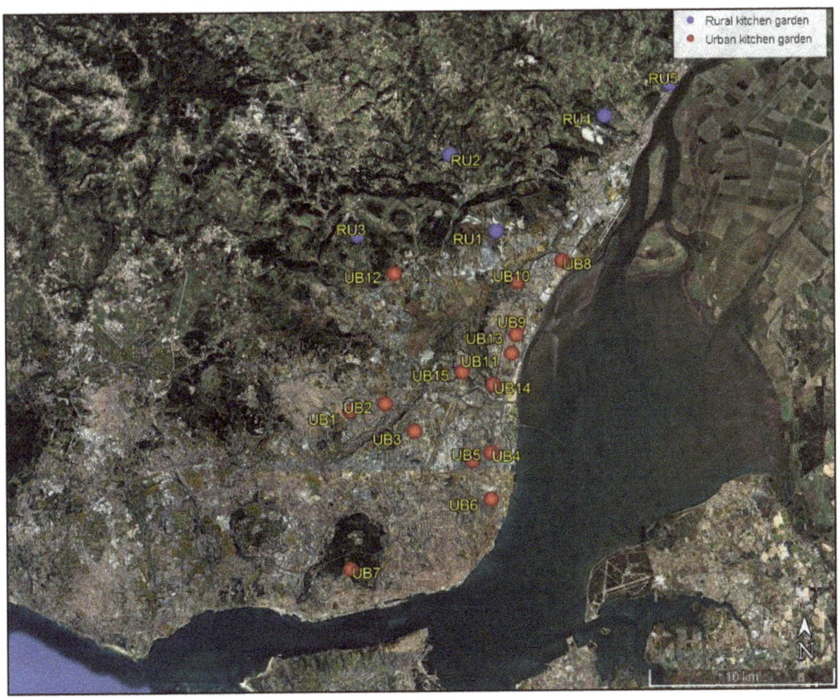

Figure 1. Location of soil sampling points in the urban kitchen gardens (red points; UB1–UB15) and rural kitchen gardens (blue points; RU1–RU5).

The soils studied have a strong anthropic influence, with intensive agricultural use and continuous irrigation. The most representative crops vary, including lettuce, various varieties of tomato, beans, pumpkins or cabbages. The soils from urban kitchen gardens can be classified as Anthrosols or Technosols [31], depending on the origin of parent materials and amounts of artefacts, while soils from rural kitchen gardens are classified as Anthrosols [31].

2.2. Soil Sampling and Physico-Chemical Analysis

Soil sampling was carried out between spring and the beginning of summer in 2023. In each sampling area, a composite soil sample was collected until 20 cm of depth using plastic tools and then stored in labelled polyethylene bags. The composite soil sample was obtained by combining several trial points (total $\approx$ 3 kg) collected across the sampling area, to obtain a representative sample.

The soil samples were homogenized, and sieved at 2 mm. In a part of each sample, we determined the moisture content while the remaining sample was air-dried for subsequent physicochemical analysis. Standard methods for characterizing soil (dry fraction < 2 mm) were used: % fine and coarse fraction, pH and electrical conductivity in water (1:2.5 *m/V*), pH in KCl (1:2.5 *m/V*) [32], total N [33], total organic C [34], labile C by pyrophosphate extraction method [35], extractable P [36]; concentration of non-acid exchangeable cations extracted with NH_4Cl 1M [37] and determined by Flame Atomic Absorption Spectroscopy (Thermo Scientific iCE™ 3300; Thermo Fisher Scientific, Waltham, MA, USA). Effective cation exchange capacity (CEC) was determined by summing the non-acid cations [38].

The multi-element concentration of the soil was determined for the following fractions: total (X-ray fluorescence spectrometry), pseudo-total and available. The element concentration in the pseudo-total fraction was determined in an international certified laboratory using the aqua regia method [39] and represents the amounts of elements associated with all soil phases except those associated with phyllosilicates. The available fraction was obtained as simulated leachates [40] and corresponds to the amounts of elements that can be taken up by plants and spread by the medium. The soil solutions from available fraction were analyzed by ICP-OES (Thermo Scientific iCAP 7000 series).

2.3. Evaluation of Environmental Risk and Contamination Level

Several indexes were calculated to evaluate the degree of soil contamination and the ecological risk to humans and the environment, based on the content of PTEs obtained.

The contamination factor (CF) is calculated as the ratio between the concentration of an element in the soil and its corresponding reference concentration (as shown in Equation (1)) [41–43]. As reference values, we considered the maximum limits proposed by Portuguese legislation for shallow soils and agricultural use [44].

$$\text{CF} = [\text{PTE soil}]/[\text{PTE reference}] \quad (1)$$

The soils are classified according to contamination levels as low contamination (CF < 1); moderate contamination (1 < CF < 3); considerable contamination (3 < CF < 6) and very high contamination (CF > 6) [42].

The degree of contamination (CD) considers the sum of the different contamination factors obtained for the PTE of a soil, described by Equation (2).

$$\text{CD} = \sum\nolimits_{n}^{i} \text{CF} \quad (2)$$

The soils are classified as having a low degree of contamination (CD < 8); moderate degree of contamination ($8 \leq$ CD < 16), considerable degree of contamination ($16 \leq$ CD < 32), or very high degree of contamination (CD $\geq$ 32) [42].

The modified contamination factor (mCF) or degree of contamination, reported by Abrahim and Parker [45], Machender et al. [46] and Rahman et al. [47], includes the normalization of the degree of contamination by introducing the total number of PTEs considered into Equation (3).

$$\text{mCF} = \sum\nolimits_{\text{n}}^{\text{i}} \text{CF}/\text{n} \tag{3}$$

This index classifies the quality of the soils as null to very low degree of contamination (mCF < 1.5); low (1.5 < mCF < 2); moderate (2 < mCF < 4); high (4 < mCF < 8); very high (8 < mCF < 16); extremely high (16 < mCF < 32); ultra-high (mCF > 32).

The pollution load index (PLI) measures the overall degree of contamination in an area or zone using Equation (4), determining areas with multi-element contamination with PLI > 1 [41,48].

$$\text{PLI} = \sqrt[n]{\text{CF1} \times \text{CF2} \ldots .\text{CFn}} \tag{4}$$

The ecological risk factor (ER) and the global potential ecological risk (GER), proposed by Hakanson [42], allow the evaluation of the possible ecological risk through the combination of the toxicity and concentration of each PTE, as well as the risk of a multi-element system. These indexes are calculated by multiplying the toxic response (Tr) of each element by its contamination factor according to Equations (5) and (6).

$$\text{ER} = \text{TR} \times \text{CF} \tag{5}$$

$$\text{GER} = \sum\nolimits_{n}^{i} \text{ER} \tag{6}$$

The TR values for Zn, Cr, Cu, As, Cd, Hg, Ni, Co, Mo, Pb and Sb were 1, 2, 5, 10, 30, 40, 5, 5, 18, 5 and 13, respectively [42,49,50]. The risk classes were established by Hakanson [42] and Al-Dahar [51].

2.4. Statistical Analysis

Data from the studied soils were analyzed to evaluate the differences in soil characteristics between Groups 1 (urban kitchen gardens) and 2 (rural kitchen gardens). A non-parametric approach was adopted due to the small size of Group 2 and the significant differences between the groups, which could lead to deviations from the normality. For each parameter, the Mann–Whitney U test (equivalent to the Wilcoxon rank-sum test) was performed to compare the distributions between the two independent groups. This test does not assume normal distribution and is appropriate for ordinal or continuous data that do not meet parametric assumptions. The analysis was performed using the dplyr, readxl, and stats packages on R software version 4.4.3.

The Spearman correlation test was used to correlate the soil characteristics (r > 0.7; $p < 0.01$ and $p < 0.05$). Quality control of the analyses was performed using analytical replicates, certified standard solutions, reference materials (OREAS 45d, 922, 907, 263, 130, 521, 620, 610, 609b), blanks and laboratory standards (Actlabs and Laboratório de Pedologia, Instituto Superior de Agronomia—Universidade de Lisboa).

3. Results and Discussion

3.1. Soil Fertility and Multi-Elemental Content

The soils from urban kitchen gardens presented a dominance of the fine fraction and a neutral reaction condition (actual and potential, determined by pH in water and KCl, respectively) with low values of electrical conductivity (Table 1), being considered as non-saline soils. With respect to C forms, the inorganic C forms represented 47 ± 15% of the total C amount, indicating the presence of considerable carbonate forms that can maintain the current pH conditions. In the organic fraction, organic C contents are considered as high [52,53], although the more easily degraded labile forms represent on average 46% (determined by labile C/organic C ratio). This fact is consistent with the low C/N values calculated (8.11 ± 2.09). These conditions, which have a high level of labile C forms and low C/N ratios, contribute to the quick degradation of organic matter and availability of nutrients to the soil–plant system. However, there is usually also a significant loss of N [53]. In fact, total N contents were correlated with organic C (r = 0.96) but the values were low rounding 2.0 g/kg (Table 1). Soils collected in urban and green areas from Lisbon also presented values of pH, electrical conductivity and total N in same range [54,55]. However, according to the same authors, the organic matter concentration varied with urban soils [54,55]. The high concentrations of organic matter obtained in some of these soils can be attributed to the lack of soil mobilization in gardens and parks, as well as the varying depths at which soil is collected. In fact, more organic matter accumulates in the top few centimetres of soil, and no-till farming can minimize its degradation.

Table 1. Main physicochemical characteristics and nutritional parameters of soils from kitchen gardens in urban and rural areas of the northern Lisbon metropolitan area (mean ± standard deviation; median). Values indicated with asterisk show significant differences between groups at $p < 0.05$.

Parameters	Unit	Urban Kitchen Garden (n = 15)	Rural Kitchen Garden (n = 5)
Coarse fraction	%	20.783 ± 9.863 22.598	23.547 ± 11.043 18.017
Fine fraction	%	79.217 ± 9.863 77.402	76.453 ± 11.043 81.983
Humidity	%	11.469 ± 10.329 8.255	22.424 ± 20.335 21.702
pH (H_2O)	-	7.728 ± 0.406 7.770	7.504 ± 0.425 7.750
pH (KCl)	-	7.109 ± 0.324 * 7.240	6.700 ± 0.370 * 6.690
EC [1]	µS/cm	482.347 ± 264.457 391.000	589.500 ± 412.829 497.100
Total C	g/kg	34.095 ± 19.937 28.230	47.314 ± 42.054 32.270
Organic C	g/kg	18.237 ± 13.937 16.853	30.580 ± 33.325 12.781
Labile C	g/kg	6.175 ± 1.864 5.824	8.927 ± 6.958 6.307
Total N	g/kg	2.110 ± 1.079 2.012	3.049 ± 1.762 2.024
Total S	g/kg	<2.500	<2.500
Total P	g/kg	1.065 ± 0.490 * 0.970	1.792 ± 0.974 * 1.430
Available P	mg/kg	166.794 ± 224.455 90.335	204.550 ± 180.629 267.210
CEC [2]	cmolc/kg	44.699 ± 18.904 46.717	65.345 ± 58.749 36.013
Exchange Ca	cmolc/kg	39.327 ± 15.842 42.523	58.234 ± 56.338 28.816
Exchange Mg	cmolc/kg	3.222 ± 3.792 1.565	4.973 ± 2.827 3.925

Table 1. *Cont.*

Parameters	Unit	Urban Kitchen Garden (n = 15)	Rural Kitchen Garden (n = 5)
Exchange Na	cmolc/kg	1.292 ± 0.728 1.405	0.904 ± 0.742 0.464
Exchange K	cmolc/kg	0.859 ± 0.420 0.760	1.234 ± 0.834 1.053

[1] Electrical conductivity; [2] effective cation exchange capacity.

The available P represents between 10 and 20% of total P in the urban soils, with very high contents in almost all samples, except for UB3 and UB10, which presented 8.69 and 10.52 mg/kg, respectively. Although the available P contents are considered as high, they showed a higher variability, reaching values of up to 897 mg/kg in UB12. No relation was obtained between organic matter and available P so, this variability can be associated with source materials of the soils and fertilization application. The presence of high levels of available P can contribute to fertility problems associated with the formation of stable solid phases with micronutrients such as Fe, Mn, or Zn [56–59], thereby reducing the concentration of these elements to the plants.

The cation exchange capacity is considered an indirect indicator of fertility. The CEC was high [52] in the soils from urban kitchen gardens (Table 1), with the exchange complex dominated by Ca (88% of the total CEC). Soils collected in urban and green areas from Lisbon also presented values of CEC in the same range [55]. The simultaneous presence of high concentration of exchangeable Ca and inorganic C can suggest the application of liming amendments based on Ca-carbonates to the soils, which is a common corrective of the pH in the agriculture sector. These conditions of Ca saturation in exchangeable complexes led to limitations in Mg uptake by the plants, with an imbalance in the Ca/Mg ratio [53].

On the other hand, although the fine fraction dominates in the soils from rural kitchen gardens (Table 1), the coarse fraction tends to increase compared to urban garden soils, varying between 12% and 35% of total. The soils from rural kitchen gardens showed a neutral current and potential reaction under low-to-medium conductivity conditions, indicating that they are non-saline soils [52,53]. Nevertheless, the pH values of KCl in rural garden soils were more acid, indicating significant differences compared to urban garden soils ($p < 0.05$). This greater difference between the actual and potential reaction (pH in water and KCl) indicates the dominance of negative charges in the soils associated with organic matter, which in this group of rural garden soils had higher average contents than in urban gardens soils (Table 1).

Concentration of inorganic C forms in rural garden soils represented a smaller proportion of the total (~35%). The labile C/Organic C ratio also showed a high presence of easily degradable forms of organic matter, which tend to quicky decomposition and mineralization. These results also agree with low C/N ratios (<10). In the studied area, the organic amendments used are manure and commercial compost in rural areas, and commercial compost or compost made from kitchen waste in situ in urban areas. All these amendments have a rapid decomposition rate. Although this type of organic amendment provides immediate benefits to the soil–plant system by quickly supplying nutrients from the exchange complex of organic matter and the organic matter itself after its decomposition, the amount of organic matter is small over the medium term, requiring frequent application [60,61].

The total P in soils from rural kitchen gardens showed contents significantly higher compared to those in urban areas, although this difference was not obtained in the available fraction (Table 1). The available P results showed a high variability (Table 1), possibly associated with chemical fertilization. Values ranged from around 13 mg/kg for RU1–RU2 to 267–398 mg/kg for RU3–RU6.

Similarly to urban kitchen soils, the CEC of the soils studied in rural areas was high [52], with mean values of 65.34 cmolc/kg and a predominance of Ca in the exchange complex (Table 1). Although high, the Ca/Mg ratios presented values lower than 10 for samples RU3 to RU5, implying that the soils may have deficiencies in Mg availability [53].

The usual soil risk assessment is based on comparing the pseudo-total concentrations of elements with the maximum values permitted by soil legislation. The concentrations of the 18 elements considered in Portuguese soil legislation [44] are presented in Table 2, grouped by soil from urban or rural kitchen gardens. These element concentrations were compared to the maximum allowed values (MAVs) established in this legislation for agriculture use. Furthermore, the contents of some representative elements (Cr, Ni, Cu, Zn, Mo, Cd, Pb and B) were determined by total, pseudo-total and available fractions. This allowed the associated risk of these elements to be assessed for each group and location (Figure 2).

Table 2. Metal and metalloid concentrations in the pseudo-total fraction of soils from kitchen gardens in urban and rural areas of the northern Lisbon metropolitan area (mean (minimum and maximum)). Values indicated with asterisk show significant differences between groups at $p < 0.05$ (*) and $p < 0.01$ (**).

Element	Unit	Urban Kitchen Garden (n = 15)	Rural Kitchen Garden (n = 5)	MAV [44]
Be	mg/kg	0.820 (0.400–1.500) **	1.560 (0.900–1.900) **	2.500
B		9.300 (4.000–16.00)	14.40 (9.000–23.00)	36.00
V		58.90 (11.00–219.0)	110.8 (21.00–284.0)	86.00
Cr		53.20 (10.00–194.0)	67.80 (20.00–228.0)	67.00
Co		15.88 (2.100–63.90)	26.96 (5.500–75.80)	19.00
Ni		42.33 (5.500–191.0)	49.30 (16.40–149.0)	37.00
Cu		35.34 (11.10–92.10) *	58.82 (27.90–75.00) *	62.00
Zn		117.9 (44.00–348.0)	131.4 (77.40–237.0)	290.0
As		5.640 (1.500–11.10)	6.080 (1.900–13.20)	11.00
Mo		0.606 (0.230–1.370)	0.786 (0.230–2.050)	2.000
Ag		0.192 (0.037–1.100)	0.073 (0.049–0.120)	0.500
Sb		0.550 (0.160–1.280)	0.426 (0.130–0.880)	1.000
Cd		0.168 (0.070–0.270)	0.166 (0.100–0.340)	1.000
Se		0.470 (0.300–0.700)	0.500 (0.300–0.700)	1.200
Ta		<0.050	<0.050	1.000
Pb		29.13 (7.800–66.20)	26.3 (7.800–55.90)	45.00
U		0.890 (0.500–1.700)	0.920 (0.700–1.300)	1.900
Hg	µg/kg	73.00 (20.00–190.0)	70.00 (30.00–100.0)	160.0

The comparative evaluation of the pseudo-total content of the two soil groups showed significant differences only in the concentrations of Be and Cu, which reached the highest values in soils from rural kitchen gardens (Table 2). The contents of Be in all soils were low and below the MAV [44], while Cu contents in samples UB12, RU3, RU4, and RU5 exceeded MAV (Table 2; Figure 2).

Although the pseudo-total concentrations of other metals and metalloids in soils from the two established groups (urban or rural) were similar, the contents showed high heterogeneity, exceeding the MAV in some samples (46% of UB and 60% of RU) for V, Cr, Co, Ni, As, Mo, Ag, Sb, Pb and Hg (Table 2).

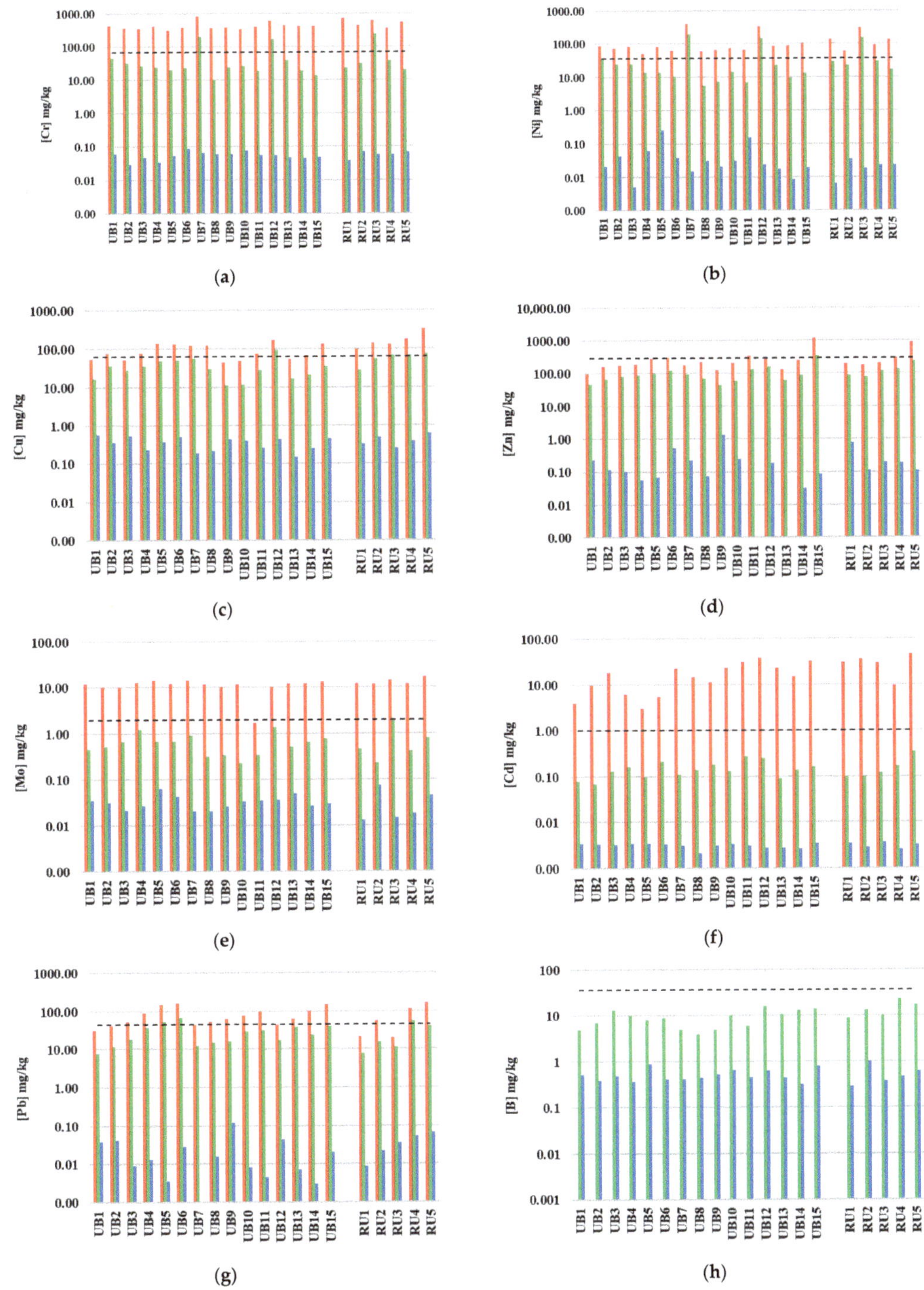

Figure 2. Distribution of total (red), pseudo-total (green) and available (blue) contents of (**a**) Cr; (**b**) Ni; (**c**) Cu, (**d**) Zn; (**e**) Mo; (**f**) Cd; (**g**) Pb and (**h**) B of soils collected from kitchen gardens in urban and rural areas of the northern Lisbon metropolitan area. The data are presented on a logarithmic scale. The black dashed line represents the value of Portuguese soil legislation for each element [44].

The pseudo-total content of V was high in some samples (UB7, UB12, RU1 and RU3), with values corresponding to double or triple those of the MAV [44]. However, these values were still within the range observed for different soils [62] suggesting that the concentrations in these samples can be associated with the local lithology. The presence of Sb above the MAV was only observed in urban kitchen garden soils (Table 2), especially in UB6 and UB15 samples. However, these values are within the ranges for basic rocks and below the phytotoxic limit [62]. The pseudo-total contents of Co exceeded MAV [44] in some samples (UB3, UB7, UB12, RU1 and RU3) with maximum values of 75.80 mg/kg in rural kitchen gardens (Table 2). The pseudo-total contents of As exceeded the MAV in the UB3 and RU4 samples, while the UB11 sample exceeded the MAV for Ag. However, Ag content of 2 mg/kg is considered the phytotoxic threshold [62].

The contents of Cr in the pseudo-total fraction exceeded the MAV by 2–3 times for samples UB7, UB12 and RU3, while the available fraction was less than 0.1 mg/kg in all samples (Figure 2). Although the soil availability of Cr is low, pH conditions would favor species associated with forms of Cr (VI), which are more soluble and toxic [63].

The pseudo-total content of Ni only exceeded the MAV in the sample UB7, which had a high concentration of 191 mg/kg (Figure 2). This is five times higher than the MAV and twice the average content of the lithosphere [63]. Nonetheless, phytotoxic concentrations for most plants and crops have been reported in the range of 40 to 260 mg/kg [62]. The available fraction of Ni in this soil sample (UB7) was low (0.01 mg/kg), similar to that in the other soils evaluated (Figure 2).

Elevated Pb concentrations in the pseudo-total fraction, greater than the MAV of 45 mg/kg, were identified in samples UB5, UB6 and RU4 (Figure 2). However, the availability of these elements is low, posing no direct risk. Sample UB5 also presented elevated pseudo-total contents of Hg.

The pseudo-total contents of B and Cd were below the MAV in all soils, indicating no risk (Table 2; Figure 2). The availability of B was low (Figure 2) corresponding to a mean value of 6.6% of pseudo-total concentrations in soils from urban kitchen gardens and 4.1% in rural gardens. B is considered the most mobile element of micronutrients but its availability in the studied soil is within normal ranges [62]. For Cd, its available fraction is lower than 0.01 mg/kg, representing between 0.9 and 5% of its pseudo-total fraction (Figure 2).

The results obtained in this study for the pseudo-total fraction, when compared to those obtained by Leitao et al. [64] for urban kitchen garden soils in Lisbon, showed similar contents of Be, Mo, Pb and Sb, higher concentrations for As, V, Cr, Cu, Ni and Zn, and the presence of additional elements such as Hg, Cd and Ag, which can pose some risk due to exceeding MAV at some sampling points. On the other hand, in soils collected in urban gardens or parks from Lisbon [54], pseudo-total concentrations of Cd were higher while Pb amount was much lower. Most of these soil samples presented concentrations of Cr and Ni in the same range [54]. Another study carried out in Lisbon's urban and green areas also found pseudo-total concentrations of several elements within the same range [55].

Strong positive correlations were found between some of the studied metals and metalloids ($p < 0.01$), as well as strong negative correlations of Se with V (for both groups) and Se with Co for urban areas (Figure 3). The relationship between these elements, as well as the significant concentrations that exceed MAV, may be due to a variety of reasons. In some cases, the geological materials present in each location may exert an influence, as is typical of volcanic rocks [54,55,64,65]. According to field observations, different remains of geological and edaphic materials from other locations (resulting from their extraction in civil work) and artefacts (e.g., building material) are used in the preparation of some kitchen gardens, being mixed with local soil. Also, the use of some organic fertilizers or manures, many of them prepared from composted sewage sludge, can increase the level of elements in soils. All these types of sources of PTEs in soils from kitchen gardens have been reported by different authors (Bidar et al. [66] and references inside). Traffic and the industrial sector can also contribute to the enrichment of certain elements in urban soils [54,55].

(a)

Positive correlation
Negative correlation

	Be	B	V	Cr	Co	Ni	Cu	Zn	As	Mo	Ag	Sb	Cd	Se	Pb	U	Hg
Be	1.0																
B	0.5	1.0															
V	0.4	0.2	1.0														
Cr	0.4	0.1	0.9	1.0													
Co	0.5	0.3	1.0	0.9	1.0												
Ni	0.4	0.1	0.9	1.0	0.9	1.0											
Cu	0.3	0.3	0.7	0.7	0.7	0.7	1.0										
Zn	−0.1	0.5	0.0	0.0	0.0	0.0	0.4	1.0									
As	0.4	0.2	−0.6	−0.6	−0.5	−0.6	−0.4	−0.1	1.0								
Mo	0.6	0.6	0.6	0.6	0.6	0.6	0.8	0.3	−0.1	1.0							
Ag	−0.2	−0.2	−0.1	−0.1	−0.2	−0.1	0.0	0.2	0.1	−0.2	1.0						
Sb	0.0	0.4	−0.5	−0.5	−0.4	−0.5	0.0	0.7	0.4	0.1	0.2	1.0					
Cd	0.0	0.2	0.1	0.1	0.1	0.0	0.4	0.4	0.1	0.2	0.6	0.4	1.0				
Se	−0.4	−0.2	−0.7	−0.6	−0.7	−0.6	−0.2	0.4	0.3	−0.3	0.4	0.7	0.4	1.0			
Pb	−0.1	0.2	−0.4	−0.4	−0.4	−0.4	0.1	0.3	0.2	0.1	0.2	0.7	0.2	0.6	1.0		
U	0.6	0.7	0.1	0.0	0.2	0.0	0.4	0.1	0.5	0.6	−0.1	0.4	0.3	0.0	0.2	1.0	
Hg	0.0	0.2	0.1	0.1	0.0	0.0	0.6	0.3	0.0	0.4	0.3	0.5	0.5	0.5	0.8	0.4	1.0

(b)

Positive correlation
Negative correlation

	Be	B	V	Cr	Co	Ni	Cu	Zn	As	Mo	Ag	Sb	Cd	Se	Pb	U	Hg
Be	1.0																
B	−0.1	1.0															
V	0.4	−0.7	1.0														
Cr	0.5	−0.4	0.8	1.0													
Co	0.5	−0.6	1.0	0.9	1.0												
Ni	0.5	−0.4	0.9	1.0	1.0	1.0											
Cu	0.0	0.6	−0.2	0.3	0.0	0.3	1.0										
Zn	−0.8	0.5	−0.4	−0.1	−0.3	−0.2	0.7	1.0									
As	0.0	1.0	−0.7	−0.5	−0.6	−0.5	0.5	0.4	1.0								
Mo	0.2	−0.4	0.8	0.9	0.9	0.9	0.4	0.1	−0.5	1.0							
Ag	−0.6	0.8	−0.7	−0.4	−0.6	−0.5	0.6	0.9	0.7	−0.2	1.0						
Sb	−0.6	0.8	−0.8	−0.5	−0.7	−0.5	0.7	0.9	0.7	−0.3	1.0	1.0					
Cd	−0.8	0.5	−0.5	−0.3	−0.4	−0.3	0.6	1.0	0.4	0.0	0.9	0.9	1.0				
Se	−0.4	0.7	−0.9	−0.6	−0.8	−0.6	0.6	0.6	0.7	−0.5	0.8	0.9	0.7	1.0			
Pb	−0.2	1.0	−0.7	−0.4	−0.5	−0.4	0.6	0.6	1.0	−0.3	0.9	0.8	0.6	0.7	1.0		
U	0.3	−0.6	0.7	0.9	0.8	0.9	0.3	0.0	−0.7	0.9	−0.4	−0.4	−0.1	−0.5	−0.5	1.0	
Hg	−0.4	0.7	−0.2	0.0	−0.1	0.0	0.7	0.8	0.6	0.2	0.8	0.7	0.7	0.4	0.8	−0.1	1.0

Figure 3. Spearman correlation matrix for studied metals and metalloids considering the soils collected from kitchen gardens in urban (**a**) and rural (**b**) areas of the northern Lisbon metropolitan area.

The distribution of the elements by different solid phases (total, pseudo-total and available) showed the existence of a variable amount associated with phyllosilicate structure (difference between total and pseudo-total concentrations) (Figure 2). These concentrations depended on elements. The evaluation of total and pseudo-total concentrations of PTE does not represent a real environmental or human risk, since only a small amount of the elements are in the available fraction, which is associated with soil solution and/or exchangeable positions associated with inorganic and organic colloids [62,67]. In general, the availability of studied elements in the soils was small, representing a small percentage of total/pseudo-total amounts. No clear relationship was obtained between available concentrations and pseudo-total and total fractions.

3.2. Pollution and Ecological Indexes

The indices established for determining the degree of contamination and ecological and environmental risks for different soils collected from kitchen gardens in urban and rural areas of the northern Lisbon metropolitan area were evaluated (Tables 3, S1 and S2).

Table 3. Results of the different indices for evaluation of environmental risk and contamination level (contamination degree indices: CD and mCD; contaminant load index: PLI and PLI zone; and the global ecological risk: GER) of the soils from kitchen gardens in urban and rural areas of the northern Lisbon metropolitan area.

	CD (n = 18)	mCD (n = 18)	PLI	PLI Zone	GER (n = 11)
UB1	5.34	0.30	0.15		23.23
UB2	6.42	0.36	0.19		30.61
UB3	8.31	0.46	0.23		44.47
UB4	7.39	0.41	0.24		44.15
UB5	7.02	0.39	0.23		38.03
UB6	8.19	0.46	0.26		46.66
UB7	16.60	0.92	0.27		67.25
UB8	3.68	0.20	0.12		26.65
UB9	4.54	0.25	0.15		27.36
UB10	5.09	0.28	0.17		20.06
UB11	7.72	0.43	0.22		28.00
UB12	18.35	1.02	0.40		24.24
UB13	6.62	0.37	0.20		35.09
UB14	6.29	0.35	0.20		77.24
UB15	7.64	0.42	0.22		80.01
Urban kitchen garden (n = 15)				0.21	
RU1	7.87	0.44	0.19		49.94
RU2	6.75	0.38	0.20		51.23
RU3	20.07	1.12	0.33		29.90
RU4	10.01	0.56	0.31		34.89
RU5	8.66	0.48	0.28		44.37
Rural kitchen garden (n = 5)				0.26	

The contamination factors (CFs) for each of the 18 PTEs included in the Portuguese soil legislation (Table S1) showed that only some soils in both groups exhibited moderate to considerable contamination for certain elements. Similar results were reported by Silva et al. [54] for some urban soils from parks and gardens. However, other urban soils were found to be highly contaminated with Cr, Ni and Pb [54].

The degree of contamination index obtained by combining these factor values (CD) varied between 3.68 and 20.07 (Table 3). The degree of contamination, as defined by Hakanson [42], enabled the identification of considerable contamination in samples UB7, UB12 and RU3, and moderate contamination in samples UB3, UB6, RU4 and RU5. This is primarily due to high levels of Ni, Co, Cr and V (Tables 2 and S1). The same degree of contamination was identified in park soils in Lisbon (moderate and considerable), but only four PTEs were considered: Cd, Cr, Ni and Pb [54].

On the other hand, determining the modified contamination level index (mCD) with 18 elements analyzed and standardized, the level of contamination in all the soils from kitchen gardens was mild to very mild (Table 3). These results were consistent with those obtained from the contamination load index (PLI), which provides a global assessment of multi-element contamination. The PLI values for studied soils, independently of the group (urban or rural soils), was less than 1, indicating the inexistence of multi-element contamination.

At the level of ecological risk associated with each element, only moderate risks were obtained for Ni in the samples UB7 and RU3 (Table S2). Overall, none of the soils from urban or rural kitchen gardens showed a global level the ecological risk (GER) (Table 3). These results did not align with those obtained for park soils in Lisbon (e.g., ER for Cd and GER for certain park and garden areas) [54].

4. Conclusions

The general quality of soils in kitchen gardens in urban and rural areas of the northern Lisbon metropolitan area is limited in terms of fertility due to neutral reaction conditions, dominated by carbonate buffering, as well as unstable organic matter, favoring rapid decomposition and mineralization and loss of nutrients especially N. The availability of carbonate forms, together with the high content of available P, can limit the availability of some essential micronutrients for vegetative growth. Similarly, the cation exchange capacity, despite having high or adequate values, is dominated by Ca under conditions of Ca/Mg ratios that can limit crop growth. Thus, improving soil fertility and crop yield requires a change in soil management.

Although the pseudo-total concentrations of some elements (e.g., Cr, Ni, Cu) exceeded the maximum allowed values for agriculture use under Portuguese soil legislation, the results indicated no environmental or human risk, since the availability of the elements in simulated leachates was low. The different indices calculated (contamination indices and associated with ecological risk) also agree with this. No clear source of enrichment or elevated pseudo-total concentrations was identified in relation to the location of the kitchen gardens. However, periodic monitoring of multi-element concentrations, especially in the available fraction, is recommended to ensure environmental and human health along the time.

Supplementary Materials: The following supporting information can be downloaded at https://www.mdpi.com/article/10.3390/toxics13080697/s1: Table S1: Contamination factor (CF) calculated for the soils collected from kitchen gardens in urban and rural areas of the northern Lisbon metropolitan area; Table S2: Ecological risk factors (ER), calculated for each sampling point.

Author Contributions: Conceptualization, D.A., O.S. and E.S.S.; methodology, D.A.; validation, D.A. and E.S.S.; formal analysis, D.A. and Y.B.; investigation, D.A., O.S., R.F.-S., Y.B. and E.S.S.; data curation, D.A.; writing—original draft preparation, D.A.; writing—review and editing, D.A., O.S., R.F.-S., Y.B. and E.S.S.; supervision, D.A.; project administration, D.A. and O.S.; funding acquisition, O.S. and E.S.S. All authors have read and agreed to the published version of the manuscript.

Funding: This work was funded by national funds through FCT—Fundação para a Ciência e a Tecnologia, I.P., under the projects UIDB/04129/2020 of LEAF-Linking Landscape, Environment, Agriculture and Food, Research Unit and LA/P/0092/2020 of Associate Laboratory TERRA.

Institutional Review Board Statement: Not applicable.

Informed Consent Statement: Not applicable.

Data Availability Statement: Data will be made available on request.

Conflicts of Interest: The authors declare no conflicts of interest.

References

1. United Nations, Department of Economic and Social Affairs, Population Division. *World Urbanization Prospects: The 2018 Revision (ST/ESA/SER.A/420)*; United Nations: New York, NY, USA, 2019; p. 125. Available online: https://www.un-ilibrary.org/content/books/9789210043144 (accessed on 2 April 2025).
2. Zhang, Z.; Zhao, M.; Zhang, Y.; Feng, Y. How does urbanization affect public health? New evidence from 175 countries worldwide. *Front. Public Health* **2023**, *10*, 1096964. [CrossRef]

3. The World Health Organization (WHO). *Urbanization and Health*; Bulletin of the World Health Organization: Geneva, Switzerland, 2010; Volume 88, p. 245. Available online: https://pmc.ncbi.nlm.nih.gov/articles/PMC2855604/ (accessed on 3 July 2025).
4. Flies, E.J.; Mavoa, S.; Zosky, G.R.; Mantzioris, E.; Williams, C.; Eri, R.; Brook, B.W.; Buettel, J.C. Urban-associated diseases: Candidate diseases, environmental risk factors, and a path forward. *Environ. Int.* **2019**, *133*, 105187. [CrossRef]
5. Pascal, M.; Corso, M.; Chanel, O.; Declercq, C.; Badaloni, C.; Cesaroni, G.; Henschel, S.; Meister, K.; Haluza, D.; Martin-Olmedo, P.; et al. Assessing the public health impacts of urban air pollution in 25 European cities: Results of the Aphekom project. *Sci. Total Environ.* **2013**, *449*, 390–400. [CrossRef] [PubMed]
6. Khomenko, S.; Cirach, M.; Pereira-Barboza, E.; Mueller, N.; Barrera-Gómez, J.; Rojas-Rueda, D.; de Hoogh, K.; Hoek, G.; Nieuwenhuijsen, M. Premature mortality due to air pollution in European cities: A health impact assessment. *Lancet Planet. Health* **2021**, *5*, e121–e134. [CrossRef] [PubMed]
7. Ballester, J.; Quijal-Zamorano, M.; Méndez Turrubiates, R.F.; Pegenaute, F.; Herrmann, F.R.; Robine, J.M.; Basagaña, X.; Tonne, C.; Antó, J.M.; Achebak, H. Heat-related mortality in Europe during the summer of 2022. *Nat. Med.* **2023**, *29*, 1857–1866. [CrossRef]
8. Pereira, B.E.; Nieuwenhuijsen, M.; Ambròs, A.; de Sá, T.H.; Mueller, N. The impact of urban environmental exposures on health: An assessment of the attributable mortality burden in Sao Paulo city, Brazil. *Sci. Total Environ.* **2022**, *831*, 154836. [CrossRef]
9. Barboza, E.P.; Cirach, M.; Khomenko, S.; Iungman, T.; Mueller, N.; Barrera-Gómez, J.; Rojas-Rueda, D.; Kondo, M.; Niewenhuijsen, M. Green space and mortality in European cities: A health impact assessment study. *Lancet Planet. Health* **2021**, *5*, e718–e730. [CrossRef]
10. WHO Regional Office for Europe. *Urban Green Spaces and Health*; WHO Regional Office for Europe: Copenhagen, Denmark, 2016.
11. Soga, M.; Gaston, K.J.; Yamaura, Y. Gardening is beneficial for health: A meta-analysis. *Prev. Med. Rep.* **2017**, *5*, 92–99. [CrossRef]
12. de Bell, S.; White, M.; Griffiths, A.; Darlow, A.; Taylor, T.; Wheeler, B.; Lovell, R. Spending time in the garden is positively associated with health and wellbeing: Results from a national survey in England. *Landsc. Urban Plan.* **2020**, *200*, 103836. [CrossRef]
13. Hartig, T.; Kahn, P.H. Living in cities, naturally. *Science* **2016**, *352*, 938–940. [CrossRef]
14. García de Jalón, S.; Chiabai, A.; Quiroga, S.; Suárez, C.; Ščasný, M.; Máca, V.; Zvěřinová, I.; Marques, S.; Craveiro, D.; Taylor, T. The influence of urban greenspaces on people's physical activity: A population-based study in Spain. *Landsc. Urban Plan.* **2021**, *215*, 104229. [CrossRef]
15. Zhang, J.; Yu, Z.; Zhao, B.; Sun, R.; Vejre, H. Links between green space and public health: A bibliometric review of global research trends and future prospects from 1901 to 2019. *Environ. Res. Lett.* **2020**, *15*, 063001. [CrossRef]
16. Lampert, T.; Costa, J.; Santos, O.; Sousa, J.; Ribeiro, T.; Freire, E. Evidence on the contribution of community gardens to promote physical and mental health and well-being of non-institutionalized individuals: A systematic review. *PLoS ONE* **2021**, *16*, e0255621. [CrossRef]
17. Bratman, G.N.; Anderson, C.B.; Berman, M.G.; Cochran, B.; de Vries, S.; Flanders, J.; Folke, C.; Frunkim, H.; Gross, J.J.; Hartig, T.; et al. Nature and mental health: An ecosystem service perspective. *Sci. Adv.* **2019**, *5*, 903–927. [CrossRef]
18. Labib, S.M.; Lindley, S.; Huck, J.J. Spatial dimensions of the influence of urban green-blue spaces on human health: A systematic review. *Environ. Res.* **2020**, *180*, 108869. [CrossRef]
19. White, M.P.; Elliott, L.R.; Grellier, J.; Economou, T.; Bell, S.; Bratman, G.N.; Cirah, M.; Gascon, M.; Lima, M.L.; Lõhmus, M.; et al. Associations between green/blue spaces and mental health across 18 countries. *Sci. Rep.* **2021**, *11*, 1–12. [CrossRef]
20. Dzhambov, A.M.; Hartig, T.; Tilov, B.; Atanasova, V.; Makakova, D.R.; Dimitrova, D.D. Residential greenspace is associated with mental health via intertwined capacity-building and capacity-restoring pathways. *Environ. Res.* **2019**, *178*, 108708. [CrossRef]
21. Soga, M.; Gaston, K.J. Extinction of experience: The loss of human–nature interactions. *Front. Ecol. Environ.* **2016**, *14*, 94–101. [CrossRef]
22. Ellen MacArthur Foundation. *Cities and Circular Economy for Food*; Ellen MacArthur Foundation: Cowes, UK, 2019.
23. FAO; IFAD; UNICEF; WFP; WHO. Transforming food systems for food security, improved nutrition and affordable healthy diets for all. In *The State of Food Security and Nutrition in the World*; Food and Agriculture Organization (FAO): Rome, Italy, 2021. [CrossRef]
24. Cobb, L.K.; Appel, L.J.; Franco, M.; Jones-Smith, J.C.; Nur, A.; Anderson, C.A.M. The relationship of the local food environment with obesity: A systematic review of methods, study quality, and results. *Obesity* **2015**, *23*, 1331–1344. [CrossRef]
25. Dinour, L.M.; Bergen, D.; Yeh, M.C. The Food Insecurity—Obesity Paradox: A Review of the Literature and the Role Food Stamps May Play. *J. Am. Diet. Assoc.* **2007**, *107*, 1952–1961. [CrossRef]
26. Rezaei, M.; Ghadamgahi, F.; Jayedi, A.; Arzhang, P.; Yekaninejad, M.S.; Azadbakht, L. The association between food insecurity and obesity, a body shape index and body roundness index among US adults. *Sci. Rep.* **2024**, *14*, 23631. [CrossRef]
27. FAO. *Framework for the Urban Food Agenda*; Food and Agriculture Organization (FAO): Rome, Italy, 2019. [CrossRef]
28. Gunapala, R.; Gangahagedara, R.; Wanasinghe, W.C.S.; Samaraweera, A.U.; Gamage, A.; Rathnayaka, C.; Hammed, Z.; Baki, Z.A.; Madhujith, T.; Merah, O. Urban agriculture: A strategic pathway to building resilience and ensuring sustainable food security in cities. *Farming Syst.* **2025**, *3*, 100150. [CrossRef]

29. Montaño-López, F.; Biswas, A. Are heavy metals in urban garden soils linked to vulnerable populations? A case study from Guelph, Canada. *Sci. Rep.* **2021**, *11*, 11286. [CrossRef]
30. IPMA—Instituto Português do Mar e da Atmosfera. Normais Climatológicas. Available online: https://www.ipma.pt/pt/oclima/normais.clima/ (accessed on 12 June 2025).
31. IUSS Working Group WRB. World Reference Base for Soil Resources. In *International Soil Classification System for Naming Soils and Creating Legends for Soil Maps*, 4th ed.; International Union of Soil Sciences (IUSS): Vienna, Austria, 2022.
32. Guitián, F.; Carballas, T. *Técnicas De Análisis De Suelos*, 2nd ed.; Pico Sacro: Santiago de Compostela, Spain, 1976; p. 288.
33. Bremner, J.M. Determination of nitrogen in soils by the Kjeldahl method. *J. Agric. Sci.* **1960**, *55*, 11–33. [CrossRef]
34. Springer, U.; Klee, J. Prüfung der Leistungsf¨ahigkeit von einigen wichtigeren Verfahren zur Bestimmung des Kohlen-stoffs mittels Chromschwefels¨ aure sowie Vorschlag einer neuen Schnellmethode. *Z. Pflanzenern Ahrung Düngung Bodenkd.* **1954**, *64*, 1–26. [CrossRef]
35. Bascomb, C.L. Distribution of pyrophosphate-extractable iron and organic carbon in soils of various groups. *Eur. J. Soil Sci.* **1968**, *19*, 251–268. [CrossRef]
36. Olsen, S. Estimation of available phosphorus in soils by extraction with sodium bicarbonate. In *Circular No. 939*; US Department of Agriculture: Washington, DC, USA, 1954; pp. 1–19.
37. Peech, M.; Alexander, L.T.; Dean, L.A.; Reed, J.F. *Methods of Soil Analysis for Soil Fertility Investigations, USDA 575*; United States Department of Agriculture: Washington, DC, USA, 1947.
38. Thomas, G.W. Exchangeable Cations. In *Methods of Soil Analysis*; Page, A.L., Buxton, R.H., Miller Keeney, D.R., Eds.; American Society of Agronomy: Madison, WI, USA, 1982; pp. 159–165.
39. *ISO 11466:1995(E)*; Soil Quality—Extraction of Trace Elements Soluble in Aqua Regia. International Organization for Standardization: Geneva, Switzerland, 1995.
40. *DIN 38414-S4*; German Standard Procedure for Water, Wastewater and Sediment Testing (Group S). Determination of Leachability by Water. Institutfür Normung: Berlin, Germany, 1984.
41. Tomlinson, D.L.; Wilson, J.G.; Harris, C.R.; Jeffrey, D.W. Problems in the assessment of heavy-metal levels in estuaries and the formation of a pollution index. *Helgol. Mar. Res.* **1980**, *33*, 566–575. [CrossRef]
42. Hakanson, L. An ecological risk index for aquatic pollution control. A sedimentological approach. *Water Res.* **1980**, *14*, 975–1001. [CrossRef]
43. Liu, W.H.; Zhao, J.Z.; Ouyang, Z.Y.; Solderland, L.; Liu, G.H. Impacts of sewage irrigation on heavy metal distribution and contamination in Beijing. *China Environ. Int.* **2005**, *32*, 805–812. [CrossRef]
44. APA—Agência Portuguesa do Ambiente. Valores de referência para solos (Revisão 3). In *Solos Contaminados—Guia Técnico*; Agência Portuguesa do Ambiente: Amadora, Portugal, 2022; p. 72.
45. Abrahim, G.M.S.; Parker, R.J. Assessment of heavy metal enrichment factors and the degree of contamination in marine sediments from Tamaki Estuary, Auckland, New Zealand. *Environ. Monit. Assess.* **2008**, *136*, 227–238. [CrossRef]
46. Machender, G.; Dhakate, R.; Prasanna, L.; Govil, P.K. Assessment of heavy metal contamination in soils around Balanagar industrial area, Hyderabad, India. *Environ. Earth Sci.* **2011**, *63*, 945–953. [CrossRef]
47. Rahman, S.H.; Khanam, D.; Adyel, T.M.; Islam, M.S.; Ahsan, M.A.; Akbor, M.A. Assessment of heavy metal contamination of agricultural soil around Dhaka Export Processing Zone (DEPZ), Bangladesh: Implication of seasonal variation and indices. *Appl. Sci.* **2012**, *2*, 584–601. [CrossRef]
48. Jorfi, S.; Maleki, R.; Jaafarzadeh, N.; Ahmadi, M. Pollution load index for heavy metals in Mian-Ab plain soil, Khuzestan, Iran. *Data Brief* **2017**, *15*, 584–590. [CrossRef]
49. Ivaneev, A.I.; Brzhezinskiy, A.S.; Karandashev, V.K.; Ermolin, M.S.; Fedotov, P.S. Assessment of sources, environmental, ecological, and health risks of potentially toxic elements in urban dust of Moscow megacity, Russia. *Chemosphere* **2023**, *321*, 138142. [CrossRef]
50. Chen, Z.; Zhao, Y.; Chen, D.; Huang, H.; Zhao, Y.; Wu, Y. Ecological risk assessment and early warning of heavy metal cumulation in the soils near the Luanchuan molybdenum polymetallic mine concentration area, Henan Province, central China. *China Geol.* **2023**, *6*, 15–26. [CrossRef]
51. Al-Dahar, R.K.; Rabee, A.M.; Mohammed, R.J. Calculation of Soil Pollution Indices with Elements in Residential Areas of Baghdad City. *Revis. Bionatura* **2023**, *8*, 43. [CrossRef]
52. De Varennes, A. *Productividade dos Solos e Ambiente*; Escolar Editora: Lisboa, Portugal, 2003; ISBN 972-592-156-9.
53. Veloso, A.; Sempiterno, C.; Calouro, F.; Rebelo, F.; Pedra, F.; Castro, I.V.; Gonçalves, M.C.; Marcelo, M.E.; Pereira, P.; Fareleira, P.; et al. *Manual de Fertilização das Culturas*, 3rd ed.; Instituto Nacional de Investigação Agrária e Veterinária, I.P.—INIAV: Oeiras, Portugal, 2022; ISBN 978-972-579-063-2.
54. Silva, H.F.; Silva, N.F.; Oliveira, C.M.; Matos, M.J. Heavy Metals Contamination of Urban Soils—A Decade Study in the City of Lisbon, Portugal. *Soil Syst.* **2021**, *5*, 27. [CrossRef]
55. Cachada, A.; Dias, A.C.; Pato, P.; Mieiro, C.; Rocha-Santos, T.; Pereira, M.E.; Ferreira da Silva, E.; Duarte, A.C. Major inputs and mobility of potentially toxic elements contamination in urban areas. *Environ. Monit. Assess.* **2013**, *185*, 279–294. [CrossRef]

56. Fan, X.; Zhou, X.; Chen, H.; Tang, M.; Xie, X. Cross-Talks Between Macro- and Micronutrient Uptake and Signaling in Plants. *Front. Plant Sci.* **2021**, *12*, 663477. [CrossRef]
57. Zhang, W.; Liu, D.Y.; Li, C.; Chen, X.P.; Zou, C.Q. Accumulation, partitioning, and bioavailability of micronutrients in summer maize as affected by phosphorus supply. *Eur. J. Agron.* **2017**, *86*, 48–59. [CrossRef]
58. Barben, S.A.; Hopkins, B.G.; Jolley, V.D.; Webb, B.L.; Nichols, B.A. Phosphorus and Manganese Interactions and their relationships with Zn in chelator-buffered solution grown russet Burbank potato. *J. Plant Nutr.* **2010**, *33*, 752–769. [CrossRef]
59. Kumpiene, J.; Lagerkvist, A.; Maurice, C. Stabilization of As, Cr, Cu, Pb and Zn in soil using amendments—A review. *Waste Manag.* **2008**, *28*, 215–225. [CrossRef] [PubMed]
60. Larney, F.J.; Angers, D.A. The role of organic amendments in soil reclamation: A review. *Can. J. Soil Sci.* **2012**, *92*, 19–38. [CrossRef]
61. Santos, E.S.; Abreu, M.M.; Macías, F.; de Varennes, A. Chemical quality of leachates and enzymatic activities in Technosols with gossan and sulfide wastes from the São Domingos mine. *J. Soil Sediments* **2016**, *16*, 1366–1382. [CrossRef]
62. Kabata-Pendias, A. *Trace Elements in Soils and Plants*, 4th ed.; CRC Press: Boca Raton, FL, USA, 2010. [CrossRef]
63. Macías, F.; Calvo de Anta, R. *Niveles Genéricos de Referencia de Metales Pesados y Otros Elementos Traza en Suelos de Galicia*; Xunta de Galicia: Santiago de Compostela, Spain, 2009; p. 232. ISBN 978-84-453-4664-8.
64. Leitão, T.E.; Cameira, M.R.; Costa, H.D.; Pacheco, J.M.; Henriques, M.J.; Martins, L.L.; Mourato, M.P. Environmental Quality in Urban Allotment Gardens: Atmospheric Deposition, Soil, Water and Vegetable Assessment at LISBON City. *Water Air Soil Pollut.* **2018**, *229*, 31. [CrossRef]
65. Moitinho de Almeida, F. *Carta Geológica do Concelho de Lisboa, Folha 1 2 3 e 4, Escala 1:10000*; Serviços Geológicos de Portugal: Lisbon, Portugal, 1986.
66. Bidar, G.; Pelfrêne, A.; Schwartz, C.; Waterlot, C.; Sahmer, K.; Marot, F.; Douay, F. Urban kitchen gardens: Effect of the soil contamination and parameters on the trace element accumulation in vegetables—A review. *Sci. Total Environ.* **2020**, *738*, 139569. [CrossRef]
67. Adriano, D. *Trace Elements in Terrestrial Environments: Biogeochemistry, Bioavailability, and Risks of Metals*, 2nd ed.; Springer: New York, NY, USA, 2001. [CrossRef]

MDPI AG
Grosspeteranlage 5
4052 Basel
Switzerland
Tel.: +41 61 683 77 34

Toxics Editorial Office
E-mail: toxics@mdpi.com
www.mdpi.com/journal/toxics

www.ingramcontent.com/pod-product-compliance
Lightning Source LLC
Chambersburg PA
CBHW040837250926
51850CB00007BA/1342
* 9 7 8 3 7 2 5 8 5 9 2 1 4 *